Satellites for Atmospheric Sciences 1

SCIENCES

Space Exploration and Technology,
Field Director – Jean-Luc Lefebvre

Space and Earth, Subject Head – Jean-Luc Lefebvre

Satellites for Atmospheric Sciences 1

Meteorology, Climate and Atmospheric Composition

Coordinated by
Thierry Phulpin
Didier Renaut
Hervé Roquet
Claude Camy-Peyret

iSTE

WILEY

First published 2023 in Great Britain and the United States by ISTE Ltd and John Wiley & Sons, Inc.

ISTE Ltd
27-37 St George's Road
London SW19 4EU
UK

www.iste.co.uk

John Wiley & Sons, Inc.
111 River Street
Hoboken, NJ 07030
USA

www.wiley.com

Library of Congress Control Number: 2023941805

British Library Cataloguing-in-Publication Data
A CIP record for this book is available from the British Library
ISBN 978-1-78945-140-5

ERC code:
PE10 Earth System Science
PE10_1 Atmospheric chemistry, atmospheric composition, air pollution
PE10_2 Meteorology, atmospheric physics and dynamics
PE10_14 Earth observations from space/remote sensing
PE9 Universe Sciences
PE9_15 Space Sciences
PE9_17 Instrumentation - telescopes, detectors and techniques

Contents

Acknowledgments . . . xi

List of Acronyms . . . xiii

Introduction . . . xxxi
Thierry PHULPIN

Part 1. Satellite Observation of the Earth's Atmosphere: International Cooperation . . . 1

Chapter 1. History of Meteorological Satellites . . . 3
Sylvain LE MOAL

1.1. The beginnings of remote sensing and the conquest of space . . . 3
1.2. It all began with Tiros-1, the first meteorological satellite . . . 6
1.3. American meteorological satellites . . . 8
1.3.1. Polar-orbiting satellites . . . 8
1.3.2. Geostationary satellites . . . 13
1.4. Russian meteorological satellites . . . 17
1.4.1. Polar-orbiting satellites . . . 17
1.4.2. Geostationary satellites . . . 20
1.5. European meteorological satellites . . . 21
1.5.1. The Meteosat saga . . . 21
1.5.2. 46 years after Tiros-1, MetOp enters the scene . . . 28
1.6. Elsewhere . . . 29
1.6.1. Japan . . . 29
1.6.2. China . . . 31
1.6.3. Korea . . . 33
1.6.4. India . . . 33
1.7. References . . . 35
1.8. Websites . . . 36

Chapter 2. Contribution of the National Oceanic and Atmospheric Administration (NOAA, USA) Meteorological Satellites Program: An Overview 37
Sid-Ahmed BOUKABARA, Mitch GOLDBERG, Timothy J. SCHMIT, Andrew HEIDINGER, Satya KALLURI, Patricia WEIR, Frank GALLAGHER, David SPENCER and Ross N. HOFFMAN

2.1. NOAA Satellite Program: historical background 38
2.1.1. Origins of NASA-NOAA Polar and Geostationary Environmental Satellite Programs 38
2.1.2. Low Earth orbit (LEO) missions. 40
2.1.3. Geostationary Earth orbit (GEO) missions 43
2.2. NOAA Current Space Constellation 45
2.2.1. The NOAA Joint Polar Satellite System (JPSS) Program 45
2.2.2. GOES-R series 49
2.2.3. Collaborative programs. 51
2.3. Applications 52
2.4. Looking ahead: designing the next-generation architecture 57
2.4.1. Factors impacting the NOAA strategy 57
2.4.2. Next-generation NOAA space architecture 59
2.5. Summary 62
2.6. Acknowledgments 62
2.7. References 63

Chapter 3. The Role of the National Aeronautics and Space Administration (NASA, USA) 67
Michael SEABLOM

3.1. The beginnings of the National Aeronautics and Space Administration (NASA) 67
3.2. The Nimbus Era (1964–1979) 68
3.3. The Earth Observing System (1982–2004) 72
3.4. The “A-train” (2004–present). 81
3.5. Decadal surveys and technological disruption (2007–present) 84
3.6. References 87

Chapter 4. The Role of the European Space Agency (ESA) 89
Paul INGMANN

4.1. Missions in geostationary Earth orbit (GEO) – ESA’s Start in Earth Observation 89
4.2. Missions in low Earth orbit (LEO) 92
4.2.1. ERS 92
4.2.2. Envisat 94
4.2.3. MetOp 95

4.2.4. The Earth Explorer and Earth Watch Concept 96
4.3. ESA's Climate Change Initiative (CCI) 113
4.4. References 114

Chapter 5. The Role of EUMETSAT (Europe). 117
François MONTAGNER

5.1. Introduction: What does EUMETSAT do? 117
5.1.1. Public service value of weather satellites 117
5.1.2. EUMETSAT, a key player in Europe 117
5.1.3. Climate and environment. 118
5.2. The organization 118
5.2.1. First steps 118
5.2.2. Stability and growth 120
5.2.3. Government 120
5.2.4. European pooling: EUMETSAT, ECMWF and EUMETNET 121
5.2.5. Global pooling by the World Meteorological Organization (WMO). . . . 122
5.3. Geostationary weather satellites: from synoptic to regional zoom 122
5.3.1. Meteosat first generation 122
5.3.2. Meteosat second generation 125
5.3.3. Agility of geostationary missions 127
5.3.4. Stabilization by rotation or on three axes: system aspects. 128
5.3.5. Meteosat Third Generation. 128
5.4. MetOp satellites, the first source for numerical weather forecasting 130
5.4.1. Synergy of observations 131
5.4.2. Continuity and innovation 132
5.4.3. The second generation of the European Polar System. 133
5.4.4. Scale economies. 134
5.4.5. Cooperation regarding the polar orbit. 135
5.5. Weather perspective and innovation 136
5.6. Climate 137
5.7. EUMETSAT and Copernicus. 137
5.7.1. A convenient partnership. 137
5.7.2. EUMETSAT and the Copernicus services 138
5.7.3. Continuity and expansion: the challenge of CO_2. 139
5.8. References 139

Chapter 6. The Role of the National Center for Space Studies (CNES, France). 141
Carole DENIEL and Pierre TABARY

6.1. The CNES and its scientific missions 141
6.2. Greenhouse gases and composition of the atmosphere 142
6.2.1. Merlin, a political French–German will. 143

6.2.2. Microcarb, a strategic and continuous project... 144
6.2.3. TRAQ, Geotrope, Mageaq, promising projects but no future developments... 146
6.3. IASI and IASI-NG, for meteorology, atmospheric composition and climate 147
6.4. Physical properties of the atmosphere 151
6.4.1. Aerosols and clouds: PARASOL, CALIPSO and the A-Train 152
6.4.2. Next: 3MI and EarthCare. 154
6.4.3. A study in the longer term: ACCP. 155
6.4.4. Megha-Tropiques and rainfall 156
6.5. Additional facilities and means of observation 157
6.6. The role of numerical models. 159
6.7. References 160

Chapter 7. A Coordinated International Effort 163
Jérôme LAFEUILLE

7.1. The challenges of international coordination 163
7.2. Multilateral coordination instances. 165
7.2.1. Overview 165
7.2.2. The World Weather Watch and its space component 165
7.2.3. CGMS 169
7.2.4. CEOS 172
7.3. The benefits of coordination 174
7.3.1. Mission continuity 174
7.3.2. Intercalibration of instruments in orbit 175
7.3.3. The climate observation strategy 177
7.3.4. Use of the radio frequency spectrum 178
7.3.5. Access to data 179
7.3.6. Bilateral cooperation 181
7.4. An extended community of space operators. 182
7.4.1. A growing number of national operational agencies. 182
7.4.2. The emergence of the private sector. 183
7.5. Conclusion 184
7.6. References 184

Part 2. The Physical Basis 187

Chapter 8. Satellite Orbits for Atmospheric Observation 189
Michel CAPDEROU

8.1. Introduction 189
8.2. Preliminaries 190
8.3. Satellites in low Earth orbit 192

8.3.1. Orbital characteristics 192
8.3.2. Sun-synchronous satellites 194
8.3.3. Non-Sun-synchronous satellites 200
8.3.4. Recurrent satellites 200
8.3.5. Spatio-temporal sampling 202
8.3.6. Collaboration with LEO satellites 208
8.4. Satellites in geostationary orbits 209
8.4.1. Orbit characteristics 209
8.4.2. Observation conditions 210
8.5. Other types of orbits used 211
8.5.1. Satellites in HEO orbits 211
8.5.2. Uses of satellites in MEO orbit 212
8.6. References 213

Chapter 9. Measurement Physics 215
Clémence PIERANGELO, Fatima KARBOU and Claude CAMY-PEYRET

9.1. Physical principles of observation of the atmosphere by satellite 215
9.1.1. Basic principles of remote sensing 215
9.1.2. Absorption, scattering, emission 218
9.1.3. Spectroscopy of gaseous species 219
9.1.4. Optical properties of particles 220
9.1.5. At the surface: reflection and emission 222
9.1.6. Spectroscopic parameter database 224
9.1.7. Aerosol and cloud databases 224
9.1.8. Atmospheric profile databases 224
9.1.9. Surface databases 225
9.2. Radiative transfer equation 225
9.2.1. Differential RTE 225
9.2.2. Integration of the RTE 226
9.2.3. Polarized RTE 228
9.2.4. Recent advances for radiative transfer 229
9.2.5. RTE analysis and implications for space-based remote sensing of the atmosphere 229
9.2.6. Example: the 4A/OP source code 232
9.3. Passive optical sensors: radiometers and spectrometers 233
9.3.1. Radiometers 234
9.3.2. Spectrometers 235
9.3.3. Level 1 processing 238
9.3.4. The sensors of the future 238
9.4. Active optical sensors: lidars 239
9.4.1. Lidar principle 239
9.4.2. Lidar equation 240

9.4.3. Different types of spatial Lidar 240
9.4.4. Comparison of optical sensors 246
9.5. Passive and active microwave sensors 247
9.5.1. Specificities of microwave sensors 247
9.5.2. Passive microwave sensors. 247
9.5.3. Active microwave sensors 249
9.5.4. List of microwave instruments. 249
9.6. References 249

Chapter 10. The Inverse Problem and Techniques for Atmospheric Variable Retrieval 253
Clémence PIERANGELO

10.1. General remarks on the inversion of atmospheric parameters 253
10.2. Matrix expression of the direct problem 254
10.2.1. Matrix expression 254
10.2.2. Linearization of the problem 255
10.2.3. Typical dimensions of the problem 255
10.3. Solutions to the inverse problem 256
10.3.1. Least squares. 256
10.3.2. Probabilistic methods 258
10.3.3. Methods with pre-calculated bases. 262
10.4. References 265

Appendices 267

Appendix 1. 269
Claude CAMY-PEYRET

Appendix 2. 277
Claude CAMY-PEYRET

Appendix 3. 287

Appendix 4. 301

Glossary 307

List of Authors 321

Index. 325

Summary of Volume 2 329

Acknowledgments

The coordinator is very grateful to Didier Renaut, Hervé Roquet and Claude Camy-Peyret for their significant help in editing this book through their advice, proofreading, corrections to chapters and help with the organization. Their support has been invaluable in bringing together quality contributions in a book that will cover the different facets of the topic *Space and Atmosphere* in a comprehensive manner. Gratitude is also expressed to Jérome Lafeuille and Jean Pailleux for their proofreading and improvements to various chapters. Finally, special thanks to the journal *La Météorologie* for having authorized the use of the glossary established for the special issue of the journal, No. 97 in 2019, devoted to satellite missions for meteorology and climate.

List of Acronyms

(3D/4D)-Var	(3/4 Dimensional) – Variational
4A/OP	Automatized Atmospheric Absorption Atlas/OPerational
AC-VC	Atmospheric Composition Virtual Constellation
ACCP	Aerosols, Clouds, Convection and Precipitation
ACX	GeoXO Atmospheric Composition
ADM	Atmospheric Dynamics Mission
AERIS	Data and Services for the Atmosphere
AERONET	AErosol RObotic NETwork
AI	Artificial Intelligence
AMV	Atmospheric Motion Vector
AOD	Aerosol Optical Depth
APHRODITE	Asian Precipitation – Highly-Resolved Observational Data Integration Towards Evaluation

APT	Automatic Picture Transmission
ARTS	Atmospheric Radiative Transfer Simulator
ASI	Agenzia Spaziale Italiana (Italian Space Agency)
AVCS	Advanced Vidicon Camera System
BC	Black Carbon
BIPM	International Bureau of Weights and Measures
BRDF	Bidirectional Reflectance Distribution Function
C3IEL	Cluster for Cloud evolution, ClImatE and Lightning
C3S	Copernicus Climate Change Service
CAMS	Copernicus Atmosphere Monitoring Service
CAPE	Convective Available Potential Energy
CALIPSO	Cloud-Aerosol Lidar and Infrared Pathfinder Satellite Observation
CATS	Cloud-Aerosol Transport System
CCI	Climate Change Initiative
CCSDS	Consultative Committee on Space Data Systems
CDR	Climate Data Record
CDRD	Cloud Dynamics and Radiation Database
CEOS	Committee on Earth Observation Satellites
CFC	ChlorFluoroCarbon
CGMS	Coordination Group for Meteorological Satellites
CI	Convection Initiation
CIMSS	Cooperative Institute for Meteorological Satellite Studies

CLW	Cloud Liquid Water
CM	Cloud Mask
CM-SAF	Climate Monitoring – Satellite Application Facility
CMA	Chinese Meteorological Agency
CMEMS	Copernicus Marine Environment Monitoring Service
CMORPH	CPC MORPHing technique
CMV	Cloud Motion Vector
CNES	Centre national d'études spatiales France
CNRM	Centre national de la recherche météorologique France
CNRS	Centre national de la recherche scientifique France
CNSA	Chinese National Space Administration
COCCON	COllaborative Carbon Column Observing Network
CoMet	Carbon Dioxide and Methane field campaign
CONTRAIL	Comprehensive Observation Network for TRace gases by AIrLiner
COP	Conference Of the Parties
CORRA	COmbined Radar-Radiometer Algorithm
CPC	Climate Prediction Center
CRM	Cloud Resolving Model
CRTM	Community Radiative Transfer Model
CTTH	Cloud Top Temperature and Height
CWDP	Commercial Weather Data Pilot

CYGNSS	CYclone Global Navigation Satellite System
DAOD	Differential AOD
DBNet	Direct Broadcast NETwork WMO-OMM
DCS	Data Collection System
DDA	Discrete Dipole Approximation
DIAL	DIfferential Absorption Lidar
DISORT	DIScrete Ordinate Radiative Transfer
DLR	Deutsches Zentrum für Luft- und Raumfahrt Allemagne
DMSP	Defense Meteorological Satellite Program
DNB	Day and Night Band
DOFS	Degrees of Freedom for Signal
DU	Dobson Unit
DVB	Digital Video Broadcast
DVB-S	Digital Video Broadcast by Satellite
EC	European Commission
ECMWF	European Centre for Medium-term Weather Forecast
ECOSTRESS	Ecostress Spectral Library Database
ECV	Essential Climate Variable
EDR	Environmental Data Record
EE	Earth Explorer
EEA	European Environment Agency
EF	Emission Factor

ELDO	European Launcher Development Organisation
ENSO	El Niño Southern Oscillation
EOS	Earth Observing System
EOSDIS	EOS Data Information System
EPS	EUMETSAT Polar System
ERA	ECMWF ReAnalysis
ERB	Earth Radiation Budget
ESA	European Space Agency
ESE	Earth Science Enterprise
ESRO	European Space Research Organisation
ESSA	Environmental Science Services Administration
ESSP	Earth System Science Pathfinder
ESTO	Earth Science Technology Office
EUMETSAT	EUropean organisation for the exploitation of METeorological SATellites
EW	Earth Watch
EXIM	EXtrapolated IMagery products
FASTEM	FAST Emissivity Model
FCDR	Fundamental Climate Data Record
FCI	Flexible Combined Imager
FCover	Fractional Cover (of vegetation)
FDR	Fundamental Data Record

FGGE	First GARP Global Experiment
FFT	Fast Fourier Transform
FGGE	First Global GARP Experiment WMO-OMM
FIR	Far IR
FOR	Field Of Regard
FOV	Field Of View
FSOI	Forecast Sensitivity Observation Impact
FSR	Forecast Sensitivity to R (the observation error covariance matrix)
FT	Fourier Transform
FTS	Fourier Transform Spectrometer
FY	Feng Yun
GADS	Global Aerosol Data Set
GARP	Global Atmospheric Research Programme WMO-OMM
GAS	GMES Atmospheric Service
GASP	GOES Aerosol/Smoke Product
GAW	Global Atmospheric Watch
GCM	Global Climate Model
GCOS	Global Climate Observing System
GDP	Gross Domestic Product
GEO	GEostationary Orbit
GEO	Group on Earth Observation
GEOS-5	Goddard Earth Observing System model – version 5

GEOSS	Global Earth Observation System of Systems
GeoXO	GEOstationary eXtended Observations
GEWEX	Global Energy and Water Cycle Experiment
GHCN	Global Historical Climatology Network
GHG	GreenHouse Gas
GLM	Geostationary Lightning Mapper
GMES	Global Monitoring for Environment and Security
GMF	Geophysical Model Function
GNP	Gross National Product
GNSS	Global Navigation Satellite System
GOCCP	GCM-Oriented CALIPSO Cloud Product (CALIPSO-GOCCP)
GPCC	Global Precipitation Climatology Center
GPCP	Global Precipitation Climatology Project
GPI	GOES Precipitation Index
GPM-CO	GPM-Core Observatory
GPROF	Goddard PROFiling algorithm
GPS	Global Positioning System
GRUAN	Global Reference Upper-Air Network GCOS
GSFC	Goddard Space Flight Center
GSICS	Global Space-based Inter-Calibration System
GSIP	GOES Surface and Insolation Project
GSMaP	Global Satellite Mapping of Precipitation

GSOD	Global Summary Of the Day
GV	Ground Validation
H-SAF	Hydrology – Satellite Application Facility
HCFC	HydroChloroFluoroCarbon
HEO	High Eccentricty Orbit
HEPAD	High Energy Proton and Alpha Detector
HFC	HydroFluoroCarbon
HLOS	Horizontal Line Of Sight
HOAPS	Hamburg Ocean Atmosphere Parameters and fluxes from Satellite data
HRIT	High Rate Image Transmission
HRPT	High Resolution Picture Transmission
HSRL	High Spectral Resolution Lidar
IAGOS	In-service Aircraft for a Global Observing System
IBTrACS	International Best Track Archive for Climate Stewardship
ICOADS	International Comprehensive Ocean-Atmosphere Data Set
ICOS	Integrated Carbon Observation System
IEC	International Electrotechnical Commission
IF	Intermediate Frequency
IFS	Integrated Forecasting System
IIP	Instrument Incubator Program

IJSP	International Joint Polar System
ILS	Instrument Line Shape
IMD	Indian Meteorological Department
IMERG	Integrated Multi-satellitE Retrievals for GPM
IMS	Ice Mapping System
IMS	Interactive Multisensor Snow
INDOEX	INDian Ocean EXperiment
INPE	Instituto Nacional de Pesquisas Espaciais Brésil
InVEST	In-space Validation of Earth Science Technology
IOC	Intergovernmental Oceanographic Commission UNESCO
IODC	Indian Ocean Data Coverage
IPCC	Intergovernmental Panel on Climate Change
IPDA	Integrated Path Differential Absorption
IPSL	Institut Pierre-Simon Laplace France
IPWG	International Precipitation Working Group CGMS
IR	Infra-Red
IROWG	International Radio Occultation Working Group CGMS
ISA	Israel Space Agency
ISCCP	International Satellite Cloud Climatology Project
ISO	International Standardization Organization
ISRF	Instantaneous Spectral Response Function
ISRO	Indian Space Research Organization

ISS	International Space Station
ITOS	Improved TIROS Operational System
ITWG	International TOVS Working Group CGMS
IUPAC	International Union of Pure and Applied Chemistry
IWP	Ice Water Path
IWWG	International Winds Working Group CGMS
JAXA	Japan Aerospace eXploration Agency
JCSDA	Joint Center for Satellite Data Assimilation
JMA	Japan Meteorological Agency
JPL	Jet Propulsion Laboratory
JPSS	Joint Polar Satellite System
JRA	Japanese ReAnalysis
KMA	Korean Meteorological Agency
LAI	Leaf Area Index
LAMP	Laboratoire de météorologie physique France
LATMOS	Laboratoire atmosphères, milieux, observations spatiales France
LEO	Low Earth Orbit
LI	Lifted Index
LIDORT	LInearized Discrete Ordinate Radiative Transfer
LMD	Laboratoire de météorologie dynamique France
LMT	Local Mean solar Time
LMT	LowerMost Troposphere

LOA	Laboratoire d'optique atmosphérique France
LOS	Line Of Sight
LRIT	Low Rate Information Transmission
LRPT	Low Resolution Picture Transmission
LRR	Laser Retro-Reflector
LSE	Land Surface Emissivity
LSI	Low Stream Approximation
LST	Land Surface Temperature
LTE	Local Thermodynamic Equilibrium
LUT	Look-Up Table
MAG	Mission Advisory Group
MAGIC	Monitoring Atmospheric composition and Greenhouse gases through multi-Instrument Campaigns
MEO	Middle Earth Orbit
MEPED	Medium Energy Proton and Electron Detector
MERRA	Modern-Era Retrospective analysis for Research and Applications
MOP	Meteosat Operational Programme
MOZAIC	Measurement of OZone and water vapour by Airbus In-service airCraft)
MSWEP	Multi-Source Weighted-Ensemble Precipitation
MTP	Meteosat Transition Programme
MW	MicroWave

MWR	MW Radiometer
NASA	National Aeronautics and Space Administration
NCEP	National Center for Environmental Prediction
NDACC	Network for the Detection of Atmospheric Composition Change
NDVI	Normalized Difference Vegetation Index
NEDT	Noise Equivalent Differential Temperature
NESDIS	National Environmental Satellite, Data, and Information Service (NOAA)
NIR	Near InfraRed (~ 1.0 to 0.7 μm)
NMVOC	Non-Methane Volatile Organic Compounds
NOAA	National Oceanic and Atmospheric Administration
NOAA/NESDIS	National Oceanic and Atmospheric Administration Satellite and Information Service
NPP	NPOESS Preparatory Program
NRT	Near Real-Time
NSOAS	National Satellite Ocean Application Service
NSOSA	NOAA Satellite Observing System Architecture study
NSSDC	NASA Space Science Data Center
NUCAPS	NOAA Unique Combined Atmospheric Processing System
NWP	Numerical Weather Prediction
NWP-SAF	Numerical Weather Prediction – Satellite Application Facility

OceanRAIN	Ocean Rain And Ice-phase precipitation measurement Network
ODS	Ozone Depleting Substances
OI	Optimal Interpolation
OLR	Outgoing Longwave Radiation
OLS	Operational Linescan System
OPAC	Optical Properties of Aerosols and Clouds
OPD	Optical Path Difference
OSCAR	Observing Systems Capability Analysis and Review tool
OSE	Observing System Experiments
OSSA	Office of Space Science Applications
OSSE	Observing System Simulation Experiments
OSTIA	Operational Sea surface Temperature and Ice Analysis
PACRAIN	Pacific Rainfall Database
PAN	PeroxyAcetyl Nitrate
PAOB	PAid OBservation
PAR	Photosynthetically Active Radiation
PATMOS-x	Pathfinder ATMOSpheres – eXtended
PERSIANN	Precipitation Estimation from Remotely Sensed Information using Artificial Neural Networks
PERSIANN-CCS	PERSIANN-Cloud Classification System
PM	Particulate Matter
PM2.5	Particulate Matter (diameter smaller than 2.5 μm)

PMW	Passive MicroWave
PNPR	Passive microwave Neural network Precipitation Retrieval
POA	Primary Organic Aerosol
POEM	Polar-Orbit Earth-Observation Mission
PPS	Precipitation Processing System
PRPS	Precipitation Retrieval and Profiling Scheme
PSC	Polar Stratospheric Cloud
QA4EO	Quality Assurance Framework for Earth Observation
QI	Quality Index
QRNN	Quantile Residual Neural Network
R&T	Research & Technology
RDT	Rapidly Developing Thunderstorm
RF	RadioFrequency
RO	Radio Occultation
ROC	Reactive Organic Carbon
ROSCOSMOS	Russian Federal Space Agency
RR	Rain Rate
RSS	Rapid Scanning Service
RT	Radiative Transfer
RTE	Radiative Transfer Equation
RTTOV	Rapid Transmission for TOVs
RTTOV-SCAT	RTTOV with SCATtering

SAF	Satellite Application Facility
SAG	Science Advisory Group
SAGE	Stratospheric Aerosol and Gas Experiment
SAM-II	Stratospheric Aerosol Measurement II
SC	Snow Cover
SCaMPR	Self-Calibrating Multivariate Precipitation Retrieval
SEM	Space Environment Monitor
SFCG	Space Frequency Coordination Group
SI	International System
SIC	Sea Ice Concentration
SNPP	Suomi National Polar-orbiting Partnership
SNR	Signal to Noise Ratio
SOA	Secondary Organic Aerosol
SORCE	SOlar Radiation and Climate Experiment
SOS	Sum Of Squares
SPARC	Stratosphere-troposphere Processes And their Role in Climate
SR	Scattering Ratio
SRSOR	Super Rapid Scan Operations for GOES-R
SSA	Single Scattering Albedo
SSEC	Space Science and Engineering Center
SST	Sea Surface Temperature
STAR	(Center for) Satellite Applications and Research

SWE	Snow Water Equivalent
SWIR	ShortWave InfraRed (~ 2.5 to 1.0 μm)
SZA	Solar Zenithal Angle
Tb	Brightness temperature
TCCON	Total Carbon Column Observing Network
TCHP	Tropical Cyclone Heat Potential
TCWV	Total Column Water Vapor
TED	Total Energy Detector
TEMPEST-D	Temporal Experiment for Storms and Tropical Systems – Demonstration
TEMPO	Tropospheric Emissions: Monitoring of Pollution
TIR	Thermal InfraRed (~ 15 to 5 μm)
TLS	Temperature of the Lower Stratosphere
TLT	Temperature of the Lower Troposphere
TMPA	TRMM Multisatellite Precipitation Analysis
TMS	Temperature of the Mid-Stratosphere
TMT	Temperature of the Mid-Troposphere
TOAR	Tropospheric Ozone Assessment Report
TOR	Tropospheric Ozone Residual
TPW	Total Precipitable Water
TTS	Temperature of the Top Stratosphere
TUS	Temperature of the Upper Stratosphere
TUT	Temperature of the Upper Troposphere

TV	TeleVision
UMORA	Unified Microwave Ocean Retrieval Algorithm
UN	United Nations
UNESCO	United Nations Educational, Scientific and Cultural Organization
UNFCCC	United Nations Framework Convention on Climate Change
UT/LS	Upper Troposphere/Lower Stratosphere
UTH	Upper Tropospheric Humidity
UV	Ultraviolet (~ 400 to 200 nm)
UV-B	Ultraviolet-B (~ 315 to 280 nm)
UVAI	Ultraviolet Aerosol Index
UVN	Ultraviolet, Visible and Near-infrared
VAAC	Volcanic Ash Advisory Centre
VarBC	Variational Bias Correction
VHI	Vegetation Health Index
VIS	VISible
VLIDORT	V-LInearized Discrete Ordinate Radiative Transfer
VOC	Volatile Organic Compound
WCRP	World Climate Research Program
WEFAX	Weather Facsimile
WF	Weighting Function

WMO	World Meteorological Organization
WWW	World Weather Watch

Introduction

Thierry PHULPIN
Retired from CNES and Météo-France, Toulouse, France

Observation of the Earth System from space

Satellite images are now in common use during television weather forecasts and are now also accessible on smartphones. Nowadays, everyone is introduced to this vision of the Earth from space at school. Thanks to satellites and their imagers, planet Earth appears to everyone as a familiar environment and is seen as a system in which all environments are interconnected. When a storm breaks out, its intensity is marked by the depth of the gyre formed by the depression and translates into very powerful winds whose impact on the ocean is very noticeable in data from oceanographic satellites. A volcanic eruption or an earthquake leaves traces in the atmosphere (ash clouds, ionospheric signals, etc.), which can be perceived and monitored. Heavy precipitation visualized using microwave instruments and droughts measured with sounders leave surface footprints that can be detected with satellites intended for monitoring continental resources. The relationship between the nature of the surface (oceans, desert, soil, bare ground, urban environment, crops, forests, etc.) and its state at a given moment (roughness, water state), in connection with atmospheric conditions, is decisive for meteorology via the mechanisms of thermal emission, or evaporation. All of these relationships can be characterized, inferred or measured by satellites, some intended for oceanography and others for monitoring land surfaces, or even for meteorology. These links between the various compartments of the Earth System began to be clearly highlighted as soon as the human or animal populations inhabiting the Earth began to move between the continents, thanks to discoveries by explorers. These discoveries, by allowing the development of a trade economy, led to the establishment of an interdependent system and had a major impact on a planetary scale, particularly on the various ecosystems. We could therefore

Satellites for Atmospheric Sciences 1,
coordinated by Thierry PHULPIN, Didier RENAUT, Hervé ROQUET and Claude CAMY-PEYRET.

already speak of an Earth System as early as the 16th century. However, it is the links between the observed natural phenomena that have made it possible to construct the notion of the climate system. Intuitively, we have perceived over the centuries that many phenomena responsible for the evolution of measurable variables were linked. These concepts can be visualized in diagrams, for example, Figure I.1. However, thanks to satellites, it has been possible to advance in the understanding of the processes (in addition to field studies), quantify certain phenomena at scales not possible from ground measurements alone, establish parametric links between certain variables and ultimately improve the physics of the models and then calibrate or validate these models.

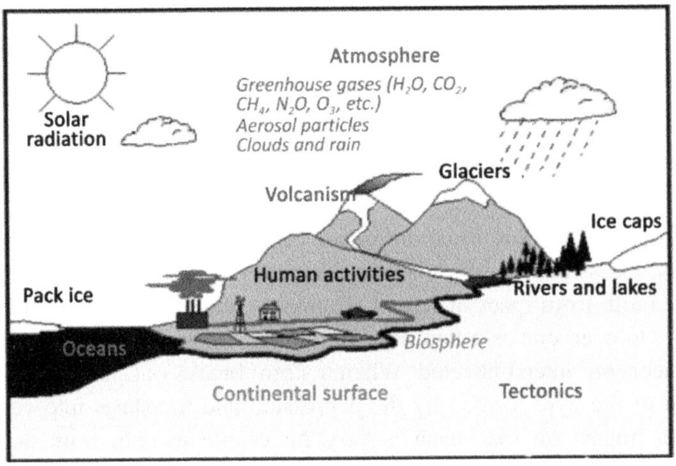

Figure I.1. *The different compartments of the Earth System, taken into account in a climate model. For a color version of this figure, see www.iste.co.uk/phulpin/satellites1.zip*

It is therefore logical to address the various subsystems of the Earth in a work devoted to space science, and to consider them in a coherent approach.

The atmosphere

This volume is devoted to satellite observation of the atmosphere and the sciences it has developed: meteorology, atmospheric composition, as well as knowledge of the climate and its evolution.

According to Blum (2019)[1], it is the mastery of meteorology that would have allowed the WWII allies to land, demonstrating the major strategic importance of investing in an observation network. Space science and technology started early on, having been boosted by know-how of German engineers who developed the V2. The first photos of the Earth were taken in 1960, which highlighted how important the images were for providing an overview of the spatial distribution of cloud cover and its movement. The US then created a program of meteorological satellites, preceding the Soviets, followed by Europe and Japan. The history of these developments will be discussed in Chapter 1. For meteorology, it is useful for the satellite measurements to be taken at the same solar time, to better follow the evolution of the thermodynamic parameters by comparing them with the readings of the previous day at the same time, hence the use of Sun-synchronous polar orbits. On the other hand, the ability to observe the Earth continuously from geostationary orbit is also very attractive for deducing information about the wind from the apparent movements of clouds and observing the development of storms or the movement of cyclones. The orbits used for these missions and their advantages are described in Chapter 8. It quickly became clear that the long-term continuity of observations under identical conditions is necessary to integrate these observations into the international network of meteorological measurements and to facilitate their use in forecasting models. We then saw a dichotomy of missions appear; on the one hand, a series of identical operational satellites for a program spanning 20 or 30 years and, on the other hand, satellites equipped with innovative instruments for research. In the US, this dichotomy is reflected in the division of responsibilities between NOAA and NASA. These agencies pioneered meteorology from space. The history of their programs to date is presented in the first chapters of this book, comprising Part 1. This part shows how the programs are developed to meet the needs of users, using the most advanced technologies whose developments are supported by the agencies that lead in this field. Following the American agencies, Chapters 4–6 are devoted to the European programs of ESA, EUMETSAT and CNES. We see in Chapter 7 the role played by the World Meteorological Organization in coordinating operational projects from the US, Europe or Asia. The chapter also discusses the issue of data acquisition anywhere on the planet for real-time use.

Part 2 of this book is devoted to the physical basis of satellite observations for atmospheric sciences. Chapter 8 presents the principles of orbital mechanics to determine how to meet the need for observations by meteorologists or orbits more suited to different issues such as monitoring precipitation in intertropical zones or air quality in the most polluted regions. The first operational satellites only offered

1. Blum, A. (2019). *The Weather Machine. How We see in the Future*. The Bodley Head, London.

observation in a visible “channel”, similar to a photograph from the sky, and another in an infrared channel giving an idea of the temperature of the objects targeted. These passive observations have allowed advances in cloud mapping and sea temperature but have proven insufficient to access other meteorological information. The range of observations was quickly extended with observations in multiple optical channels, and the observation domain was extended to the microwave domain, allowing for, due to the digitization of measurements, the transition to the quantitative use of observations. Measurement physics, developed to propose passive or active measurement techniques and better use the resulting data, is presented in Chapter 9. In Chapter 10, we present the problem of finding atmospheric variables of interest from the observations and the main methods used for this purpose.

Part 1 of Volume 2 is more specifically devoted to meteorology. With regard to the three-dimensional distribution of temperature, humidity and wind, the main data useful for meteorology has long been provided by radiosondes. However, these observations are rare over the oceans and are therefore supplemented with satellite measurements of vertical distributions of temperature and humidity. The technique used is based on the selective absorption of carbon dioxide or oxygen, which is scarce but evenly distributed in the atmosphere (at least to the first order), and of water vapor and other greenhouse gases, which strongly absorb infrared or microwave radiation. This method, first tested by the Americans with the Infrared Interferometer Spectrometer (IRIS*) on the Nimbus IV satellite, then gave rise to the TIROS Operational Vertical Sounder (TOVS*) radiometric sounders, before technically evolving towards much more precise spectrometric measurements with the Atmospheric Infrared Sounder (AIRS*) and Infrared Atmospheric Sounding Interferometer (IASI*). Chapters 1 to 3 of Volume 2 present the restitution of the main variables characterizing the state of the atmosphere, or that of the surface variables directly useful or necessary for the inversion of the atmospheric characteristics. Chapters 4 to 6 of Volume 2 focus on the major applications of these observations: numerical weather forecasting, short-term forecasting and cyclone tracking.

Atmospheric sounding missions using spectrometry are successful by analyzing minority atmospheric components that are important for the planet or for human health, such as stratospheric ozone, gases and particles that are involved in air quality or the greenhouse effect.

The use of space observations and the benefits derived from them are presented in Part 2 of Volume 2: detection of polluting species dangerous to health (Chapters 7

---

*See Glossary.

and 8, Volume 2), observation of clouds of desert dust and elements emitted by bush or forests. Chapter 11 of Volume 2 is devoted to stratospheric chemistry, better known and understood thanks to the many satellite observations devoted to monitoring the hole in the ozone layer.

Because they offer precise and well-targeted observations with global spatial coverage, satellites have contributed significantly to better understanding the processes at play in the Earth's climate. In addition, the continuity of long-term observations (more than 30 years of data!) makes it possible to monitor the evolution of the climate and to attribute certain phenomena to this evolution. Part 3 of Volume 2 concerns the use of satellite observations for the analysis of climatic processes and the monitoring of climatic variables. In its introduction, Chapter 12 of Volume 2 recalls the importance of well-calibrated data, the need to reprocess the data acquired over time in a homogeneous manner and to repeat the reprocessing of the products as and when more efficient algorithms and processes become available. Chapter 13 of Volume 2 is devoted to the restitution of the greenhouse gases CO_2 and CH_4 and to the location of the sinks and sources of these gases, which play a major role in the warming of the atmosphere. Clouds and water vapor also play an essential role. However, these variables have been monitored from the very first satellites. Analytical techniques have been refined and new data produced with more precise instruments have arrived, making it possible to more precisely estimate the spatio-temporal distribution of atmospheric water (Chapter 14 of Volume 2). Clouds and water vapor are also involved in precipitation: the distribution and intensity of which are regularly monitored (Chapter 15 of Volume 2).

In addition to an index, this book also includes a glossary to clarify the definition of certain concepts, appendices including a compilation of the names of instruments and satellites in question (to be updated as and when new missions occur), as well as a list of acronyms.

PART 1

Satellite Observation of the Earth's Atmosphere: International Cooperation

1

History of Meteorological Satellites

Sylvain LE MOAL
Centre de météorologie spatiale, Météo-France, Lannion, France

1.1. The beginnings of remote sensing and the conquest of space

For centuries, the atmosphere has remained enigmatic, its movements are difficult to decipher from the ground. Scientists interested in weather, and therefore in meteorology, sought to unravel its mysteries. At first, they could only gain altitude by climbing mountains, loaded with measuring instruments. Later, they operated flying instruments (balloons, kites, planes, rockets, etc.), measuring and recording atmospheric parameters: temperature, pressure, humidity, wind direction and speed. Thanks to on board cameras, it was finally possible to see the Earth from the sky. Among so many other scholars, Pascal, Gay-Lussac, Mendeleïev, Teisserenc de Bort, Bjerknes and Idrac were part of this quest.

On October 23, 1858, Félix Tournachon, known as Nadar, a great photographer and ballooning enthusiast, took the first aerial photograph of Paris from a tethered balloon 80 m high. He can be considered the father of remote sensing. Today, it is unfortunately impossible to get your hands on this shot. In Nadar's footsteps, two years later, on October 13, 1860, James Wallace Black photographed part of the city of Boston from a hot air balloon. This first aerial view taken in America is kept at the Metropolitan Museum of Art in New York. The work is titled "Boston, as seen by eagles and wild geese".

For a color version of all the figures in this chapter, see www.iste.co.uk/phulpin/satellites1.zip.

Satellites for Atmospheric Sciences 1,
coordinated by Thierry PHULPIN, Didier RENAUT, Hervé ROQUET and Claude CAMY-PEYRET.

It was not until October 24, 1946 that a photograph of the Earth seen from more than 100 km above sea level was obtained using a 35 mm camera on board a German-made V2 rocket at the end of World War II. Taking off from the White Sands Missile Range (New Mexico), it captured the roundness of the Earth and, for the first time, showed cloud cover from space.

Figure 1.1. *October 24, 1946 – first photo taken from space (source: Wikimedia commons)*

From 1947 to 1950, experiments of this type multiplied in the United States thanks to the V2 rockets, then Aerobee and Viking. They gave rise to a classified report by the Rand Corporation in April 1951 entitled "Inquiry into the feasibility of weather reconnaissance from a satellite vehicle". This report shows the benefit that future satellites would have in facilitating the recognition of clouds, to delimit air masses, to estimate temperatures and winds, and to help in the drawing and analysis of weather maps at the synoptic scale*.

In 1954, photographs taken from a U.S. Navy Aerobee rocket confirmed the value of images produced from space. The structure of the storm coming from the Gulf of Mexico and the distribution of the associated rains were deciphered, whereas with conventional methods this is not possible.

The International Geophysical Year of 1957–1958 marked the beginning of the Space Age. As early as July 1955, the United States and the Soviet Union announced that they were each preparing an artificial satellite for this event. On the

*See Glossary.

American side, the competition was fierce. The U.S. Navy was a candidate, with its Aerobee and Viking rockets, as well as the U.S. Air Force and the U.S. Army. The U.S. Army had in its ranks Wernher von Braun and his team of German engineers who designed the V2s, who moved to the West at the end of the war. Von Braun outlined his plan to launch the first artificial satellite using the Redstone missile as a launcher, directly derived from the German V2. On this basis, the U.S. Army launched the Orbiter project. As President Eisenhower did not want the first launch of an American satellite to be entrusted to a predominantly German team, the Orbiter project was abandoned in favor of the U.S. Navy's project Vanguard, which was less accomplished than its competitor, and the von Braun team was banned from any attempt to launch a satellite. Vanguard was a fiasco, with nine failures on 14 attempts (including the first two test launches).

Figure 1.2. *October 5, 1954 – composite image of a storm (source: Wikimedia commons)*

On the Soviet side, Sergei Korolev was developing missiles with increasing capabilities from the V2. On August 21, 1957, the intercontinental ballistic missile R-7 Semiorka made its first successful flight. It was chosen to place the first Soviet satellite in orbit. The launch of Sputnik took place on October 4, 1957 from what would become the Baikonur Cosmodrome (Kazakhstan). Sputnik was an aluminum sphere 58 cm in diameter, equipped with four antennas and weighing approximately 84 kg. Its payload is a radio transmitter whose signal can be picked up by radios around the world. The Americans found this launch traumatic. According to them, it highlighted that the Soviet Union had the technology to attack the United States with a nuclear projectile. The Sputnik crisis triggered a space race between the two parties. The dog Laïka went aboard Sputnik-2 on November 3, 1957. The United States then entrusted the Jet Propulsion Laboratory (JPL) and the U.S. Army, therefore von Braun, with the mission to make up for the gap between them and the Soviets. Three months was enough. Launched from Cape Canaveral (Florida) by a

Juno-I rocket on January 31, 1958, Explorer-1 was the first American artificial satellite. Its data led to the discovery of Van Allen belts (torus-shaped regions in space, very dense in energetic particles). In order to bring order to the various initiatives and to pilot space programs and aeronautical research in the United States, the National Aeronautics and Space Administration (NASA) was founded on July 29, 1958.

Verner Suomi and Robert Parent, both from the University of Wisconsin, managed to convince von Braun to board Explorer-7 with the radiometer they had developed. Launched on October 13, 1959 from Cape Canaveral, it was the first satellite associated with meteorology. The radiometer consisted of five bolometers that converted the incident electromagnetic energy flux into heat. Its objective was to measure the radiated power by separating the reflected solar radiation from the radiation emitted by the Earth and its atmosphere. It was the first satellite to provide information on the Earth's radiation budget, which is a still current concern.

1.2. It all began with Tiros-1, the first meteorological satellite

In April 1958, following President Eisenhower's announcement to create NASA, the Department of Defense (DoD) founded the Advanced Research Projects Agency (ARPA), which began to develop a meteorological satellite, before NASA took over its cruising speed. After a historic meeting at the Pentagon, ARPA formed a committee chaired by William Kellogg to define the content of the payload of future satellites. A camera made of a newly developed Vidicon tube was chosen. The main challenge was to be able to take photographs with this camera attached to a rotation-stabilized satellite; the solution was to mount it on the rotation axis. Originally, Kellogg wanted to have three cameras, from the wide-angle camera to the most precise. The latter was refused because of its horizontal resolution of 100 m, which was too detailed and therefore militarily sensitive. This was the beginning of the Television Infrared Observation Satellite (TIROS) program. TIROS was finally transferred to NASA in early 1959.

Tiros-1 thus marked the beginning of the space age in meteorology. It was launched on April 1, 1960, early in the morning, from the Cape Canaveral base in Florida, by a Thor-Able rocket. This 122 kg satellite, 56 cm high and 107 cm in diameter, revolved around the Earth at an altitude of 720 km in an orbit inclined at 48.4° from the plane of the equator. It took pictures at scheduled intervals. Its two cameras targeted the same region with different field widths. The first had a wide angle and covered an area of 1,200 km by 1,200 km with a spatial resolution of approximately 3 km; the second was equipped with a zoom for an area of 120 km by 120 km with a resolution of 300 m to 800 m. Each camera had its own tape recorder and could store up to 32 images. The recorded images were transmitted when Tiros-1 passed over one of the two receiving stations. At the same time, the image being created was sent in real time. The first station was the U.S. Air Force Tracking Station at Kaena Point (Hawaii), the second being the U.S. Army Signal Research and Development Laboratories at Fort Monmouth (New Jersey). At both stations, the images were recorded on film and photographed for immediate processing by meteorologists who carried out nephanalyses. These were forwarded to the Weather Bureau by fax. They generated little enthusiasm among forecasters, since their reading required a little effort to draw a meteorological interpretation. On April 9, 1960, Tiros-1 transmitted the image of a hurricane in the Coral Sea, off the northeast coast of Australia. This was the first of a long series, the hurricanes no longer escaping the eye of the satellites. Tiros-1 only worked for 78 days, falling victim to a power failure after having transmitted, in two and a half months, 22,952 images leading to 333 nephanalyses. On April 27, 1961, Francis Reichelderfer, head of the Weather Bureau declared: "For the first time, Man has a complete view of weather conditions over a large part of the Earth and the data collected by Tiros-1 would have required thousands of weather ships scattered over all the oceans".

Figure 1.3. *April 1, 1960 – first image of Tiros-1 – Coasts of Maine and Maritime Provinces of Canada (source: NASA)*

Nephanalysis is the graphical interpretation of cloud data on a map. At the very beginning, the interpretation involved analyzing photographs using two basic criteria, the degree of cloud brilliance and cloud shape, in order to plot on a map the outline of the cloudy sectors and the symbols corresponding to the cloud classification and cloudiness. This analysis was significantly improved in 1970 by the use of infrared images, in addition to images of the visible channel. See example in Figure 1.5.

1.3. American meteorological satellites

1.3.1. *Polar-orbiting satellites*

1.3.1.1. *Tiros – 1st generation*

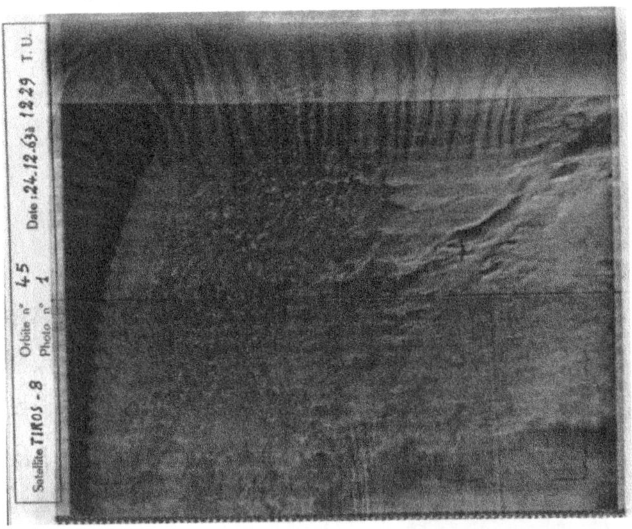

Figure 1.4. *December 24, 1963 – first image received in Europe from a meteorological satellite, at the CEMS in Lannion (source: Météo-France)*

Ten satellites made up the first series of Tiros, whose launches were spread over five years. Tiros-8 was equipped with automatic picture transmission (APT), real-time image broadcasting, which allowed many meteorological services to receive the images. The cost of a receiving station was low, which allowed many countries to have them. This was the case of the Center for Space Meteorological Studies (CEMS in French), newly created in Lannion by the French National Meteorological Department. Good signal reception assumes that the antenna is pointed towards the satellite and follows it as it moves. In practice, a readjustment every 30 seconds is sufficient but the tracking is manual by following the position of the satellite using the ephemeris provided by NASA. As a nice Christmas present, the first image of a portion of orbit no. 45 of Tiros-8 was received on December 24, 1963 in Lannion around 12:30 UTC. The CEMS is the first center in Europe to receive an image from a meteorological satellite. The interpretation is tricky and requires a bit of imagination to extract meteorological information! Tiros-9 and Tiros-10, which complete the program, were the first to be positioned in a sun-synchronous orbit.

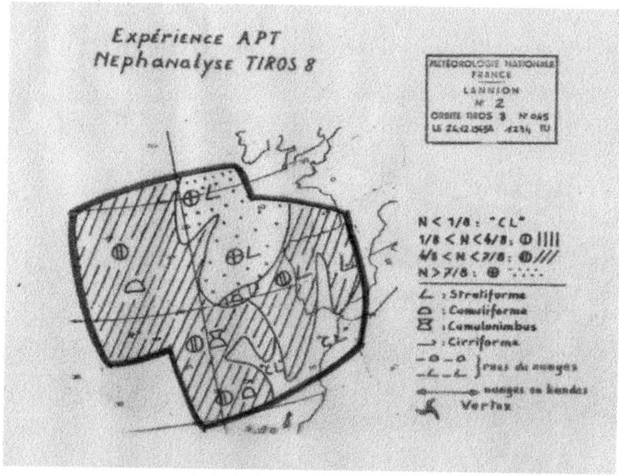

Figure 1.5. *December 24, 1963 – nephanalysis performed by CEMS meteorologists in Lannion (source: Météo-France)*

1.3.1.2. *Nimbus*

The family of Nimbus satellites, all Sun-synchronous, took over from Tiros. These were seven experimental satellites, designed by NASA, launched between 1964 and 1978. They made it possible to test many instruments, in particular for vertical probing of the atmosphere. The instruments on board varied according to the satellites. In addition to conventional imagers were the high-resolution infrared sounder HIRS**, the infrared interferometer IRIS**, the limb infrared sounder LRIR**, microwave radiometers ESMR** and SMMR**, the spectrometer dedicated to ozone TOMS** and the coastal zone color scanner CZCS**. These instruments were then deployed on operational satellites.

**See Appendix 3.

1.3.1.3. *Tiros operational satellite (TOS) – 2nd generation*

The TOS program included nine satellites named ESSA, named after the administration in charge of the program: Environmental Science Services Administration (predecessor of the National Oceanic and Atmospheric Administration – NOAA). ESSA-1 was put into orbit on February 3, 1966 from Cape Canaveral and carried the same type of Vidicon camera as the last Tiros. ESSA-9, the last of the series, was launched on February 26, 1969.

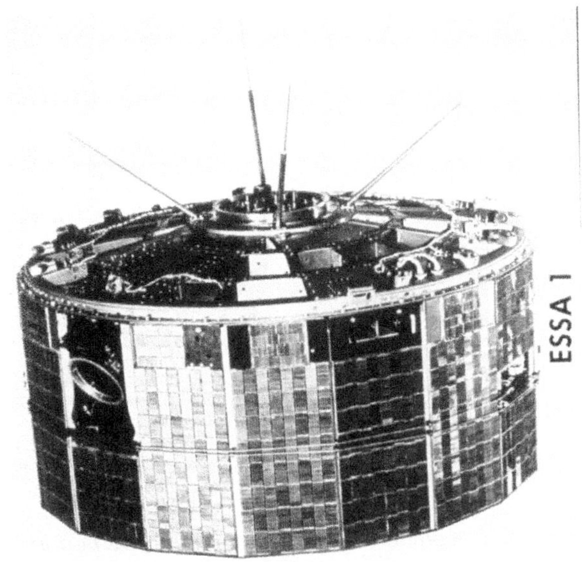

Figure 1.6. *ESSA-1 satellite (source: Wikimedia commons)*

1.3.1.4. *Improved Tiros Operational System (ITOS) – 3rd generation*

Ensuring the continuity of TOS, ITOS was a series of six satellites in flight between 1970 and 1979. ITOS-1, also called TIROS-M, was the first. It was succeeded by NOAA-1. From NOAA-2 onwards, the Vidicon cameras were replaced by a two-channel imaging radiometer, one in the visible spectrum, the second in the thermal infrared (VHRR**). Thanks to this infrared channel, images could therefore be used in the nocturnal part of the Earth. NOAA-2 also carried an eight-channel infrared sounder (VTPR**) whose objective was to measure temperatures at different levels in the atmosphere and the integrated water vapor content.

**See Appendix 3.

1.3.1.5. *From Tiros-N to NOAA-14 – 4th generation*

Tiros-N was the first of a series of 10 satellites (Tiros-N, then the satellites from NOAA-6 to NOAA-14, whose launches span from 1978 to 1994). Launched on October 13, 1978, it carried instruments inherited from the Nimbus, which allowed a major advance in the measurement of atmospheric parameters: a five-channel visible and infrared imager (AVHRR**), an infrared sounder for measuring temperature and humidity (HIRS/2), a microwave sounder (MSU**) and a stratospheric sounder (SSU**). These last three instruments constitute the Tiros Operational Vertical Sounder (TOVS**). All of these instruments are described in Chapter 2.

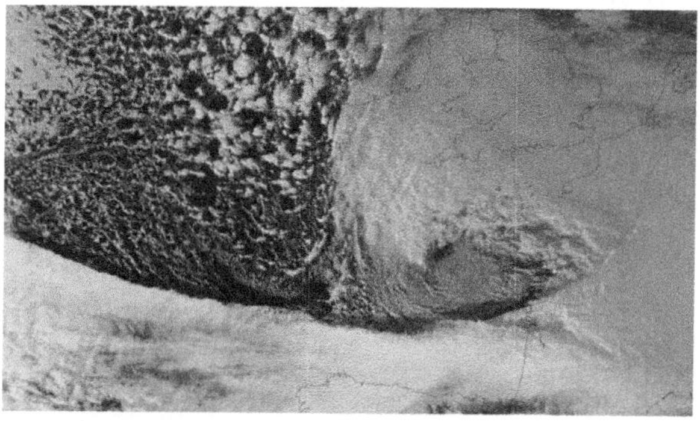

Figure 1.7. *December 27, 1999 at 16:00 UTC – NOAA-14 – image of storm Martin (source: Météo-France). For a color version of this figure, see www.iste.co.uk/phulpin/satellites1.zip*

1.3.1.6. *From NOAA-15 to NOAA-19 – 5th generation*

The objectives of this fifth generation of satellites known as Polar Operational Environmental Satellites (POES) were to provide an uninterrupted flow of imagery data, atmospheric sounding, hydrological and surface information, and ozone observations; to establish a long series of data for climate monitoring and a significant contribution to space weather (with its main objective being to understand the influence of the Sun on the Earth's environment). It was a program under the responsibility of NOAA in partnership with NASA.

With a take-off weight of 2,232 kg, NOAA-15 was put into orbit on May 13, 1998. It carried more efficient instruments than those of the previous generation: the AVHRR/3 imager, the HIRS/3 infrared sounder, AMSU-A** and AMSU-B microwave sounders.

**See Appendix 3.

Thanks to APT and High Resolution Picture Transmission (HRPT) data transmission systems, the fifth-generation NOAA satellites were received by thousands of users around the world. NOAA-19, launched on February 6, 2009, completed the program. NOAA-18 and NOAA-19 were part of a new partnership with EUMETSAT to form the Initial Joint Polar System (IJPS) in which the United States provided afternoon orbit coverage, while Europe ensured the morning orbit with its MetOp satellites.

1.3.1.7. *Joint Polar Satellite System (JPSS)*

JPSS was born from the ashes of the National Polar-orbiting Operational Environmental Satellite System (NPOESS) program. The objective of the NPOESS program was to provide, using polar satellites, meteorological, oceanographic, climatic and space data that met the operational requirements of American civilian and military communities. It was canceled in 2010 for budgetary reasons. The civil and military programs are now separate.

Figure 1.8. *Verner Suomi on January 25, 2012 (source: University of Wisconsin)*

JPSS is a series of four satellites and an NPP precursor satellite. The acronym NPP, intended for the NPOESS Preparatory Project, was kept for JPSS under the name National Polar-orbiting Partnership, to which was added Verner Suomi, who was considered the founder of satellite meteorology. This is most often referred to as S-NPP. Scheduled to operate until around 2040, the JPSS satellites are dedicated to the afternoon orbit (see Chapter 8), while the European MetOp satellites carry out the morning orbit.

S-NPP was launched on October 28, 2011 by a Delta II rocket from Vandenberg Air Force Base. Its payload is composed of new instruments, with a 22-channel visible and infrared imager (VIIRS**), a hyperspectral infrared sounder (CrIS**), a microwave sounder for measuring temperature and humidity (ATMS**) and a spectrometer for measuring ozone (OMPS**). These instruments are described in Chapter 2.

Figure 1.9. *May 11, 2020 at 01:29 UTC – Suomi-NPP – day–night channel image (source: Météo-France). For a color version of this figure, see www.iste.co.uk/phulpin/satellites1.zip*

1.3.2. *Geostationary satellites*

1.3.2.1. *Synchronous Meteorological Satellite (SMS)*

SMS was the first operational series of geostationary meteorological satellites. SMS benefitted from experiments carried out by the Applications Technology Satellite (ATS) program developed by NASA in the 1960s. ATS includes technological and instrumental experiments relating to meteorology. ATS-1, ATS-3,

**See Appendix 3.

and then ATS-6, respectively, carried SSCC**, MSSCC** and VHRR imagers. SMS-1 was launched on May 17, 1974 and was placed at 45° W longitude. It was a 2.30 m long cylinder with a diameter of 1.90 m. It was stabilized by a rotation at approximately 100 rpm. Its main instrument was the VISSR** radiometer with two channels (visible and infrared), which produced complete disk images at a frequency of once every 30 min, with a horizontal resolution of 900 m at nadir for the visible channel. It was followed by the SMS-2, which was launched on February 6, 1975 and was placed at 115° W longitude. These geostationary satellites were a real improvement for meteorologists, owing to their imagery. Each satellite could see almost 40% of the Earth's surface and delivered high-frequency images, enabling quality animations essential for monitoring large-scale phenomena in real time.

Figure 1.10. *Artist's impression of the SMS satellite (source: Wikimedia commons). For a color version of this figure, see www.iste.co.uk/phulpin/satellites1.zip*

**See Appendix 3.

1.3.2.2. *Geostationary Operational Environmental Satellite (GOES) – 1st generation*

The first generation of GOES corresponds to the satellites GOES-1 to GOES-7. VISSR was the main instrument of the satellites GOES-1, GOES-2 and GOES-3. GOES-1, launched on October 16, 1975 and initially placed at 55° W, was moved several times, changing in longitude. It was even entrusted to the European Space Agency (ESA) on December 1, 1978 and was relocated to the Indian Ocean for a year to participate in the Global Atmospheric Research Program (GARP).

The VAS** instrument replaced the VISSR in and after GOES-4, which was launched on September 9, 1980. VAS has the particularity of combining atmospheric imagery and sounding of temperature and humidity. It was the first sounding instrument on board a geostationary satellite.

To meet the needs of the United States, the GOES system used two satellites simultaneously, one covering the Pacific Ocean (GOES-West – longitude ~135° W), the other monitoring the Atlantic Ocean (GOES-East – longitude ~75° W).

1.3.2.3. *From GOES-8 to GOES-15 – 2nd generation*

Compared to the previous generation, the stabilization of the platform was now ensured by a three-axis system and the imaging and sounding missions were performed using separate instruments. The imager had five channels and the sounder 19 channels.

Figure 1.11. *August 29, 2005 at 15:00 UTC – GOES-12 – Hurricane Katrina (source: Météo-France). For a color version of this figure, see www.iste.co.uk/phulpin/satellites1.zip*

**See Appendix 3.

GOES-9, launched on May 23, 1995, experienced technical problems from 1997. It was kept in orbit as a backup satellite and replaced by GOES-10. Reactivated in 2003, following the failed launch of the Japanese satellite MTSAT-1, it was moved to 155° E to fill the gap in the geostationary belt before the launch of MTSAT-1R.

1.3.2.4. *GOES – 3rd generation*

Today, it is a new generation satellite that observes the Americas. GOES-R was successfully launched from Cape Canaveral on November 19, 2016. It was declared operational on December 18, 2017 as GOES-16 at 75.2° W longitude. This new-generation program includes four satellites to ensure an observation mission until around 2036. GOES-16 is equipped with the new ABI** imager, which makes it possible to observe the Earth in 16 different spectral bands (compared to five for the previous generation of GOES satellites). The spatial resolution of the images is between 500 m and 1 km for the visible band channels and between 1 and 2 km for the infrared channels. The temporal frequency of the images is increased with full-discs transmitted every 10 min. It even has the ability to scan a targeted 1,000 km x 1,000 km area every 30 seconds to monitor fast-developing hazards. GOES-16 is also equipped with a GLM** instrument, which is an important innovation. This is, in fact, the first operational lightning imager on a geostationary satellite. It complements ground lightning detection networks and anticipates the risk of severe weather, with a rapid increase in the number of lightning flashes being a signal of a sudden deterioration in meteorological conditions.

GOES-17 is located at longitude 137.2° W above the Pacific Ocean. It carries the same payload as GOES-16 but has some cooling issues. As a result, in the infrared, images can be degraded or absent for a few hours at times of the year close to the equinoxes. GOES-18 and -19 will complete the series in future years.

**See Appendix 3.

Figure 1.12. *September 12, 2018 at 12:00 UTC – GOES-16 – Geo-color image (source: Météo-France). For a color version of this figure, see www.iste.co.uk/phulpin/satellites1.zip*

1.4. Russian meteorological satellites

1.4.1. *Polar-orbiting satellites*

In the USSR, the development of the first meteorological satellites was initiated at the request of the military who, during the Cold War, sought to know meteorological conditions across the entire planet for their intercontinental bombers and their war fleet. Soviet strategists do not forget to remind their leaders of all the efforts made by the United States. Convinced, the Soviet government ordered, on October 30, 1961, the development of the experimental network of meteorological satellites, Meteor, assigned to the OKB-586 design bureau in Dnepropetrovsk. In its initial version, the satellite was light enough to be launched by the Cosmos rocket (type 65S3) under development within the same design office. But the OKB-586 played a crucial role in the strategic intercontinental missile program, and its manager Mikhail Yangel decided to transfer the Meteor program, which was considered to be less prestigious, to the Moscow-based VNIIEM research institute. This transfer became effective in May 1962.

On April 13, 1963, VNIIEM launched Cosmos-14, an experimental satellite and test platform for on board electrotechnical systems that allowed the orientation and stabilization of future meteorological satellites. The satellite was a cylinder 1.8 m long and 1.2 m in diameter, with two hemispherical ends. The stabilization system consisted of flywheels driven by electric motors. The kinetic energy of the flywheels was damped by electromagnets interacting with the Earth's magnetic field. This system provided a three-axis-type stabilization and oriented the satellite towards the center of the Earth. A similar test flight was carried out almost eight months later with Cosmos-23 and then by the Cosmos-44 prototype. These three satellites, as well as the flights of Cosmos-58, Cosmos-100 and Cosmos-118, were the first stage of development and led to the first Soviet experimental meteorological satellite, Cosmos-122. The latter was launched on June 25, 1966 and transmitted its images to three receiving stations on the ground. These missions and then those of the following satellites (Cosmos-144, Cosmos-156, etc.) paved the way for the first operational meteorological satellite, Meteor-1-1. Weighing 2,200 kg, Meteor-1-1 was put into orbit by a Vostok-2M launcher on March 26, 1969 in an orbit inclined at 81.2°. Its payload consisted of a radiation budget sensor, a television camera and an infrared instrument. This was the first in a series ending with Meteor-1-27, launched into orbit on April 5, 1977.

Figure 1.13. *Model of the Cosmos-122 satellite at the Air and Space Museum, Paris-Le Bourget (source: Wikimedia commons). For a color version of this figure, see www.iste.co.uk/phulpin/satellites1.zip*

The early 1970s saw the gradual demilitarization of Soviet meteorological satellites. In December 1971, the government decreed that space resources could also be used to meet civilian needs. This decision paved the way for the development of the Meteor-Priroda series of satellites. The first specimen, Meteor-P.-1 (also called Meteor-1-18) was launched on July 9, 1974 from the Plesetsk cosmodrome. It was a three-axis stabilized satellite weighing 3,800 kg. The third satellite in the Priroda series, Meteor-P.-3, put into orbit on June 29, 1977, became the first Sun-synchronous satellite, at an altitude of 630 km and an inclination of 98°. It mainly included two multispectral scanners working in the visible and near-infrared domains for imaging, an infrared interferometer and a microwave sounder to obtain temperatures at different levels in the atmosphere. The sixth and final satellite in the series was launched on July 10, 1981.

At the end of the 1960s, the second generation of Meteor satellites was considered. Despite the technical problems encountered, it resulted in the launch of Meteor-2-1 on July 11, 1975, the first of a series of 21 satellites. Among the four instruments on board, the payload of the Meteor-2 satellites included an instrument consisting of three television camera-type sensors that took measurements in the visible and infrared spectra. Cloud cover information was thus obtained during the day and at night. An infrared sounder was used to reconstruct atmospheric temperature and humidity profiles. Two or three satellites, whose passage times were spaced out, made up the nominal Meteor-2 network, guaranteeing high-frequency observations. Hurricanes, storm cells and vast forest fires no longer escaped forecasters. The satellites stored and then released their data as they passed over the main stations in Moscow, Novosibirsk and Khabarovsk, but data could also be received in real time by autonomous receiving stations. There were about 50 of them across Soviet territory, allied countries and even at sea. In May–June 1978, the icebreaker Sibir, equipped with such a receiver, attempted a pioneering voyage across the ice-covered Arctic Ocean, from Murmansk to the Bering Strait. Meteor imagery was crucial in locating areas where the ice was thinnest, allowing the ship to travel at speeds of up to 22 knots. The last in the Meteor-2 series was put into orbit on August 31, 1993 from the Plesetsk site using a Tsiklon-3 launcher.

The development of the Meteor-3 program began in the early 1980s. It was a series of meteorological–environmental satellites of the USSR RosHydroMet agency. The system comprised six polar-orbiting satellites. The main objectives of the Meteor-3 program were to obtain information about:

– cloud cover, snow on the ground and sea ice;

– the vertical distribution of temperature and humidity in the atmosphere;

– radiation in near-Earth space.

Meteor-3-1 was placed in orbit on October 24, 1985. With some additional instruments, Meteor-3-6 took off from the Plesetsk Cosmodrome on January 25, 1994.

Delayed due to funding problems, the Meteor-3M polar-orbiting satellite program spans the period 2001–2033. Overall responsibility for the program is given to RosHydroMet, with Roscosmos being responsible for the space segment. They are Sun-synchronous satellites equipped with numerous instruments. Among these are (on the Meteor-M N2, launched July 8, 2014):

– two radiometers for a high spatial resolution imaging;

– a 2,670-channel infrared interferometer for atmospheric temperature and humidity sounding, ozone profiling and greenhouse gas detection;

– a microwave radiometer to reconstruct temperature and humidity profiles, even in the presence of clouds, and to collect information on precipitation.

On November 28, 2017, the launch of Meteor-M N2-1 ended in failure. After taking off from the Vostochny Cosmodrome near Tsiolkovsky in Siberia, the Soyuz-2.1b rocket encountered a problem during the separation of its third stage, leading to the loss of the satellite which probably crashed into the North Atlantic Ocean. According to Roscosmos, it was due to a bug in the algorithm of the navigation software. Some blame it on human error: a poor configuration that does not take into account the location of the new Vostochny Cosmodrome instead of Baikonur. The situation caused a commotion and made it to the highest level in the government.

Meteor-M N2-2, launched on July 5, 2019, did not experience the same setbacks and reached its Sun-synchronous orbit at an altitude of 821 km.

1.4.2. *Geostationary satellites*

As far as geostationary satellites are concerned, GOMS-1 (also called Electro-1) was the first Russian experimental satellite. The program began in the early 1970s for a planned launch in 1979. Various problems related to the payload and software development, and then to the break-up of the Soviet Union, delayed the program. Fifteen years later, GOMS-1 was finally launched on October 31, 1994 from the Baikonur Cosmodrome in Kazakhstan. It carried a main instrument: the imaging radiometer with two channels (visible and thermal infrared). It sent images every 30 min from longitude 76.5° E.

Following the GOMS-1 experiment, the Russian government entrusted the company NPO Lavochkin with the development of a new family of geostationary satellites called Electro-L, who inaugurated a new platform. These were three-axis stabilized satellites with a 10-channel imager as the main instrument. Their spatial resolution was 1 km at nadir for the visible spectrum channels and 4 km for the infrared channels. The first satellite in the series was put into orbit on January 20, 2011 at 76° E. It was followed by two other satellites in the series, launched in 2015 and 2019.

Figure 1.14. *Artistic view of the satellite Electro-L (source: ESA). For a color version of this figure, see www.iste.co.uk/phulpin/satellites1.zip*

1.5. European meteorological satellites

1.5.1. *The Meteosat saga*

In the 1960s, France, the United Kingdom, the Federal Republic of Germany, Italy and other European nations embarked, on their own initiative and in a dispersed manner, on space activities. Thanks to its National Center for Space Studies (*Centre national d'études spatiales*, CNES), created on December 19, 1961, France became the third world power, after the Soviet Union and the United States, to have access to space by putting into orbit, with its own launcher, its first satellite on November 26, 1965. Asterix was launched from Hammaguir, Algeria, using a Diamant-A rocket. Little by little, the idea of a European collaboration germinated and gave birth to two organizations with complementary objectives: the European Space Research Organization (ESRO) and the European Launcher Development Organization (ELDO), which were operational from 1964. However, it was not until the creation of the European Space Agency (ESA) that European space activities really took off. The convention creating ESA was signed on May 30, 1975 by 11 Member States. As the successor to the ELDO, ESA supports the development of European launchers. Ariane-1 made its first flight on December 24, 1979.

Pierre Morel had the idea for the Météosat project around Christmas 1968 when he and his colleagues from the CNRS Dynamic Meteorology Laboratory were participating in a balloon launch campaign in the middle of the Australian desert. He remembered the ATS-1 geostationary satellite and Verner Suomi's idea to place a 15 cm aperture telescope there. Like its operational successors, ATS-1 provided an animated view of atmospheric circulation from its position nearly 36,000 km away, fixed relative to Earth. Taking up Pierre Morel's idea, CNES is enthusiastically studying the possibility of launching a geostationary meteorological satellite as part of the World Weather Watch and as part of the GARP of the World Meteorological Organization (WMO) to complete the belt of geostationary satellites.

Figure 1.15. *Professor Pierre Morel (source: ESA–CNES)*

For both the French government and Jean-François Denisse, president of CNES, this research mission of the French space agency was not compatible with the implementation of an observation system, such as Meteosat. It was desirable that the National Meteorological Department as well as other European meteorological services actively participated in it. The project was then submitted to ESRO but it immediately came up against a competing project inspired by British meteorologists, whose satellites had to operate in a polar orbit. After some attempts, the proposal of a coherent Earth observation system took preference over multiple smaller regional initiatives. Meteosat was integrated into the ESRO program on March 23, 1972, before being taken over and continued by ESA. Eight Member States decided to fund the program: Belgium, Denmark, France, Italy, the Federal Republic of Germany, the United Kingdom, Sweden and Switzerland. But what name should this "Europeanized" project bear? MetSat, as proposed by ESRO officials, Geostationary European Meteorological Satellite (GEMS) or Meteosat (anglicized, without the accent but close to the initial name)? It was this third option that was ultimately chosen, delighting the French designers.

The technical design of the Meteosat satellite and its MVIRI** instrument was almost complete when CNES handed over control to the European agency's project team. It was a three-channel imager (a visible channel at 0.7 μm, a thermal infrared channel at 11.5 μm and an infrared channel in the water vapor absorption band centered at 6.4 μm). The spatial resolution at nadir is 2.5 km for the visible channel and 5 km for the other two channels. The Earth's disk is scanned every 30 minutes from a satellite stabilized in rotation at 100 rpm. Meteosat is a cylinder 3.2 m high with a diameter of 2.1 m, weighing 626 kg. On this module are the solar panels. It is surmounted by a second cylinder, which is smaller and equipped with antennas.

Produced by the European industrial consortium Cosmos, under the project management of Aerospatiale together with Matra, the Meteosat system comprised three satellites. The first, Meteosat-1, was launched on November 23, 1977 from Cape Canaveral by an American Delta rocket. Its mission was to monitor weather conditions on the Earth's disk comprising the Atlantic Ocean, Europe and Africa. However, certain technical uncertainties remained, in particular that of the cooling of sensitive parts of the telescope. In order to ensure favorable operating conditions, it was planned to flip the satellite at each equinox. The satellite was oriented upside down when it was put into orbit and the image was taken from south to north. A few months later, with the cooling system functioning nominally and with a sufficient margin, the decision was made to continue without flipping the satellite; an operation that could prove to be tricky. The following Meteosat satellites remained oriented upside down. At present, with the second-generation Meteosat satellites (MSG), the scanning of the disk is always done from south to north, which is a tenacious European peculiarity!

**See Appendix 3.

Figure 1.16. *First-generation Meteosat satellite (source: ESA). For a color version of this figure, see www.iste.co.uk/phulpin/satellites1.zip*

Figure 1.17. *December 9, 1977 – Meteosat-1 – first image (source: ESA)*

The Meteosat program was qualified as pre-operational. The technical design of the first three flight models was adapted to the provision of an operational service, but no provision guaranteed the continuity of service. This is what triggered the creation of a new European organization. During a long transition period, the solution was found. The provision of a meteorological service from space and its financing were accepted by Europe. The convention establishing the European Organization for the Exploitation of Meteorological Satellites (EUMETSAT) was adopted in Geneva on May 24, 1983. It was not until June 19, 1986 and the ratification of the national parliaments that the new organization was actually constituted, with 16 Member States (in 2020 EUMETSAT had grown to 30 Member States). EUMETSAT's main objective was to set up, maintain and run European systems of operational meteorological satellites, taking into account as far as possible the recommendations of the World Meteorological Organization. Meanwhile, Meteosat-2 was launched on June 19, 1981 from Kourou (French Guiana) using an Ariane-1 launcher. It took over from the first Meteosat, which broke down prematurely on November 24, 1979. EUMETSAT entrusted ESA with carrying out an initial operational program of three Meteosats on behalf of its Member States as well as for various countries in Africa and the Middle East. ESA also operated the Meteosat-1 and Meteosat-2 satellites, as well as Meteosat-3, put into orbit on June 15, 1988.

Figure 1.18. *February 25, 2007 at 08:00 UTC – Meteosat-5 from its position over the Indian Ocean at 67.5° E longitude – color composition image (source: Météo-France). For a color version of this figure, see www.iste.co.uk/phulpin/satellites1.zip*

Meteosat-4 was therefore the first model of the Meteosat Operational Program (MOP). It was launched on March 6, 1989. Its main instrument, as for all first-generation Meteosat satellites, was the MVIRI imager. This was followed by Meteosat-5 (March 2, 1991) and Meteosat-6 (November 20, 1993). In 1990, in order to ensure continuity between the MOP and Meteosat second-generation (MSG) programs, EUMETSAT adopted the Meteosat Transition Program (MTP). The MTP included the production of the Meteosat-7 satellite by ESA using spare parts from the MOP, the development of a new ground segment, and the operation of the system by EUMETSAT from 1995. Meteosat-7 was launched on September 2, 1997 to ensure the continuity of Earth's observation from the vertical of the Gulf of Guinea, the main position of the Meteosat satellites, generally ensured by the satellite most recently placed in orbit. It should be noted that Meteosat-3, a satellite that became a backup, was moved to 50° W longitude on August 1, 1991 then to 75° W longitude on February 1, 1993 to fill the gap left by NOAA and the delay in the program of second-generation GOES satellites. As of 1998, Meteosat-5, -6 and -7 were successively moved over the Indian Ocean. EUMETSAT therefore simultaneously carried out two operational missions from two orbital positions. The lifetime of Meteosat-7 was exceptionally long (20 years). The last image it transmitted was on March 31, 2017 at 12:00 UTC. On April 3, 2017, Meteosat-7 set off for a final mission, being placed definitively in a so-called graveyard orbit, about 300 km above the geostationary orbit, where satellites are transferred at the end of their working life. This maneuver, which takes about a week, was carried out successfully.

The first model of a second-generation Meteosat satellite, MSG-1, was placed in orbit by the European launcher Ariane-5, which took off from the Kourou Space Center in French Guiana on August 28, 2002. This satellite became operational on January 9, 2004 and was then named Meteosat-8. Four satellites developed by ESA on behalf of EUMETSAT formed the MSG program. Two and a half times larger than their predecessors, they were cylindrical satellites 3.22 m in diameter and 3.74 m long. On take-off, their mass reached two tons, almost half of which is made up of the propellants necessary to put them and keep them in position for the seven years, the time that their mission theoretically lasted. MSG satellites were equipped with new, more powerful and more precise sensors for the continuous observation of the Earth's atmosphere. The SEVIRI** imager was the main instrument of their payload. It was a 12-channel imager in the visible and infrared spectra with a horizontal resolution of 3 km at nadir. One channel, the high-resolution visible (HRV), even achieved a resolution of 1 km. It transmitted images of the Earth's disk every 15 min. When Meteosat-9 became the operational satellite, Meteosat-8 provided a backup, as well as the rapid scanning service (RSS) mission: it scanned the northern third of the disk, therefore mainly Europe, every 5 min. As for first-generation Meteosats, an MSG satellite was moved above the Indian Ocean, at 41.5° E.

The third generation of Meteosat then emerged. This new program includes six geostationary satellites: four for imaging, MTG-I, and two for sounding, MTG-S, which will be placed between longitudes 10° E and 10° W. In addition to state-of-the-art imagery thanks to the FCI** imager, the MTG satellites have new infrared sounding and lightning detection capabilities onboard. MTG-S satellites also carry Sentinel-4**. This is an instrument dedicated to monitoring air quality within the framework of the European Copernicus program, which aims to bring together all the data obtained from environmental satellites and in situ measuring instruments, in order to produce a global and complete view of the state of our planet. This gives Meteosat satellites a new ability to analyze atmospheric chemistry and identify gas concentrations. The launch of the first MTG satellite will be a major milestone for European meteorology.

Figure 1.19. *December 6, 2011 at 12:00 UTC – Meteosat-9 – color composite image (source: Météo-France). For a color version of this figure, see www.iste.co.uk/phulpin/satellites1.zip*

**See Appendix 3.

1.5.2. *46 years after Tiros-1, MetOp enters the scene*

Originally, the MetOp satellites were part of an ESA satellite program called the Polar-Orbit Earth Observation Mission (POEM) intended to succeed ERS-1 and -2. The payload must cover the provision of meteorological data and, more broadly, environmental data. In 1992, the ESA Council decided to split the POEM mission into two programs: MetOp for meteorology and Envisat for the environment. MetOp is a family of three satellites. They form the space segment of the EUMETSAT Polar System (EPS) and constitute the European part of a polar meteorological observation system shared with the United States, the Initial Joint Polar Satellite System (IJPS), whose agreement was signed in November 1998. The MetOp satellites occupy the morning orbit, whereas the NOAA satellites occupy the afternoon orbit. This complementarity allows better coverage of the globe.

It was not until October 19, 2006 that MetOp-A took off from the Baikonur Cosmodrome (Kazakhstan) aboard a Soyuz rocket. Once in orbit, MetOp-A came under the control of ESA's European Space Operations Center (ESOC), which placed it in its final orbit, deploying its solar panel and antennas. EUMETSAT then took control of the satellite for the commissioning phase. Lasting six months, it covered satellite verification operations and the scientific validation of the products generated by its instruments. Once this phase was completed, MetOp-A was declared operational and took over the morning service, replacing the American satellite NOAA-17.

The MetOp satellites travel approximately 830 km from the Earth in a quasi-circular, heliosynchronous orbit, known as the "morning orbit" because they cross the equatorial plane at 9:30 am in the descending direction (9:30 pm in the ascending direction). They collect data on all areas of the globe, even at high latitudes, which geostationary satellites over the equator cannot achieve. The images of the polar zones mixing water and ice are spectacular under the low sun. Low orbit is also used to probe the atmosphere in detail in order to collect temperature and humidity profiles, in particular thanks to the IASI** sounder, the result of a collaboration between CNES and EUMETSAT. IASI is an infrared interferometer of unprecedented precision that also collects information on the chemical components of the atmosphere, including greenhouse gases. The MetOps carry numerous instruments including a radar scatterometer, whose main function is to reconstruct wind measurements at the surface of the oceans. With MetOp-A, then MetOp-B and then MetOp-C, the volume of data that fed numerical weather prediction models increased exponentially.

**See Appendix 3.

Figure 1.20. *Artistic view of the satellite MetOp (source: ESA). For a color version of this figure, see www.iste.co.uk/phulpin/satellites1.zip*

Approved by the EUMETSAT Council in June 2010, EPS-SG (SG for Second Generation) is on the horizon, with the planned launch of the first MetOp-SG-A1 in 2023. It is a program comprising six satellites, three "A" platforms which will carry a suite of passive instruments (imagers and sounders) and three "B" platforms with additional instruments, a radar scatterometer, and microwave imagers in particular.

1.6. Elsewhere

1.6.1. *Japan*

Japan only operates geostationary satellites. Three generations of Himawari satellites (sunflower in Japanese) have succeeded each other since 1977. The first generation, called Geostationary Meteorological Satellite (GMS), included five satellites. It was the result of a partnership between the Japan Meteorological Agency (JMA) and the National Space Development Agency of Japan (NASDA). NASDA was created on October 1, 1969 to develop space technologies and promote the use of space for peaceful purposes. GMS-1 was launched on July 14, 1977 using an American Delta rocket, a few months before the European satellite Meteosat-1. The GMS are spin-stabilized satellites and carry the American VISSR imager with two channels as their main instrument. Like its successors, GMS-1 was positioned at 140° E longitude. The GMS-2 to -5 satellites were successively placed in orbit by a Japanese launcher. GMS-5 saw its VISSR imager have two additional channels (the addition of a channel in the water vapor absorption band and another in the thermal infrared). It was put into orbit on March 18, 1995 and was scheduled to be replaced in 1999 by a second-generation satellite.

The second generation of Japanese meteorological satellites was called Multi-functional Transport SATellites (MTSAT). On May 15, 1999, MTSAT-1 was destroyed during its launch from Tanegashima Space Center. The JMA immediately decided to build a replacement satellite (MTSAT-1R) to extend the life of GMS-5 and negotiated with NOAA on the loan of a satellite. In April 2003, GOES-9 reached its new position at 155° E to replace GMS-5 at end-of-life, while waiting for MTSAT-1R. MTSAT-1R was launched on February 26, 2005. It was a three-axis stabilized satellite with a more efficient imager (JAMI**), equipped with five channels to produce images every 30 min. Its successor MTSAT-2, the second and last satellite of the MTSAT program, launched on February 18, 2006, became an operational satellite in 2010 and captured its last image in 2016.

Himawari-8 and Himawari-9 form the third generation of Japanese geostationary satellites. Himawari-8 was launched on October 7, 2014 and has a new 16-channel AHI** imager. With blue, green and red channels, it makes it possible to produce true color images, which is a notable improvement in particular for the detection of aerosols. It provides imaging in 10 minute intervals up to a resolution of 500 m for its channel at 0.645 μm.

Figure 1.21. *March 6, 2020 at 06:00 UTC – Himawari-8 – Pacific Ocean agitated by a beautiful storm (source: Météo-France). For a color version of this figure, see www.iste.co.uk/phulpin/satellites1.zip*

**See Appendix 3.

1.6.2. *China*

As early as 1971, the National Satellite Meteorological Center (NSMC) was created within the China Meteorological Administration (CMA) to manage research and operations related to Fengyun (wind and cloud) satellites. Fengyun, abbreviated as FY, refers to all weather satellites developed by the People's Republic of China. FY programs carry odd numbers to denote polar-orbiting satellites and even numbers for geostationary satellites.

1.6.2.1. *Polar-orbiting satellites*

The FY-1 orbiting satellite program, which is the first series of Chinese meteorological satellites, was founded in 1977. The satellites were built by the Shanghai Institute of Satellite Engineering and their payload by the Shanghai Institute of Technical Physics. FY-1A, an experimental satellite, was launched on September 7, 1988 from Taiyuan base. It suffered from several malfunctions: humidity contaminated the detectors altering the infrared measurement, then major attitude control problems led to the loss of the satellite after 39 days in orbit. Its successor, FY-1B, had the same attitude control difficulties but these could be corrected from the ground during the first few months. FY-1B delivered images for about a year thanks to its MVISR** imager with five channels. FY-1C and FY-1D followed with respective launches on May 10, 1999 and May 15, 2002. They benefited from the experience gained and the technical improvements, notably by carrying a new 10-channel imager with a spatial resolution of 1.1 km.

The second generation of Chinese polar-orbiting satellites was called FY-3. This program was the result of collaboration between the CMA and the China National Space Administration (CNSA). Approved in 1998, its development really began in 1999; eight satellites made up the series. The first two were considered experimental. Regular launches, every two years, were planned with equatorial crossing times of orbits around 06:00 (FY-3 E and H), 10:00 (FY-3A, C and F) and 14:00 (FY-3B, D and G). The experimental satellites FY-3A and B were launched on May 27, 2008 and November 4, 2010, respectively. Satellites B, C and D are currently operational. Although their payload was different depending on whether they were morning or afternoon satellites, the improvements are significant compared to FY-1, with increased capacities concerning in particular:

- VIS/IR imagery with spatial resolutions from 250 m to 1 km;

- atmospheric soundings;

- microwave imaging.

1.6.2.2. *Geostationary satellites*

FY-2 was the first generation of Chinese geostationary satellites. The development of the program began in the 1980s with the first ready in 1994. On April 2, 1994, while the satellite's fuel tank was being filled, a fire broke out, followed by an explosion that killed a technician from the Xichang Satellite Launch Center and injured 31 others. It took three years to build a new satellite with a modified tank. It was therefore on June 10, 1997 that FY-2A was launched. Like Meteosat, it was a spin-stabilized satellite. Its main instrument was the three-channel S-VISSR** imager. It operated in a nominal way until April 1998 when it began to experience transmission problems. Its capabilities were then limited to producing six images per day. The second experimental satellite, FY-2B, was put into orbit on June 26, 2000. These two experimental satellites were followed by six satellites, until FY-2H, which was launched on March 27, 2017 carrying a third-generation S-VISSR imager with five channels.

FY-4 is the next program, consisting of seven satellites. The performance of the FY-4s improved considerably. FY-4A was put into orbit on December 10, 2016. It was considered the pre-operational satellite of the series. It had a 14-channel imager, an infrared interferometer, as well as a lightning imager. The FY-4s are expected to be operational until around 2040.

Figure 1.22. *February 20, 2020 at 05:15 UTC – FY-4A – first color composition image (source: China National Space Administration). For a color version of this figure, see www.iste.co.uk/phulpin/satellites1.zip*

**See Appendix 3.

1.6.3. *Korea*

Figure 1.23. *Artistic view of the satellite Kompsat-5 (source: Wikimedia commons). For a color version of this figure, see www.iste.co.uk/phulpin/satellites1.zip*

South Korea began to develop its space program in the late 1980s. On October 10, 1989, the South Korean space agency, Korea Aerospace Research Institute (KARI), was born. The first of the Kompsat polar-orbiting Earth observation satellites was put into orbit on December 20, 1999. It was not until June 26, 2010 that COMS, the first geostationary satellite, dedicated, among other things, to meteorology, was launched. It had a five-channel imager as well as a second imager dedicated to the color of the ocean with eight channels. It was positioned at 128.2° E longitude and produced images every half hour. It was followed by the GEO-KOMPSAT-2A geostationary satellite in 2018.

1.6.4. *India*

The Indian National Satellite (INSAT) program was created in the 1970s. Four geostationary satellites were planned, positioned around longitudes 74° E and 93° E. These satellites carried both a domestic telecommunications system including television and a weather observation system. Built by the American company Ford Aerospace Corporation, their VHRR imager had two channels. INSAT-1A, the first of the series, was launched on April 10, 1982. The last satellite of the program, INSAT-1D, was deactivated on May 14, 2002 after 12 years of good and loyal service.

Figure 1.24. *Insat-1B in a clean room (source: Wikimedia commons). For a color version of this figure, see www.iste.co.uk/phulpin/satellites1.zip*

The second generation of satellites, INSAT-2, was designed and built entirely by India. Like INSAT-1, the INSAT-2 program had multiple missions: the satellites were used for both telecommunications and Earth observation. It was a program of five satellites. They carried a more efficient imager than that of INSAT-1 (the spatial resolution of the visible channel drops from 2.75 km to 2 km and that of the infrared channel from 11 km to 8 km). INSAT-2A was put into orbit on July 9, 1992 by an Ariane-4 launcher. INSAT-2C was totally devoted to telecommunications. The INSAT-2E payload included a new imager that had a third channel dedicated to water vapor.

At the end of the 1990s, the Indian Space Research Organization (ISRO) signed agreements with NASA and NOAA, and then with EUMETSAT for real-time data exchange. At the beginning of 2000, in collaboration with EUMETSAT, a reception station was installed at the Indian Meteorological Department (IMD) to receive high resolution data from Meteosat-5, located at 63° E longitude.

MetSat-1 was launched on September 12, 2002. It was the first Indian satellite dedicated solely to meteorology. It was positioned at 74.1° E longitude. On February 6, 2003, it was renamed Kalpana-1 in tribute to Kalpana Chawla, a female Indian astronaut who died in the destruction of the American space shuttle Columbia during its re-entry into the atmosphere. Kalpana-1 had a three-channel imager with a spatial resolution of 2 km in the visible and 8 km in the infrared.

Like its predecessors, INSAT-3A, put into orbit on April 9, 2003, was a satellite dedicated to both telecommunications and meteorology. It carried the same instruments as Kalpana-1 and was positioned at 93.5° E longitude.

We have to wait for INSAT-3D to deal with the new generation of Indian satellites. It was put into orbit on July 25, 2013 from Kourou by an Ariane-5 and was positioned at 82° E longitude. It carries both an imager and a sounder. The imager, which offers full disks every 30 min, has six channels, the visible and very near-infrared channels of which achieve a spatial resolution of one kilometer. As for the 19-channel sounder, its purpose is to reconstruct temperature and humidity profiles.

1.7. References

Ashby, J.H. (1964). *A Preliminary History of the Evolution of the Tiros Weather Satellite Program (HHN-45)*. NASA/Goddard Space Flight Center, Greenbelt.

Butler, H. (1964). Meteorological satellite data systems. Report, Goddard Space Flight Center, Greenbelt.

Courteix, S. (1983). Eumetsat ou l'Europe de la météorologie par satellites. *Annuaire français de droit international*, 29, 624–638.

CS (1983). Convention portant création d'une Organisation européenne pour l'exploitation de satellites météorologiques ("Eumetsat"). Confédération Suisse.

Greenfield, S.M. and Kellogg, W.W. (1960). Inquiry into the feasibility of weather reconnaissance from a satellite vehicle. Unclassified edition of USAF Project RAND Report R-218.

Hamon, J. (1995). Trente années de météorologie spatiale. *La Météorologie 8e série – numéro spécial*, 101–106.

Lebeau, A. (2001). Météosat, une expérience sur les structures spatiales européennes. *La Météorologie*, 33, 61–67.

Morel, P. (2003). Le projet Météosat, le CNES et l'Europe. *La Météorologie*, 41, 38–42.

Moulin, H. (2006). *La France dans l'espace, 1959–1979. Contribution à l'effort spatial européen*. European Space Agency, Paris.

Ratier, A., Roquet, H., Schmetz, J., Sénési, S. (2003). L'imagerie géostationnaire et son évolution. *La Météorologie*, 40, 74–79.

Renaut, D. (2004). Les satellites météorologiques. *La Météorologie*, 45, 33–37.

Rochas, M.J. (2010). Les débuts de la météorologie spatiale. *La Météorologie*, 70, 45–50.

Vaughan, W.W., Greenfield, S.M., Schnapf, A., Haas, I.S., Shapiro, R., Rivers Jr., J.W., Arnold, C.P., Fermelia, L.R., Mitchell, K.L., Wilson, G.S. et al. (1982). Meteorological satellites – Past, present and future. NASA Conference publication 2227, Washington, DC.

Vaughan, W.W. and Johnson, D.L. (1994). Meteorological satellites – The very early years, prior to launch of Tiros-1. *Bulletin of the American Meteorological Society*, 75(12), 2295–2302.

1.8. Websites

http://www.russianspaceweb.com/meteor.html

https://www.wmo-sat.info/oscar/spacecapabilities

https://directory.eoportal.org/

https://nssdc.gsfc.nasa.gov/nmc/

http://www.astronautix.com/

2

Contribution of the National Oceanic and Atmospheric Administration (NOAA, USA) Meteorological Satellites Program: An Overview

Sid-Ahmed BOUKABARA[1], Mitch GOLDBERG[1], Timothy J. SCHMIT[1], Andrew HEIDINGER[1], Satya KALLURI[1], Patricia WEIR[1], Frank GALLAGHER[1], David SPENCER[1] and Ross N. HOFFMAN[1,2]

[1] *NOAA National Environmental Satellite, Data and Information Services (NESDIS), Maryland and Wisconsin, USA*

[2] *Cooperative Institute for Satellite Earth System Studies (CISESS), University of Maryland, USA*

Over the past few decades, global monitoring of the Earth's environment has improved considerably (Boukabara et al. 2021a). Simultaneously, we are witnessing significant departures from expected conditions, including larger and more intense wildfires, reduction of polar ice, retreating glaciers and more intense storms. Earth observations play a significant role in monitoring these changes, and as input to models to predict future change on a wide range of time scales – from weather nowcasting and short-term forecasting to seasonal climate predictions and much

---

For a color version of all the figures in this chapter, see www.iste.co.uk/phulpin/satellites1.zip.

Satellites for Atmospheric Sciences 1,
coordinated by Thierry PHULPIN, Didier RENAUT, Hervé ROQUET and Claude CAMY-PEYRET.

longer-term climate projections. Observing and predicting the Earth's environment makes it possible to protect lives and properties and prepare for future impacts.

This chapter presents an overview of the United States operational Earth-observing satellite program, led by the National Oceanic and Atmospheric Administration (NOAA). We will cover the history of the NOAA program dating back to the 1960s and summarize the evolution of its capabilities in terms of diversity and quality of the measurements, on both polar-orbiting and geostationary platforms. Two important classes of sensors that will be discussed are imagers and sounders. Both passively measure the Earth System radiation. Imagers provide images of the Earth System in different frequency bands, while sounders are designed to determine the vertical profiles of atmospheric variables, such as temperature and humidity, from a number of narrow frequency bands, often referred to as channels. We will discuss the past (section 2.1), present (section 2.2) and expected upcoming portfolio of NOAA satellites (section 2.4), as well as the products and services that are or will be derived from the NOAA satellites (section 2.3). This chapter has been written at a time of intense discussions within NOAA on the future of the satellite program and its renewal and enhancement in the next few decades. We will discuss the major challenges and opportunities driving the design and evolution of this next-generation space architecture. We will also provide general information as to what shape this evolution will take.

2.1. NOAA Satellite Program: historical background

The overall evolution of the Earth-observing satellite constellation is outlined in terms of early beginnings, the NOAA LEO satellite program including the International Joint Polar System (IJSP) partnership with the European Organisation for the Exploitation of Meteorological Satellites (EUMETSAT), and the geostationary-based GOES satellite program. For the context of how the United States' contributions fit into the overall international effort, see Chapter 1. Here, the focus is on the evolution of sensor capabilities and the measurements.

2.1.1. *Origins of NASA-NOAA Polar and Geostationary Environmental Satellite Programs*

The first Television Infrared Observation Satellite (TIROS-1) provided forecasters with the first visualization from space of cloud formations as they developed and moved across the Earth's surface, and demonstrated the utility of satellites in the study of the Earth's weather. TIROS-1 was launched by National Aeronautics and Space Administration (NASA) on April 1, 1960, carried two TV cameras and two video recorders, and was operational for 78 days. NASA launched 10 TIROS satellites during 1960–1965 and seven Nimbus satellites during 1964–1978 (see Chapter 3). These NASA experimental missions developed new technologies for NOAA's operational polar systems – first, the TIROS Operational Satellite (TOS) program, and then the Improved TIROS Operational System (ITOS). The many instrument types and orbital configurations of the early TIROS and Nimbus NASA satellites demonstrated the feasibility required for an operational system of weather satellites and provided the trade space for the development of more sophisticated meteorological observation satellites.

Figure 2.1. *The first visible image from GOES-1, from October 25, 1975 (source: NOAA, public domain)*

The NOAA Geostationary Program has its beginning in the NASA Applications Technology Satellite (ATS) Series (1966–1974). These satellites proved the concept of imaging the Earth from a stable geostationary platform. The ATS Series was followed on by the experimental Synchronous Meteorological Satellites (SMS). While SMS was a NASA project, NOAA cooperated with NASA to make operational use of the data for meteorological applications. SMS-1 and -2 were the first geostationary satellites to occupy what became the traditional GOES East and West positions. Following the successful collaboration of NASA and NOAA on SMS, the model of NASA designing, procuring and launching, and NOAA operating geostationary satellites has become the standard procedure. The first Geostationary Operational Environmental Satellite (GOES-1) was launched in 1975. (NOAA satellites are assigned letters on Earth and numbers after being commissioned on orbit.) Figure 2.1 shows the first image from GOES-1.

2.1.2. *Low Earth orbit (LEO) missions*

The ITOS Series of satellites was the second generation of operational polar-orbiting satellite. The first of the improved satellites (ITOS-1) was launched on January 23, 1970. Over the next six years, NOAA-1 (ITOS-A) through NOAA-5 (ITOS-H) were launched. The evolution of the ITOS Series provided improved observations, higher resolution data and expanded operational capabilities.

By the mid-1970s, NOAA and NASA agreed to produce the Polar Operational Environmental Satellites (POES) Series operationally based on the TIROS-N generation of satellites. Figure 2.2 shows the timeline of the POES Series. TIROS-N, a research and development spacecraft, served as a prototype for the operational follow-on series, through NOAA-19. TIROS-N, launched on October 13, 1978, introduced the third-generation operational, polar-orbiting, environmental satellite system.

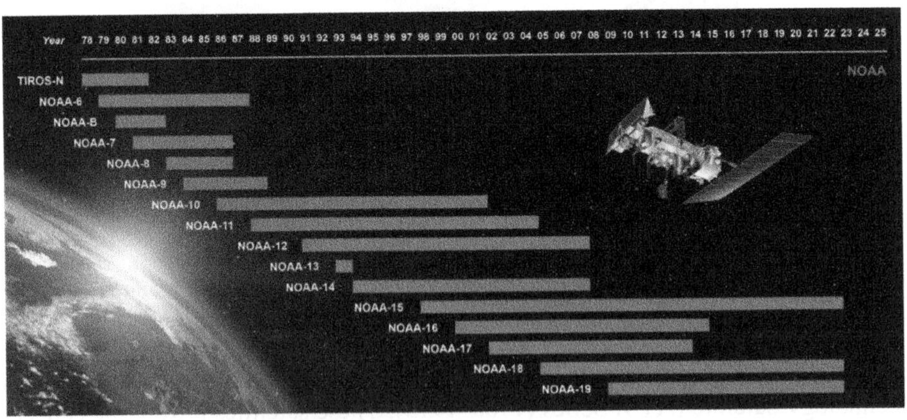

Figure 2.2. *Timeline of POES satellites from TIROS-N to NOAA-19 (source: NOAA, public domain)*

TIROS-N was a cooperative effort of the United States (NOAA and NASA), the United Kingdom and France. TIROS-N and the following operational satellites, NOAA-A through NOAA-G satellites, had the following sensors.

– The TIROS Operational Vertical Sounder (TOVS) system consisted of three sensors:

- the High-Resolution Infrared Radiation Sounder (HIRS/2) – a 20-channel instrument that made measurements primarily in the infrared (IR) region of the spectrum;

- the Stratospheric Sounding Unit (SSU) provided by the United Kingdom, through its Meteorological Office (Met Office), Ministry of Defense that observed radiances in the 15 µm carbon dioxide band, with three channels peaking at about 29, 35 and 45 km in the atmosphere;

- the Microwave Sounding Unit (MSU) – a four-channel Dicke radiometer, which made passive measurements in the 5.5 mm oxygen band.

– The Advanced Very High-Resolution Radiometer (AVHRR) – a four-channel scanning radiometer (five channels on later satellites of this series) sensitive in the visible, near-infrared and infrared window regions.

– The Space Environment Monitor (SEM) consisted of three separate data instruments and a data processing unit:

- the Total Energy Detector (TED) – measured a broad range of energetic particles from 0.3 KeV to 20 KeV in 11 bands;

- the Medium Energy Proton and Electron Detector (MEPED) – sensed protons, electrons and ions with energies from 30 Kev to several tens of MeV;

- the High Energy Proton and Alpha Detector (HEPAD) – sensed protons and alpha particles from a few hundred MeV up through relativistic energies above 840 MeV.

– The Data Collection System (DCS) – a random-access system to acquire data from fixed and free-floating terrestrial and atmospheric platforms. This was built by *Centre national d'études spatiales* (CNES) in France.

NOAA's POES historically had a two-satellite constellation, flying in two orbital planes, referred to as an "AM" or "morning" satellite and a "PM" or "afternoon" satellite. When launched, the AM satellites had an equatorial crossing time in the descending node of 0730 and the PM satellites had an equatorial crossing time in the ascending node of 1330.

Enhancements were made to the TIROS follow-on missions to accommodate new sensors. NOAA-8 (NOAA-E) was the first of the Advanced TIROS-N (ATN) satellites and included a stretched structure to provide growth capability. Two new sensors were added to NOAA-9: the Earth Radiation Budget Experiment (ERBE) designed to measure the diurnal variation of shortwave and longwave fluxes; and the Solar Backscatter Ultraviolet Radiometer (SBUV/2) for measuring ozone concentrations globally. ERBE sensors flew on NOAA-9 and -10, and SBUV/2 flew on NOAA-9 through NOAA-19, excluding NOAA-12 and NOAA-15.

In the spring of 1998, a new series of NOAA POES commenced with the launch of NOAA-K (NOAA-15) and ended with NOAA-N Prime (NOAA-19). This series represented an improvement over the previous series of NOAA polar-orbiting satellites beginning with TIROS-N (launched in October 1978), and continuing with NOAA-6 through NOAA-14 (launched in December 1994).

The AVHRR/3 on the NOAA "K, L, M" Series (NOAA-15, -16, -17) is a six-channel scanning radiometer providing three solar channels in the visible–near-infrared region and three thermal infrared channels. However, only five are transmitted to the ground at any one time. Channel 3A is time-shared with the 3.7 μm channel, designated 3B, to provide five channels of continuous data. Operationally, channel 3A is active during the daytime part of each orbit and 3B is active during the nighttime part of each orbit.

The Advanced TOVS (ATOVS) introduced on the K, L, M Series included the High-Resolution Infrared Sounder/3 (HIRS/3), the Advanced Microwave Sounding Unit-A (AMSU-A) and the Advanced Microwave Sounding Unit-B (AMSU-B). The HIRS/3 has 20 spectral bands, 19 in the IR band and one in the visible band. This instrument is basically the same as the instrument flown on earlier spacecraft, except for five spectral band changes to improve sounding parameter accuracy. The AMSU consists of two functionally independent units, the 15-channel AMSU-A for temperature sounding in the troposphere and stratosphere and the five-channel AMSU-B for atmospheric water vapor sounding in the troposphere. AMSU-A relies primarily on channels in the oxygen absorption band that peaks at 57 GHz (50.3, 52.8, 53.6, 54.4, 54.94, and 55.5 GHz, plus six centered at the 57.29 GHz resonance), while AMSU-B relies primarily on channels in the water vapor absorption band that peaks at the 183 GHz resonance. Additional channels (at 23.8, 31.4 and 89 GHz for AMSU-A and at 89.9 and 150 GHz for AMSU-B) help to remove the effects of the surface and condensed phases of water (snow, clouds). AMSU-B was replaced by the Microwave Humidity Sounder (MHS) from EUMETSAT starting with NOAA-18, and on following MetOp satellites.

2.1.3. *Geostationary Earth orbit (GEO) missions*

The GOES spatial, spectral and temporal resolutions have improved through four generations or series, from the experimental satellites in the late 1960s to early 1970s to the operational era beginning in 1975, and continued through the GOES-13/14/15 Series of spacecraft (Menzel 2020) and into the GOES-R series era. Figure 2.3 shows the evolution of the GOES Series imagery. In addition to resolution improvements, there have been calibration, image navigation and other improvements. The first two GOES Series (GOES-1 through -3, followed by GOES-4 through -7) achieved orbital stability by rotating parallel to the axis of rotation of the Earth. Such spinning GEO satellites viewed the Earth only 5% of the time. The next GOES Series, GOES-8 through -15, employed three-axis stabilization to stare at the Earth continuously. Water vapor imagery improved starting with GOES-4, with further improvements in subsequent series. The current GOES Series is GOES-R, which will fly as GOES-16 through -19. At the time of writing, GOES-16 and GOES-18 are the operational GOES East and West, respectively, with GOES-17 serving as an on-orbit standby for the operational GOES constellation.

Figure 2.3. *Noon visible images from each GOES Imager (from GOES-1 to GOES-18) (source: T. Schmit)*

Each of the geostationary instruments hosted an imager. Figure 2.3 shows a sample visible image from each of the GOES satellites that successfully reached orbit. The imager has undergone incremental improvements over each generation. Table 2.1 captures this evolution of the imager from ATS-1 to GOES-18. The first imager on ATS was the Spin Scan Cloud Camera (SSCC). The first SSCC instruments measured only single visible bands (in the green part of the visible spectrum) while the Multicolor SSCC (MSSCC) on ATS-3 measured blue, green and red bands which provided the first true color imagery from GEO. The ATS-3 true color images remain the only true color images made from a geostationary imager over the Western Hemisphere. On the SMS satellites, the SSCC was replaced by the Visible Infrared Spin Scan Radiometer (VISSR). The VISSR provided a single visible and a single infrared window band on GOES-1 through -3. Infrared channels provide nighttime as well as daytime GEO imagery. GOES-4 through -7 added 3.9 µm and 6.7 µm IR bands to the VISSR. The Imager on the GOES-8 through -11 satellites further extended geostationary imagery to five channels (0.65, 3.9, 6.7, 11 and 12 µm). The GOES-12 through -15 Imagers saw the replacement of the 12 micrometer channel by the 13.3 micrometer channel. Finer spatial and temporal resolutions have fueled numerous new ways that imager information has been used, including at NOAA's Storm Prediction Center to better monitor on fine time and space scales the mesoscale environment (over-shooting tops, thermal couplets, cloud-top cooling, anvil expansion, cloud-level divergence, above-anvil cirrus plumes and other satellite imagery signatures associated with severe weather) and to derive motion vectors of hurricanes.

		Number of channels		Resolution		
Time	Spacecraft (Instruments)	VIS/NIR	IR	Temporal (min)	VIS/NIR Spatial (km)	IR Spatial (km)
1966+	ATS-1/2 (SSCC)	1	N/A	20	4	N/A
1967+	ATS-3 (MSSCC)	3	N/A	24	4	N/A
1974	ATS-6 (GVHRR)	1	1	25	6	11
1974/ 1975	SMS-1/2 (VISSR)	1	1	20	1	8
1975+	GOES-1/3 (VISSR)	1	1	30	1	8
1980+	GOES-4/7 (VISSR)	1	3	30	1	7–14
1994+	GOES-8/12 (Imager)	1	4	7–30	1	4–8
2010+	GOES-13/15 (Imager)	1	4	7–30	1	4
2017+	ABI	6	10	0.5–10	0.5–1	2

Table 2.1. *Evolution of the imagers on the NASA/NOAA Geostationary Satellites. The spatial resolutions are the approximate values at the sub-point locations, without factoring in any over-sampling. GVHRR: Geosynchronous Very High-Resolution Radiometer; VIS: visible*

The GOES pioneered the use of IR sounders in geostationary orbit beginning with the experimental VISSR Atmospheric Sounder (VAS) on GOES-4. The VAS added 12 IR spectral bands to the VISSR to allow for sounding of temperature and moisture profiles in the atmosphere. Starting with GOES-8, a separate, operational Sounder was added, which measured 18 IR bands and a single visible band. The Sounder remained unchanged through GOES-15. The products derived from the GOES Sounder radiances included temperature and moisture profiles; total precipitable water vapor (TPW); atmospheric stability indices, such as convective available potential energy (CAPE) and lifted index (LI); cloud-top properties; total column ozone; and mid-level atmospheric motion vectors (Schmit et al. 2002). Plans to replace the low spectral resolution Sounder with the GOES High-Resolution Interferometer Sounder (GHIS) on the later GOES-I/M spacecraft were not carried out. Unfortunately, the planned Hyperspectral Environmental Sounder (HES), first known as the Advanced Baseline Sounder, was dropped during the development of GOES-R in 2006, predominantly because of budgetary pressure. As a result, there are no sounders on the GOES-R Series. Currently, the only operational hyperspectral sounder in geostationary orbit is the Geosynchronous Interferometric Infrared Sounder (GIIRS) on board the FengYun-4A. Radiances from GIIRS have been shown to be valuable to derive four-dimensional wind fields (Ma et al. 2021) and to improve NWP forecasts of tropical cyclones. Also, as described in more detail in Chapter 5, EUMETSAT's Meteosat Third Generation (MTG) will include two operational advanced IR sounding (MTG-S) geostationary platforms beginning in the mid-2020s. The first of these, MTG-I, was recently launched on December 22, 2022 and is going through the commissioning phase at the time of writing.

2.2. NOAA Current Space Constellation

This section describes the different components of the current NOAA Earth-Observing satellites, stratified by types: the Joint Polar Satellite Systems (JPSS) and the current geostationary satellites (the GOES-R Series). Brief mention is also made of other select missions developed in partnership with national and international institutions.

2.2.1. *The NOAA Joint Polar Satellite System (JPSS) Program*

The Joint Polar Satellite System (JPSS) is NOAA's current generation polar-orbiting operational environmental satellite system and is the successor to the legacy POES program (Goldberg et al. 2013). JPSS is a collaborative program between NOAA and NASA. As the backbone of the global observing system, JPSS polar satellites circle the Earth from pole-to-pole and cross the equator about 14 times daily in each direction (i.e. ascending from south to north and descending from north to south), providing full global coverage twice a day. It is expected that JPSS will provide continuity of critical, global observations of Earth's atmosphere, oceans and land through at least 2038.

The NPOESS Preparatory Project (NPP), launched in October 2011, is the predecessor to the JPSS Series spacecraft. The NPP was renamed the Suomi National Polar-orbiting Partnership (Suomi NPP) in honor of Verner E. Suomi, a University of Wisconsin meteorologist widely recognized as the "Father of Satellite Meteorology", who designed the very first satellite meteorological sensor – the Earth radiation budget instrument that was launched October 13, 1959 on Explorer 7.

Suomi NPP was constructed with a design life of five years (although it is still functioning normally) and carries five state-of-the-art instruments: (1) Visible Infrared Imaging Radiometer Suite (VIIRS), (2) Cross-track Infrared Sounder (CrIS), (3) Advanced Technology Microwave Sounder (ATMS), (4) Ozone Mapping and Profiler Suite (OMPS) and (5) Clouds and the Earth's Radiant Energy System (CERES). Apart from CERES, which was only carried on JPSS-1, and some variations in the configuration of OMPS (section 2.2.1.4), the JPSS program consists of four identical satellites – JPSS-1, -2, -3 and -4 – that all carry the same instruments as Suomi NPP. The JPSS instruments are described in the following sections. Suomi NPP and JPSS provide users with several land, atmosphere, ocean and cryosphere products, operational for several applications besides severe weather (Zhou et al. 2019) (Figure 2.4). JPSS-1 became NOAA-20 after it was launched on November 18, 2017. JPSS-2 became NOAA-21 after it was launched on November 10, 2022. NOAA-20 and NOAA-21 each have a design life of seven years.

2.2.1.1. *VIIRS*

VIIRS extends and improves upon a series of measurements initiated by the AVHRR, the Moderate Resolution Imaging Spectroradiometer (MODIS) and the Operational Linescan System (OLS). VIIRS provides a number of products (Figure 2.4), has better spectral and spatial resolution with a larger swath compared with legacy POES imagers. VIIRS provides operational and research users with:

– spectral coverage from 412 nm to 12 μm in 22 bands;

– five high resolution bands ("I" bands) at ~375 m nadir resolution;

– 16 moderate resolution bands ("M" Bands) at ~750 m nadir resolution;

– and a panchromatic day and night band (DNB) at ~750 m nadir resolution with low, medium and high gain settings capable of observing the Earth during both day and night.

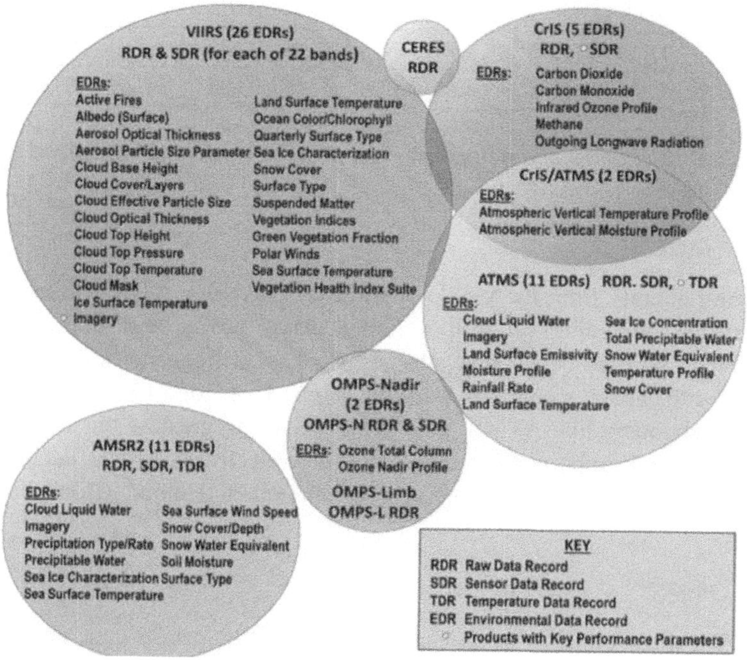

Figure 2.4. *JPSS data products (source: NOAA[1])*

2.2.1.2. *CrIS*

The Cross-track Infrared Sounder (CrIS), the first advanced operational hyperspectral sounder on NOAA satellites, provides more accurate, detailed atmospheric temperature and moisture profiles for weather and climate applications compared to legacy POES sensors. It is a Fourier Transform Spectrometer (FTS) providing high resolution IR spectra similar to IASI on the MetOp spacecraft. It has 2211 spectral channels at full spectral resolution (FSR) mode, with a resolution at 0.625 cm^{-1} for all three bands (Table 2.2). Each CrIS Field of Regard (FOR) has a 3x3 array of Fields of View (FOV) where each FOV is 14 km in diameter at nadir. A large number of products are derived from CrIS by the NOAA unique combined atmospheric processing system (NUCAPS).

1. https://www.star.nesdis.noaa.gov/jpss/JPSS_products.php.

	Spectral range (cm^{-1})	Spectral range (μm)	Spectral resolution (cm^{-1})	Number of in-band (unapodized) channels
Longwave	650–1,095	15.38–9.14	0.625	713* (717)
Mid-wave	1,210–1,750	8.26–5.71	0.625	865* (869)
Shortwave	2,155–2,550	4.64–3.92	0.625	633* (637)

Table 2.2. *CrIS channel characteristics. The asterisk (*) denotes apodized channels*

2.2.1.3. *ATMS*

Drawing directly from its AMSU-A and MHS heritage, ATMS offers improved characteristics, including reduced volume, mass, power and improved spatial coverage with no gaps between swaths. ATMS has 22 frequencies from 23 GHz to 183 GHz with nadir resolution varying from 15.8 to 74.8 km. Historically, microwave sounders like ATMS and AMSU have the greatest impact on forecast accuracy. Several important products are derived from ATMS data including profiles of temperature, water vapor, precipitation and cloud (Boukabara et al. 2011).

2.2.1.4. *OMPS*

OMPS collects total column and vertical profile ozone data and continues the daily global data produced by previous ozone monitoring systems but with higher fidelity and larger swaths. OMPS consists of three spectrometers: a downward-looking nadir mapper, a nadir profiler and a limb profiler that provide a number of products (Flynn et al. 2014). The entire OMPS suite, OMPS-Nadir (OMPS-N) and OMPS-Limb (OMPS-L), currently flies on board the Suomi NPP spacecraft and is scheduled to fly on the JPSS-2 satellite mission. The nadir total column spectrometer provides total column ozone data with a 50 × 50 km resolution at nadir. The nadir profile spectrometer provides ozone profiles in a single ground pixel of 250 × 250 km at nadir from Suomi NPP, and with five ground pixels of 50 × 50 km at nadir from JPSS-1 (NOAA-20). The limb sounder provides 1 km vertical sampling with spectral coverage from 290 to 1000 nm using a prism spectrometer enabling ozone profile retrieval on Suomi NPP and JPSS-2 spacecraft (but not on NOAA-20).

2.2.1.5. *CERES*

CERES on Suomi NPP and NOAA-20 provide continuity with earlier CERES instruments launched on NASA's Terra, Aqua and Tropical Rainfall Mapping

Mission (TRMM). These sensors have been providing observations of Earth's radiation budget (ERB) since 2000. CERES measures radiances in three channels: 0.3–5 μm (shortwave); 8–12 μm (window IR); and 0.3 to > 50 μm (broadband) at 20 km spatial resolution. CERES data are used along with data from other imagers on polar (e.g. VIIRS, MODIS) and geostationary satellites to produce highly accurate ERB estimates.

2.2.2. *GOES-R series*

GOES-R followed the GOES-NOP series and represented a major advancement in capabilities in terms of resolutions and new measurement capabilities. While a new hyperspectral IR sounder was eliminated during its development, GOES-R did innovate the Advanced Baseline Imager (ABI) and the Geostationary Lightning Mapper (GLM). Goodman et al. (2020) review the GOES-R sensors and applications. As described below, the ABI was a significant improvement in performance (Figure 2.5) and the GLM provided the first geostationary observations of lightning.

Figure 2.5. *The "local noon" true color composite image with both GOES-17 and GOES-16 ABI data on September 4, 2019. Note the clearly visible eyes of Hurricane Juliette southwest of Baja California and Hurricane Dorian near Florida (source: R. Kohrs, UW/SSEC)*

While the GOES-R Series was a "game-changer" in many respects, the follow-on Series (GeoXO, described in section 2.4.2.2) has the potential to offer many other firsts as well, for example, operational GEO hyperspectral IR sounders over the Western Hemisphere, plus sensors related to ocean color and atmospheric composition.

2.2.2.1. *ABI*

As Table 2.1 shows, there was a step function of capability moving from the GOES-13/15 imagers to the GOES-R Series with the Advanced Baseline Imagers (Schmit et al. 2018). For example, the single spectral band in the visible/near-IR spectral region increased to a total of six spectral bands, and four infrared bands increased to 10 bands. This improved spectral coverage was accompanied by a factor of four increase in spatial resolution, a factor of five increase in temporal resolution, improved signal-to-noise-ratio (SNR) and improved image navigation.

From the spectral bands of the ABI (Figure 2.6), many quantitative products are being operationally generated (Table 2.3) (Schmit et al. 2018). This includes products monitoring the atmosphere, ocean, land and other hazards. Much more on the current GOES constellation is detailed in the book *The GOES-R Series: A New Generation of Geostationary Environmental Satellites* (Goodman et al. 2020).

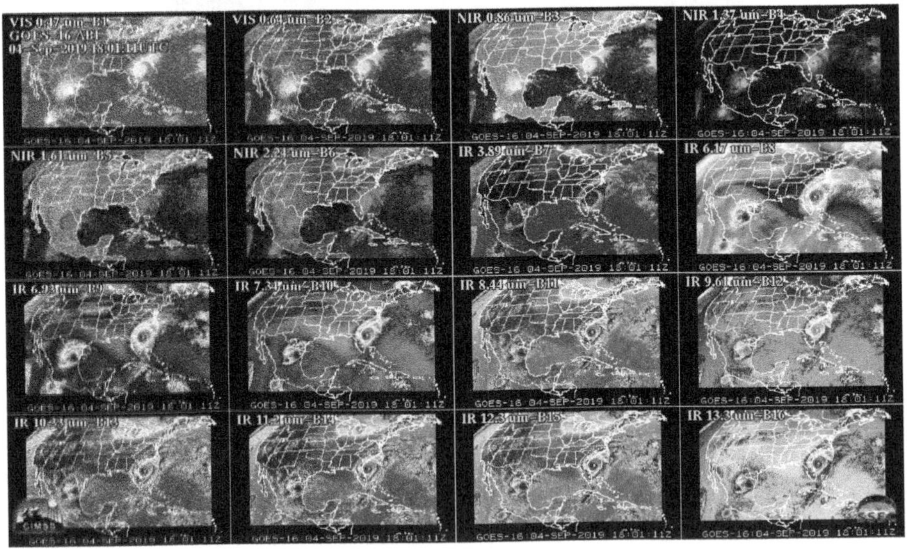

Figure 2.6. *The 16 spectral bands of the GOES-16 ABI from September 4, 2019 (source: Mat Gunshor, UW/CIMSS)*

Product	Acronym meaning	Product	Acronym meaning
ACHA	Cloud top height	DSI	Derived stability indices
ACHT	Cloud top temperature	DSR	Downward shortwave radiation: surface
ACM	Clear sky masks	FDC	Fire/hotspot characterization
ACTP	Cloud top phase	FLS	Fog and low stratus
ADP	Aerosol optical depth	FSC	Snow cover
CMIP	Cloud and moisture imagery	LST	Land surface (skin) temperature
MCMIP	Multiband cloud and moisture imagery	LVMP	Legacy vertical moisture profile
COD	Cloud optical depth	LVTP	Legacy vertical temperature profile
CPS	Cloud particle size distribution	RRQPE	Rainfall rate/QPE
CTP	Cloud top pressure	RSR	Reflected shortwave radiation: TOA
DMW	Derived motion winds	SST	Sea surface (skin) temperature
DMWV	Derived motion winds for ABI band 8	TPW	Total precipitable water

Table 2.3. *A list of GOES-R Series products from the ABI. QPE: quantitative precipitation estimate; TOA: top of atmosphere*

2.2.2.2. *Geostationary Lightning Mapper (GLM)*

The GOES-R Geostationary Lightning Mapper (GLM) observes lightning throughout much of the Western Hemisphere. The continuous sampling of the GLM provides insights over a shorter period of time (i.e. months vs. years) compared to similar data from low Earth orbiting sensors. GLM applications include severe storm monitoring and diagnosis, lightning safety, aviation planning and fire weather forecasting. A suite of gridded GLM products have been developed to help forecasters characterize the convective scene. The GLM has contributed to a significant advance in the observation of lightning. The continuous availability of spatially extensive total lightning data will spur more rapid progress towards the synthesis of these observations with other meteorological datasets and forecasting tools. Wide-ranging efforts continue towards: (1) documenting, diagnosing and communicating the GLM data quality, (2) developing and implementing the most useful operational products and (3) promoting the proper use of the GLM data within data assimilation systems and among a broad range of operational and research users. Many examples of the products and applications of GLM are provided by Rudlosky et al. (2020).

2.2.3. *Collaborative programs*

Today, NOAA's National Environmental Satellite, Data, and Information Service (NESDIS) has ongoing interagency and international partnerships with over

25 countries. NOAA's successes depend on scientific and data exchange with our partners, which has been built on years of cooperation and collaboration. An extremely important example is the continuing collaboration in which the United States afternoon polar-orbiting satellites carry EUMETSAT provided instruments and the EUMETSAT morning polar-orbiting satellites carry NOAA instruments. (Here, afternoon and morning indicate satellites crossing the Equator northbound in the afternoon and southbound in the morning, respectively.) Details about this collaboration are provided in Chapters 5 and 7.

Another significant example of NOAA's international and interagency partnerships is the Constellation Observing System for Meteorology, Ionosphere and Climate (COSMIC)-2/FORMOSAT-7 constellation of global navigation satellite system (GNSS) radio occultation satellites. The GNSS radio occultation technique measures the time delay in the transmitted GNSS radio signal as it is refracted through the atmosphere, allowing the derivation of profiles of temperatures and moisture in the troposphere and lower stratosphere and electron density in the ionosphere because the relatively fast-moving low Earth orbiting receiving satellites trace out different paths through the atmosphere. COSMIC-2/FORMOSAT-7 and its predecessor COSMIC-1/FORMOSAT-3 have significantly advanced numerical weather prediction (NWP), space weather monitoring and climate monitoring (Ho et al. 2020). For NWP, COSMIC-2 provides accurate and high vertical resolution atmospheric measurements. These data improve forecasting skill and more specifically produce positive impacts on the prediction of track and intensity of tropical cyclones from genesis to landfall. More recently, the NOAA Commercial Weather Data Pilot (CWDP), a technology demonstration project, was established in 2016 to evaluate commercial data to demonstrate the quality of the data and its impact on weather forecast models, and to inform NOAA's process for acquiring, evaluating and utilizing commercial data in the future. NOAA concluded that, based on the results of CWDP Round 2, the commercial sector is able to provide radio occultation data that can support NOAA's operational products and services. NOAA is currently purchasing and operationally assimilating commercial radio occultation data.

2.3. Applications

The legacy and the new generation satellites provide us with a long and continuous record of Earth observations. While it is difficult to capture all of the contributions of meteorological satellites in this chapter, a few applications will be highlighted. The observations from meteorological satellites are also valuable in building extensive multi-year environmental data records that allow us to study changing climate patterns over the past few decades.

The data from the LEO satellites, for example, have long been assimilated into weather prediction models in countries around the world (Eyre et al. 2020). IR and microwave (MW) sounder data from NOAA and MetOp satellites continue to be the backbone of NWP (Menzel et al. 2018). Currently, the primary mission objectives of JPSS include: increased timeliness and accuracy of severe weather event forecasts; providing advanced atmospheric temperature and moisture profiles from space; providing advanced imaging capability to analyze fires, volcanoes, oil spills and other natural hazards; and maintaining continuity of climate observations and critical environmental data from the polar orbit.

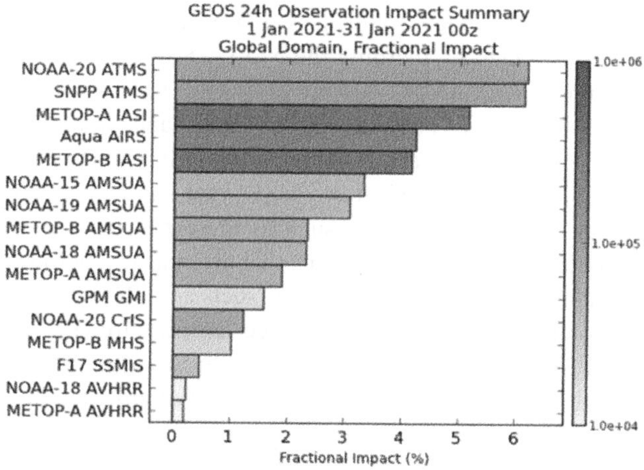

Figure 2.7. *Forecast sensitivity-based observation impact of various satellite observations on GEOS-5 atmospheric data assimilation system (source: Gelaro et al. (2010))*[2]

Figure 2.7 shows the results of the forecast sensitivity-based observation impact (FSOI) method applied to experiments undertaken by NASA. In this study, MW sounders have the highest impact on NWP models followed by IR sounders. This technique is one of the techniques employed by NWP centers to assess the value of observation (in particular satellite data). Other techniques employed include observing system experiments (OSEs) and observing system simulation experiments (OSSEs). These assessment techniques are complementary and provide useful information for space agencies about the impact of their data on global NWP systems. They also provide guidance on how these systems should evolve in the future. The results shown in Figure 2.7 are consistent with the findings from multiple agencies and meteorological centers highlighting the importance of

2. https://gmao.gsfc.nasa.gov/forecasts/systems/fp/obs_impact/.

microwave, infrared and other sensors tracking atmospheric winds from space. Besides their utility in initializing NWP through data assimilation, the MW and IR sounders are also used for the retrieval of atmospheric parameters such as TPW, profiles of temperature, water vapor and trace gases, such as those derived using the Microwave Integrated Retrieval System (Boukabara et al. 2011) and NUCAPS (Nalli et al. 2018).

In addition to NWP and real-time soundings, the polar satellites also provide valuable Climate Data Records (CDR) for monitoring climate change. Figure 2.8 shows monthly global mean anomaly time series and trends for layer temperatures of mid-stratosphere (TMS), upper-stratosphere (TUS) and top-stratosphere (TTS). This time series is constructed by merging the NOAA SSU from 1979 to 2006 with SSU equivalent channels from AMSU-A from 2001 to present (Zou and Qian 2016). Similar monthly global mean anomaly time series and trends for the layer temperatures of mid-troposphere (TMT), upper-troposphere (TUT) and lower-stratosphere (TLS) constructed from MSU and AMSU-A have also been developed (Zou et al. 2018). Time series such as these provide a global picture of climate trends compared to in situ observations that are sparse. Microwave soundings also provide important hydrological parameters such as TPW, cloud liquid water (CLW), sea–ice concentration (SIC), land surface temperature (LST), land surface emissivity (LSE) for 23, 31, 50 GHz, rain rate (RR), snow cover (SC), ice water path (IWP) and snow water equivalent (SWE) (Figure 2.9) (Ferraro et al. 2018).

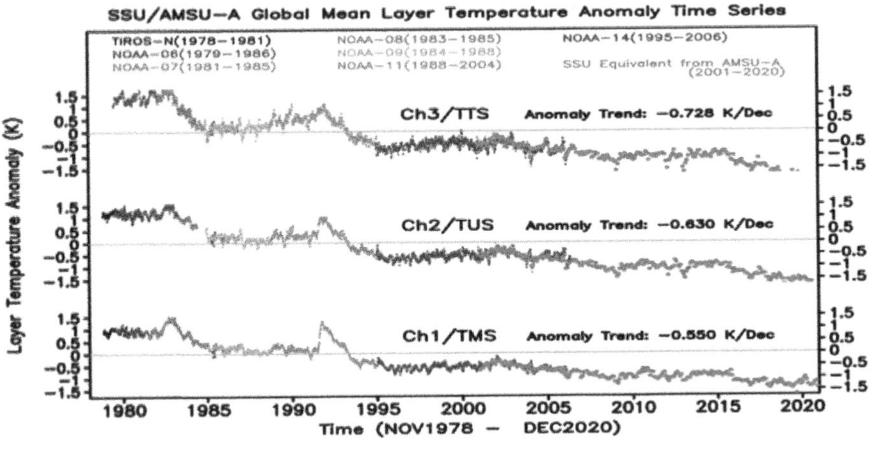

Figure 2.8. *Monthly global mean anomaly time series and trends for layer temperatures of mid-stratosphere (TMS), upper-stratosphere (TUS) and top-stratosphere (TTS) from NOAA satellites (source: NOAA/NESDIS/STAR)*

Multipurpose imaging radiometers such as AVHRR and the follow-on VIIRS not only continue to provide global imagery providing synoptic views of land, clouds and weather patterns for operational meteorologists, but are also used to produce several ocean, land and atmospheric products such as sea surface temperatures (SST), fire detection and aerosols. AVHRR sensors have been flying on POES satellites since 1978 and MetOp-C carries the last AVHRR. Several operational products continue to be produced, such as SST and vegetation health indices (Figure 2.10) (Kogan 1997), that are used in operational meteorology and drought monitoring, respectively. Temporal composites of vegetation indices computed from AVHRR bands 1 and 2 are one of the most extensively used products for a wide variety of applications such as modeling biogeochemical cycles, land use Land cover change, drought monitoring, vector-borne disease modeling, hydrological modeling and climate change (Kalluri 2021). About 30 operational products are currently produced from VIIRS by NOAA which are vital for operational meteorology and disaster management (Zhou et al. 2016) (Figure 2.11).

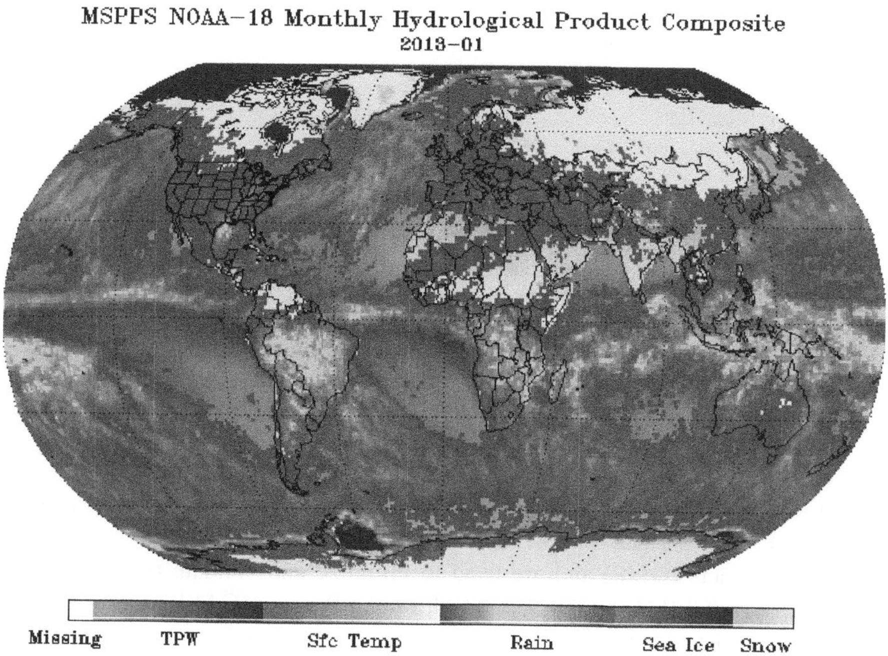

Figure 2.9. *Monthly composite of hydrological parameters derived from AMSU on NOAA-18 for January 2013. TPW: total precipitable water; Sfc Temp: surface temperature (source: Ralph Ferraro)*

GOES Imager and Sounder data have been applied to a number of applications by several institutions and agencies over many years. Only selected applications are mentioned here. With the successful demonstration of stable imagers from the ATS SSCC in the 1960s, meteorologists began using the data to study storm development. Professor Suomi at the University of Wisconsin–Madison quantified cloud-drift wind vectors using a set of three images to derive the speed and direction of the movement of clouds. Once IR observations were available, estimates of the cloud height were possible and could be used to vertically place the cloud-drift winds. As the number of IR channels increased, methods such as CO_2 slicing and water vapor intercept were developed and applied to GOES Imager and Sounder data and resulted in increasing cloud height accuracies (Heidinger et al. 2020). With the advent of IR water bands, the cloud-drift motion algorithms were applied to clear-sky water features. The GOES Sounder data were used to derive temperature and moisture profiles (Schmit et al. 2002), as well as atmospheric stability indices such as the lifted index. The Sounder was also used to estimate cloud properties, total precipitable water and total ozone.

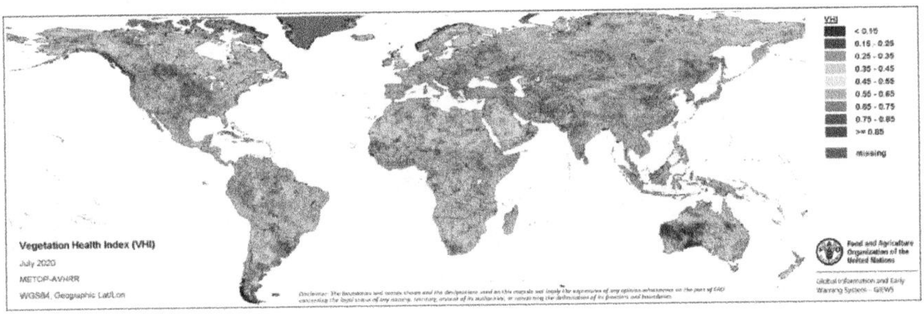

Figure 2.10. *Vegetation Health Index (VHI) produced by AVHRR on MetOp-C satellite for July 2020 (source: Food and Agricultural Organization[3])*

The visible channel on the Geostationary Imagers has also been used for many applications. Currently, cloud optical depths, estimates of solar insolation at the surface and land surface temperature are included in the GOES Surface and Insolation Products (GSIP) and are used to initialize and validate the land surface model used by NCEP. The GOES Aerosol/Smoke Product (GASP) uses the visible channel to estimate smoke and aerosol optical depth. GOES Imagers support fire applications. The operational Hydro-Estimator satellite rainfall estimates are based on available geostationary data, including the GOES Imagers.

3. http://www.fao.org/giews/earthobservation/asis/index_2.jsp?lang=en.

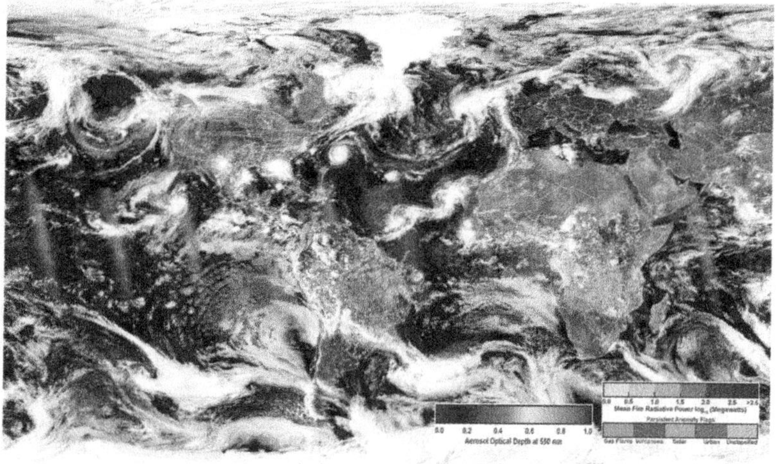

Figure 2.11. *VIIRS true color image from NOAA-20 collected on September 14, 2020, overlaid by fire detection map and aerosol optical thickness (source: NOAA, public domain)*

The use of GOES data for climate monitoring is growing as new techniques to improve the radiometric quality of GOES data are developed. GOES data contributed to the Global Energy Water Exchanges (GEWEX) suite of products including the International Satellite Cloud Climatology Project (ISCCP) and the Global Precipitation Climatology Program (GPCP).

2.4. Looking ahead: designing the next-generation architecture

The NOAA satellite program led by NOAA's NESDIS has a vision going forward to provide a truly integrated digital understanding of our Earth environment that can evolve quickly to meet changing user expectations.

2.4.1. *Factors impacting the NOAA strategy*

There are many factors impacting the design and evolution of the future NOAA sensors and constellation architecture. Three major factors, which we will describe in more detail below, are: (1) new remote sensing and space mission technologies including small satellites, (2) evolving user requirements and (3) trends in space access affordability.

Technology: technological advances are expected to result in new variables observed as well as increases in the spatial, vertical and temporal resolutions compared to the antecedent observations. Of particular interest is the ability to develop compact payloads hosted by affordable yet entirely capable small satellites. It is worth noting that advances in remote sensing technology are leading to a data volume explosion (termed Big Data) which will lead to challenges in exploiting all these data – challenges in terms of processing, distribution, archiving, accessing and securing these data. Innovations in sensors and payload hosting solutions will likely result in improved signal-to-noise ratios in sensors, added diversity in launch options and increased cost-effectiveness in both sensors and satellites. We also note that potential spectrum interference poses a challenge on microwave sensors that needs to be tackled in the design of the future constellation.

In addition to incremental advances in sensor hardware technology, we anticipate new data processing techniques and capabilities to accompany these advances, new products with higher information content, new business models and partnerships for acquiring environmental space-based data, and entirely new sensors. New data processing approaches will include an increased role for Artificial Intelligence (AI) (Boukabara et al. 2021b) and an increased dependence on data fusion techniques resulting in integrated, blended and fused data products with multiple layers of information content, which will facilitate generating added-value products and services for users. There will likely be an increasing role for commercial and other sources (including the Internet of Things (IoT) and citizen data) of Earth observations both as a complement to traditional observing systems and as a validation source for satellite data. Innovative new sensors will evolve following recent demonstration missions (such as GNSS reflectometry (CYGNSS) and Doppler wind lidars (Aeolus)) that increased their potential for operational implementation. NOAA is considering all of these developments in hardware technology, ground processing systems, business models and innovative missions, in designing and evolving the space architecture of the future.

Requirements: technical advances match to first order the increased need for a variety of satellite data at increased resolutions from a variety of applications. Indeed, requirements from several environmental applications are evolving due to increasing public demand for environmental data to support a wide range of services. Activities to map user requirements to current and expected satellite remote sensing capabilities occur at the national level (e.g. the NOAA Satellite Observing System Architecture (NSOSA) study described below) and at the international level, notably in the "Vision for a World Meteorological Organization (WMO) Integrated Global Observing System in 2040" (WMO 2019) and the WMO Observing Systems Capability Analysis and Review

(OSCAR) tool[4]. NOAA is considering the evolution of these requirements in its effort to design and evolve the space architecture of the future.

Space access: current trends in access to space are leading to more affordable and frequent launch opportunities, especially for small satellite missions, which can be used for technology validation, Earth observation and space exploration. Advances in small satellite platforms, the availability of low-cost launches for small satellites and the miniaturization of instruments can help enable new Earth-observation missions which can benefit from the use of constellations of small satellites. Due to its enormous potential for cost reductions, reusability will, in the long-term, likely become a key determinant for the competitiveness of commercial launch providers. Small vehicles tailored to smaller satellites are also showing enormous potential, including the potential for "disaggregation" – the ability to accomplish certain goals in space technologies with multiple small units rather than a single large one. Multiple small satellites also reduce the chance of failure, therefore increasing robustness and resilience.

2.4.2. *Next-generation NOAA space architecture*

In 2016, NOAA undertook the NSOSA study (Anthes et al. 2019) to assess the optimal way to design the next-generation space architecture. The principal goal of the NSOSA study has been to recommend one (or possibly more than one) alternative satellite observing system architecture for the 2028–2050 time period with no preconceived notions of what the future architecture should be. The goal was to recommend a preferred satellite architecture, not a single specific alternative constellation. Assessing the relative value of the constellation alternatives to the NOAA mission required revisiting and evaluating NOAA observational requirements and polling the existing space technology capabilities.

An unsurprising yet essential finding of the NSOSA study is that the preferred architecture depends on the funding level considered. The NSOSA study has been structured to consider a range of alternatives, showing how benefits and architectural choices vary with the funding available. Because of overall cost–benefit issues and other strategic considerations, the "Hybrid" architecture can be considered the recommended architecture regardless of the budget level. This is the primary strategic decision resulting from the NSOSA study.

The key criterion for selecting an alternative space architecture is its cost–benefit at a given average annual cost. But other non-cost factors, such as risks, technology

4. https://www.wmo-sat.info/oscar/spacecapabilities/.

maturity, launch cadence considerations, selection of capabilities and partnerships, are also appropriate criteria to use in selecting alternative architecture candidates. The details of the configurations considered are being determined in current and future formulation activities at the time of writing. However, more information is given in the sections below for the LEO and GEO components of the future constellation.

2.4.2.1. *The future of LEO*

The LEO component of the NOAA space constellation has an opportunity to utilize SmallSats in a cost-effective manner. This has been enabled by the innovative commercial and academic applications of the CubeSat standards to reduce the size of components by using commercial parts and cutting-edge technology where practical. This has subsequently created the demand from commercial entities and academia to find inexpensive access to space for their SmallSats, creating a market for smaller launch vehicles and rideshares. These developments have driven down the cost of building and launching small satellites with small sensors to the point where it is feasible to launch a constellation of a dozen or so satellites for the cost of a single large satellite a decade ago.

NOAA is taking advantage of this current situation in the SmallSat industry by applying the disaggregated SmallSat constellation approach to meet NOAA's current and future observational needs while, at the same time, enabling resilience and adaptability. An added benefit is obviously increased temporal and spatial resolution resulting from a higher density of low-flying small satellites. NOAA is performing conceptual design studies with industry in the early 2020s to determine the best method to meet its operational requirements. NOAA hopes these studies prove that this concept is of value, thus enabling new technologies and businesses to move weather observations into an adaptive and evolutionary future.

2.4.2.2. *The future of GEO*

The Geostationary Extended Observations (GeoXO) program will follow the GOES-R Program with a first launch tentatively scheduled in the early 2030s. At this time, GeoXO is still being formulated. The final approval for GeoXO will occur a few years later. Until that time, the sensor complement and the GeoXO constellation are preliminary. Unlike past GOES programs, the space weather component has been separated from GeoXO and will not be described here. The material given here represents the current thinking and may be subject to change[5].

5. Readers interested in tracking updates to the evolution of GeoXO should visit: https://www.nesdis.noaa.gov/GeoXO.

Since the launch of GOES-R with its advanced imager and pioneering lightning mapper, the amount and type of observations from geostationary sensors launched by various countries have and will continue to increase during the period before the GeoXO launch. NOAA has the potential to leverage these technologies for use in the GeoXO program. Further, working group and user engagement workshops provided NOAA with a set of observational priorities. The current set of GeoXO sensors being considered includes the following:

– GeoXO Imager (GXI) – an incremental improvement on the GOES-R Advanced Baseline Imager (ABI);

– GeoXO Lightning Mapper (LMX) – an incremental improvement on the GOES-R Geostationary Lightning Mapper (GLM);

– GeoXO Infrared Sounder (GXS) – a new hyperspectral infrared sounder;

– GeoXO Ocean Color Imager (OCX) – a new ocean color imager;

– GeoXO Atmospheric Composition (ACX) – a new atmospheric composition instrument;

– GeoXO Day/Night Band Imager (GXL) – a new low-light imaging instrument.

NOAA also considered microwave sounding from geostationary orbit. While the scientific value of such observations is clear, the cost of this observation was deemed too large during formulation and GeoXO is not currently considering the inclusion of microwave observations.

GeoXO should usher in an era of synergistic applications that use multiple instruments. For example, the fine spatial and temporal observations from the GXI should offer value to all of the other GeoXO sensors in terms of cloud detection and sub-pixel characterization. The GXI is expected to provide sub-pixel variability in cloud mask, cloud height and radiances to support the use of the GXS applications such as radiance assimilation. GXS can provide information on atmospheric profiles that will aid in the GXI applications of fire and smoke detection, cloud and aerosol products and high-spatial resolution water vapor and stability products.

In addition to the instrument complement and the performance specifications of each instrument, GeoXO must also concern itself with the number, size and location of the satellites in the GeoXO constellation. Due to anticipated additional instruments, GeoXO is expected to expand beyond the traditional East and West satellites at 75W and 135W, respectively. The number, the size, the location and the ownership of the GeoXO constellation are still to be determined at this time. GeoXO will learn from the NASA experiences in hosting its TEMPO and GEOCARB instruments on commercial geostationary satellites.

In summary, NOAA's Geostationary Extended Observations (GeoXO) satellite system is planned by NOAA for launch in the early 2030s to follow the current GOES-R Series. GeoXO will continue the Imager and Lightning Mapper records from GOES-R and is planning for observations of ocean color and atmospheric composition, low-light imagery and high-resolution infrared hyperspectral sounding.

2.5. Summary

NOAA has pioneered operational meteorological satellites for decades in both LEO and GEO, following the first successful NASA missions to deploy meteorological satellites. The sensors on these satellites have evolved gradually but significantly over this time period in terms of accuracy, spatial, vertical and spectral resolutions, temporal refresh and other capabilities. This was done both in response to gradual increases in user requirements, as well as to leverage new sensing technologies. Due to space limitations, this chapter is not comprehensive but has outlined the evolution of these sensors and their capabilities as well as the diversity of the applications where they play a major role. The JPSS and GOES programs are the two foundational pillars in the NOAA Earth-observing space program and will continue to be so in the next two decades. The applications that depend on these satellites, and our understanding and techniques used to exploit these data, have improved and are expected to continue to improve over the years. The emergence of the commercial data sector and the CWDP in NOAA will likely play an increasingly important role in NOAA's future observing systems, initially as a complement to the critical backbone systems in LEO and GEO. All of these government and commercial players are expected to increasingly take advantage of the advent of smaller satellites and maturing sensor technologies for easier access to space, faster development cycles, more robustness and overall cost-effectiveness in a disaggregated hybrid system. Flexibility and agility and leveraging new technology through rapid infusion will characterize the evolution of the next-generation architecture of the NOAA space assets which will therefore be characterized by robustness, resiliency, quality, continuity and continual improvement.

2.6. Acknowledgments

The authors would like to extend their gratitude to the many scientists, engineers, managers, collaborators, partners and support staff in the government, academic and private sectors who have contributed over the past several decades to the NOAA satellite program. The authors also extend their thanks to all those who are currently working on the design and evolution of the future NOAA satellite program to secure its long-term sustainability and improvement. The scientific results and conclusions as well as any views or opinions expressed herein are those of the author(s) and do not necessarily reflect those of NOAA or the U.S. Department of Commerce or the U.S. Government.

2.7. References

Anthes, R.A., Maier, M.W., Ackerman, S., Atlas, R., Callahan, L.W., Dittberner, G., Edwing, R., Emch, P.G., Ford, M., Gail, W.B. et al. (2019). Developing priority observational requirements from space using multi-attribute utility theory. *Bull. Am. Meteorol. Soc.*, 100(9), 1753–1774. https://doi.org/10.1175/bams-d-18-0180.1.

Boukabara, S.-A., Garrett, K., Chen, W., Iturbide-Sanchez, F., Grassotti, C., Kongoli, C., Chen, R., Liu, Q., Yan, B., Weng, F. et al. (2011). MiRS: An all-weather 1DVAR satellite data assimilation and retrieval system. *IEEE Trans. Geosci. Remote Sens.*, 49(9), 3249–3272. https://doi.org/10.1109/tgrs.2011.2158438.

Boukabara, S.-A., Eyre, J., Anthes, R.A., Holmlund, K., St. Germain, K., Hoffman, R.N. (2021a). The earth-observing satellite constellation: A review from a meteorological perspective of a complex, interconnected global system with extensive applications. *IEEE Geosci. Remote Sens. Magazine*, 9(3), 26–42. https://doi.org/10.1109/MGRS.2021.3070248.

Boukabara, S.-A., Krasnopolsky, V.M., Penny, S.G., Stewart, J.Q., McGovern, A., Hall, D., Ten Hoeve, J.E., Hickey, J., Huang, H.-L.A., Williams, J. et al. (2021b). Outlook for exploiting artificial intelligence in the earth and environmental sciences. *Bull. Am. Meteorol. Soc.*, 102(5), E1016–E1032. https://doi.org/10.1175/bams-d-20-0031.1.

Eyre, J.R., English, S.J., Forsythe, M. (2020). Assimilation of satellite data in numerical weather prediction. Part I: The early years, *Q.J.R. Meteorol. Soc.*, 146(726), 49–68. https://doi.org/10.1002/qj.3654.

Ferraro, R.R., Nelson, B.R., Smith, T., Prat, O.P. (2018). The AMSU-based hydrological bundle climate data record – Description and comparison with other data sets. *Remote Sens.*, 10(10), 1640. https://doi.org/10.3390/rs10101640.

Flynn, L., Long, C., Wu, X., Evans, R., Beck, C.T., Petropavlovskikh, I., McConville, G., Yu, W., Zhang, Z., Niu, J. et al. (2014). Performance of the ozone mapping and profiler suite (OMPS) products. *J. Geophys. Res. Atmos.*, 119(10), 6181–6195. https://doi.org/10.1002/2013jd020467.

Gelaro, R., Langland, R.H., Pellerin, S., Todling, R. (2010). The THORPEX observation impact intercomparison experiment. *Mon. Weather Rev.*, 138(11), 4009–4025. https://doi.org/10.1175/2010mwr3393.1.

Goldberg, M.D., Kilcoyne, H., Cikanek, H., Mehta, A. (2013). Joint polar satellite system: The United States next generation civilian polar orbiting environmental satellite system. *J. Geophys. Res. Atmos.*, 118(24), 13463–13475. https://doi.org/10.1002/2013jd020389.

Goodman, S.J., Schmit, T.J., Daniels, J., Redmon, R.J. (eds) (2020). *The GOES-R Series: A New Generation of Geostationary Environmental Satellites*. Elsevier, Amsterdam.

Heidinger, A.K., Pavolonis, M.J., Calvert, C., Hoffman, J., Nebuda, S., Straka, W., Walther, A., Wanzong, S. (2020). Chapter 6 – ABI cloud products from the GOES-R series. In *The GOES-R Series*, Goodman, S.J., Schmit, T.J., Daniels, J., Redmon, R.J. (eds). Elsevier, Amsterdam.

Ho, S.-P., Anthes, R.A., Ao, C.O., Healy, S., Horanyi, A., Hunt, D., Mannucci, A.J., Pedatella, N., Randel, W.J., Simmons, A. et al. (2020). The COSMIC/FORMOSAT–3 radio occultation mission after 12 years: Accomplishments, remaining challenges, and potential impacts of COSMIC–2. *Bull. Am. Meteorol. Soc.*, 101(7), E1107–E1136. https://doi.org/10.1175/bams-d-18-0290.1.

Kalluri, S., Cao, C., Heidinger, A., Ignatov, A., Key, J., Smith, T. (2021). The advanced very high resolution radiometer: Contributing to earth observations for over forty years. *Bull. Am. Meteorol. Soc.*, 102(2), E351–E366. https://doi.org/10.1175/bams-d-20-0088.1.

Kogan, F.N. (1997). Global drought watch from space. *Bull. Am. Meteorol. Soc.*, 78(4), 621–636. https://doi.org/10.1175/1520-0477(1997)078<0621:gdwfs>2.0.co;2.

Ma, Z., Li, J., Han, W., Li, Z., Zeng, Q., Menzel, W.P., Schmit, T.J., Di, D., Liu, C.-Y. (2021). Four-dimensional wind fields from geostationary hyperspectral infrared sounder radiance measurements with high temporal resolution. *Geophys. Res. Lett.*, 48(4), e2021GL093794. https://doi.org/10.1029/2021GL093794.

Menzel, W.P. (2020). Chapter 2 – History of geostationary weather satellites. In *The GOES-R Series*, Goodman, S.J., Schmit, T.J., Daniels, J., Redmon, R.J. (eds). Elsevier, Amsterdam.

Menzel, W.P., Schmit, T.J., Zhang, P., Li, J. (2018). Satellite-based atmospheric infrared sounder development and applications. *Bull. Am. Meteorol. Soc.*, 99(3), 583–603. https://doi.org/10.1175/bams-d-16-0293.1.

Nalli, N.R., Gambacorta, A., Liu, Q., Barnet, C.D., Tan, C., Iturbide-Sanchez, F., Reale, T., Sun, B., Wilson, M., Borg, L. et al. (2018). Validation of atmospheric profile retrievals from the SNPP NOAA-Unique combined atmospheric processing system. Part 1: Temperature and moisture. *IEEE Trans. Geosci. Remote Sens.*, 56(1), 180–190. https://doi.org/10.1109/tgrs.2017.2744558.

Rudlosky, S.D., Goodman, S.J., Virts, K.S. (2020). Chapter 16 – Lightning detection: GOES-R series geostationary lightning mapper. In *The GOES-R Series*, Goodman, S.J., Schmit, T.J., Daniels, J., Redmon, R.J. (eds). Elsevier, Amsterdam.

Schmit, T.J., Feltz, W.F., Menzel, W.P., Jung, J., Noel, A.P., Heil, J.N., Nelson III, J.P., Wade, G.S. (2002). Validation and use of GOES sounder moisture information. *Weather Forecast.*, 17(1), 139–154. https://doi.org/10.1175/1520-0434(2002)017<0139:VAUOGS>2.0.CO;2.

Schmit, T.J., Lindstrom, S.S., Gerth, J.J., Gunshor, M.M. (2018). Applications of the 16 spectral bands on the Advanced Baseline Imager (ABI). *J. Operational Meteor.*, 6(4), 33–46. https://doi.org/10.15191/nwajom.2018.0604.

WMO (2019). Space-based capabilities (OSCAR/Space), Technical report, World Meteorological Organization, Geneva [Online]. Available at: https://www.wmo-sat.info/oscar/spacecapabilities/.

Zhou, L., Divakarla, M., Liu, X. (2016). An overview of the joint polar satellite system (JPSS) science data product calibration and validation. *Remote Sens.*, 8(2), 1–13. https://doi.org/10.3390/rs8020139.

Zhou, L., Divakarla, M., Liu, X., Layns, A., Goldberg, M. (2019). An overview of the science performances and calibration/validation of Joint Polar Satellite System operational products. *Remote Sens.*, 11(6), 698. https://doi.org/10.3390/rs11060698.

Zou, C.-Z. and Qian, H. (2016). Stratospheric temperature climate data record from merged SSU and AMSU-A observations. *J. Atmos. Oceanic Technol.*, 33(9), 1967–1984. https://doi.org/10.1175/jtech-d-16-0018.1.

Zou, C.-Z., Goldberg, M.D., Hao, X. (2018). New generation of U.S. satellite microwave sounder achieves high radiometric stability performance for reliable climate change detection. *Sci. Adv.*, 4(10), eaau0049. https://doi.org/10.1126/sciadv.aau0049.

3

The Role of the National Aeronautics and Space Administration (NASA, USA)

Michael SEABLOM
NASA Earth Science Technology Office, Science Mission Directorate, Washington, DC, USA

3.1. The beginnings of the National Aeronautics and Space Administration (NASA)

In the United States, focused efforts on observing the Earth from space began in earnest with the creation of the National Aeronautics and Space Administration (NASA) in 1959. Under the Office of Space Science Applications (OSSA), the new agency began work on a research and satellite technology development program, Nimbus, that would produce new insights and understanding of the Earth's weather, oceans, atmosphere, land surface and ecosystems. The program was designed to complement data and insights provided by operational satellites operated by the United States Weather Bureau (later known as the Environmental Science Services Administration, or ESSA), primarily under the Television Infrared Observation Satellite (TIROS) program. The Nimbus program was designed to provide the second generation of such observations. It launched a total of seven missions between 1964 and 1978 and was notable for its successful development of new instrument technologies and for expanding the capabilities of low Earth orbiting

For a color version of all the figures in this chapter, see www.iste.co.uk/phulpin/satellites1.zip.

Satellites for Atmospheric Sciences 1,
coordinated by Thierry PHULPIN, Didier RENAUT, Hervé ROQUET and Claude CAMY-PEYRET.

spacecraft. The knowledge gained from Nimbus (NSSDC 2003) helped to blaze the trail for nearly all future missions, including the Landsat series, the satellites that would comprise the NASA Earth Observing System, and beyond. The technology and standards that were developed were transferred to the operational weather agencies in the US and adopted globally.

Figure 3.1. *Nimbus satellite (source: NASA)*

3.2. The Nimbus Era (1964–1979)

The 1964 launch of Nimbus-1 included the Advanced Vidicon Camera System (AVCS), the Automatic Picture Transmission (APT) camera and the High-Resolution Infrared Radiometer, enabling real-time imaging of cloud cover and the measurement of nighttime radiative temperatures of cloud tops and of surface terrain. The AVCS was designed to image an area of nearly 1,300 km^2, transmitted using a slow-scan (four lines per second) system. A new stabilization and control system allowed for a precise, three-axis pointing of the spacecraft, which was 3.5 m in height and nearly 3 m across when the solar panels were extended. The mission lasted for approximately four months.

Subsequent missions employed increasingly advanced technologies for both the instruments and for the spacecraft platforms. In 1966, Nimbus-2 was launched with an instrument suite similar to its predecessor; however, the camera now had the capability of generating an image of 720 x 3,400 km. Nimbus-2 also carried the Medium-Resolution Infrared Radiometer (MRIR), which measured the Earth's radiation balance in four spectral bands covering water vapor absorption, long-wave infrared energy, carbon dioxide absorption and Earth's albedo, along with one visible band. Although some real-time data were collected as late as 1969, most of the data archives are limited to the last six months of 1966 due to failures of the magnetic tape-recording system. In April 1969, the agency launched Nimbus-3 (also referred to as "Nimbus B"), which contained a satellite infrared spectrometer (SIRS), an infrared interferometer spectrometer (IRIS), the MRIR used by Nimbus-2, a high-resolution infrared radiometer (HRIR), an instrument to measure solar ultraviolet radiation (MUSE) and an image dissector camera system (IDCS) for providing daytime cloud-cover pictures along with advanced power and communication systems. The Nimbus-3 mission ended in September 1970, when the attitude control system of the spacecraft failed. Nimbus-4 was launched in 1970 and contained all predecessor instruments along with the Backscatter Ultraviolet (BUV) detector for measuring the vertical distribution of atmospheric ozone, and the selective chopper radiometer (SCR) that determined the vertical atmospheric temperature profile by measuring atmospheric absorption in the 15 μm band. The mission operated nominally for a year when attitude control problems once again appeared. Some data continued to be collected until the late 1980s.

Passive microwave instruments for remote sensing made their first appearance in the program with the launch of Nimbus-5 (or "Nimbus E") in 1972. The Nimbus E microwave spectrometer (NEMS) measured thermal radiation in five spectral bands between 0.51 and 1.35 cm, enabling atmospheric temperature profiles even in cloudy conditions. A second microwave instrument, the three-channel electrically scanning microwave radiometer (ESMR), measured horizontally polarized radiation at a wavelength of 1.55 cm. The ESMR data were successfully used in numerous sea ice studies, for example Gurney et al. (1993), and the instrument was a precursor for more capable microwave sensors in the 1990s. The mission also included the infrared temperature profile radiometer (ITPR) for obtaining vertical profiles of temperature and moisture at a resolution of 32 km, the SCR used by Nimbus-4, and the temperature/humidity infrared radiometer (THIR), which measured cloud-top temperatures and stratospheric water vapor. Also included with Nimbus-5 was the three-channel surface composition mapping radiometer (SCMR), which had the capability to measure radiation from the visible to the infrared spectrum suitable for studying surface composition and estimating sea surface temperature. While the SCMR failed shortly after launch, the remaining instruments operated until the end of the mission in 1973 (Staelin et al. 1973).

The ESMR was flown again on Nimbus-6, which launched in 1975. Seven other instruments were also included: the Earth radiation budget (ERB) scanner, the high-resolution infrared radiation sounder (HIRS), the limb radiance inversion radiometer (LRIR), the pressure modulated radiometer (PMR), the scanning microwave spectrometer (SCAMS) and the temperature humidity infrared radiometer (THIR). The ERB measured incoming solar radiation in the spectral range of 0.2–50 μm, and terrestrial radiation in three sub-intervals of that range. Vertical temperature profiles and water vapor distribution were provided by HIRS, while the THIR provided a window channel that provided images of clouds and temperatures of the cloud tops and land surfaces, along with water vapor distribution near the tropopause and in the lower stratosphere. The PMR extended temperature measurements to the stratopause and lower mesosphere. In addition to the ESMR, the SCAMS provided microwave radiance measurements in five bands, which resulted in data products that included tropospheric temperature profiles, water vapor and cloud water content. Nimbus-6 operated until 1983.

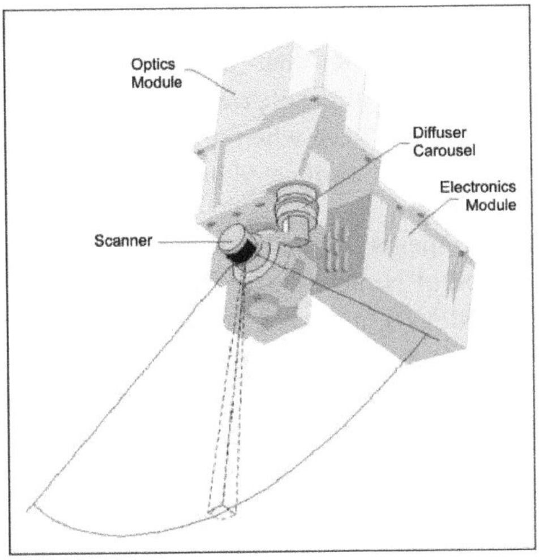

Figure 3.2. *Illustration of the Nimbus TOMS instrument*

In 1978, the last mission of the series, Nimbus-7, was launched with eight instruments on board. Two of the instruments – the ERB and the THIR – were re-flights from the previous missions. The new instruments included the coastal zone color scanner (CZCS), the limb infrared monitor of the stratosphere (LIMS), the scanning multispectral microwave radiometer (SMMR), the solar backscatter ultraviolet/total ozone mapping spectrometer (SBUV/TOMS), the stratospheric aerosol measurement II (SAM-II) and the stratospheric and mesospheric sounder (SAMS). The goal of the CZCS was to make a variety of measurements of the ocean, including chlorophyll concentration, salinity and temperature of ocean currents. The LIMS instrument, a follow-on to the LRIR of Nimbus-6, measured stratospheric ozone concentration, water vapor, nitrogen dioxide and nitric acid. SMMR indirectly measured ocean surface winds, water vapor, cloud properties and sea ice. SBUV measured the solar ultraviolet irradiance and ultraviolet radiation backscattered from the surface, while TOMS measured Earth radiance at six wavelengths in the band 0.31–0.38 μm. SAM-II, a solar photometer, provided information on the vertical distribution of stratospheric aerosols in the Earth's polar regions, measuring solar radiation at 1.0 μm. The SAMS radiometer measured temperature and vertical concentrations of water, nitrous oxide, methane, carbon monoxide and nitric oxide. The mission operated until 1994.

Numbers of Instruments / Type	Nimbus Mission Number						
	1	2	3	4	5	6	7
Visible Imaging	1	1	1	1			
Infrared Imaging	1	2			1	1	1
Microwave Imaging					1	1	1
Sounders			1	2	2	4	2
Radiometers			2	1	1	1	3
Other Spectrometers				1			1
	1964	1966	1969	1970	1972	1975	1978

Table 3.1. *Instrument types and count for Nimbus missions*

Programmatic changes occurred in 1979 which relocated the management of Nimbus to the Operational Satellite Improvement Program, through which the U.S. National Oceanic and Atmospheric Administration (NOAA) tasked NASA's science programs with developing technologies and demonstrating them with satellite missions. The multi-agency approach was difficult to implement, and NASA made the decision to terminate the Nimbus program and to begin defining the next generation of Earth observation missions.

The Nimbus program enabled a leap in our scientific understanding of the Earth's radiation budget, and its atmospheric temperature and moisture content characteristics. The assimilation of microwave and infrared profiles to operational numerical weather forecasts produced remarkable gains in predictive skill. The direct measurement of both incoming and outgoing solar radiation helped to establish a benchmark for establishing accurate measurement trends necessary for studying climate change, and aided in the development of improved climate models, most notably the processes that produce and maintain the protective ozone layer. During the 1980s, Nimbus-7 measurements confirmed evidence from ground-based measurements that indicated ozone depletion over the Antarctic during winter. The global scientific community took notice, and a series of field campaigns were conducted to better understand the severity of the problem, culminating with the Montreal Protocol that phased out the use of ozone-depleting chemicals.

3.3. The Earth Observing System (1982–2004)

In response to growing international concern over ozone depleting substances and changes in climate, studies led by the National Research Council began considering deployment of a wide range of capabilities in low altitude Earth orbit to make comprehensive measurements needed for the various disciplines of Earth Science, including climate, atmospheric chemistry, oceanography and land remote sensing. Scientists from around the world were engaged to develop a long-term strategy that would define the measurement requirements and the related needs for predictive modeling and operational weather forecasting. The results of the study, which was led by Dr. Francis Bretherton (National Research Council 1988), called for a program that would advance the scientific understanding of the Earth System, describe how the component parts and their interactions have evolved, and how they might change in the future. This was the beginning of NASA's Earth Observing System (EOS; personal communication with Dr. Dixon Butler, January 2021).

Figure 3.3. *The Upper Atmosphere Research Satellite*

During the formulation stages of EOS, the period between the end of the Nimbus program and EOS implementation that began in 1997, there were several notable missions that contributed significantly to the fields of meteorology and climate science. The Upper Atmosphere Research Satellite (UARS), launched in 1991, probed what had been the sparsely observed altitudes of the atmosphere between 75 km and 350 km. UARS consisted of 10 instruments that measured atmospheric chemistry, particularly concentrations of gases that contribute to ozone depletion.

The total ozone mapping spectrometer (TOMS), launched in 1996, continued total ozone column measurements that had begun with Nimbus-7. In 1997, the Sea-viewing Wide Field-of-view Sensor (SeaWiFS) mission, implemented through a commercial provider, flew with an optical scanner to obtain highly accurate ocean color data to quantify the characteristics of chlorophyll and production by marine phytoplankton. The SeaWiFS instrument represented a technological evolution of the Nimbus-7 CZCS.

NASA's planning for a space station started with a broad needs-assessment of Earth observations from low-altitude Earth orbit, which included planned deployment of EOS instruments. This resulted in two polar platforms being added to the initial concepts, and brought EOS into the context of a partnership among NASA and some of the International Space Station (ISS) Partner organizations, including the European Space Agency (ESA), the Japanese Aerospace Exploration Agency (JAXA) and the Canadian Space Agency (CSA). In addition to the space agencies, each of the four parties also included their operational weather agencies.

For the observing constellation, the ISS Partners designed a solution in which NASA satellites would provide coverage at the afternoon equator crossing time and ESA satellites would comprise the morning orbit for meteorological observations. JAXA would provide a microwave surface imager to address measurement requirements over oceans, cryosphere and atmosphere, and a high spatial resolution imager with visible, near-infrared and thermal infrared channels that would address measurement requirements over land, ice and for the study of clouds. In addition, CSA would provide an L-band synthetic aperture radar (SAR), which it would launch in partnership with NASA to address measurement needs over oceans, cryosphere and land surface. Under the original configuration, NASA EOS satellites would be serviceable by the U.S. Space Shuttle program.

In the early 1990s, budget realities intervened to help craft the final configuration in which EOS would consist of three separate satellites, one optimized for land surface observations in a morning orbit, one optimized for meteorological observations in an afternoon orbit and one optimized for atmospheric chemistry observations in a noon orbit. The initial names "AM", "PM" and "Chem" were eventually replaced with the more familiar "Terra", "Aqua" and "Aura". Because of these changes, EUMETSAT took over the responsibility for the morning orbit operational meteorological satellite, reducing costs for the US that had resulted from providing NOAA satellites in both morning and afternoon orbits.

Figure 3.4. *Terra satellite (source: NASA)*

To fully implement EOS, NASA instituted a series of programmatic changes, both at its Headquarters in Washington and at nearby Goddard Space Flight Center, one of the Agency's primary Earth Science facilities. In 1993, EOS management was placed under the new Earth Science Enterprise (ESE), replacing the OSSA role, and an Earth Science Directorate was established at Goddard. Several new mission support offices were set up, most notably for data management and technology. The EOS Data Information System (EOSDIS) was established to address, for that time, an unprecedented volume of data retrieved from satellites (the volume was predicted to reach 2 terabytes per day by the turn of the century), along with the need for processing and delivery of data products to hundreds of thousands of users. The task was particularly challenging in the late 1990s, when the nascent Internet was in a rapid state of evolution. The Earth Science Technology Office (ESTO) was established, which would identify and manage advanced technology projects deemed necessary by the ESE, would balance mission-specific, short-term technology solutions with a consolidated program that enables investment priorities to be based on Enterprise priorities rather than on local optimizations, and would reduce cost, schedule uncertainties and risks of performance by developing and maturing technologies at adequate levels of readiness prior to implementation into a flight mission.

The full EOS instrument set required several orbits that were not covered by the three Sun-synchronous orbit satellites, resulting in additional missions that operated parallel to the EOS flagships. The Clouds and the Earth's Radiant Energy System (CERES) instruments, which measure Earth's radiation budget, required three orbits with one in a non-polar inclination to make observations at all times of day. The EOS scatterometer, used to measure sea surface winds, became part of the joint NASA-JAXA Advanced Earth Observing Satellite (ADEOS) mission that was launched in 1996.

Figure 3.5. *TRMM satellite (source: NASA)*

The Tropical Rainfall Measuring Mission (TRMM), a joint mission between NASA and JAXA launched the following year, hosted CERES in support of the program with its primary instruments consisting of a precipitation radar, a microwave imager, a visible-to-infrared scanner and a lightning imaging sensor. With the first precipitation radar to fly in space, TRMM was a breakthrough mission for both climate research and for operational weather forecasting (National Research Council 2006). For the first time, a definitive description of the diurnal cycle of precipitation over the oceans and tropical land masses was made possible. Operationally, TRMM data were used for tropical cyclone monitoring, real-time rainfall estimation, assimilation of passive microwave information for improved numerical weather prediction, monitoring of precipitation and sea surface temperature.

The first EOS flagship satellite mission, Terra, was launched in 1999. Although its primary focus was on land surface type, condition, processes and change, it provides observations that are crucial to our understanding of the atmosphere, oceans, cryosphere and solar radiation. On board are five primary instruments.

– ASTER (Advanced Spaceborne Thermal Emission and Reflection Radiometer) – provided through a partnership with Japan – collects data in 14 spectral bands from the visible to the thermal infrared at spatial resolutions from 15 to 90 km.

– CERES (Clouds and the Earth's Radiant Energy System) comprises two broadband instruments, each with three spectral bands (shortwave, longwave and total radiance), measuring emitted and reflected radiations from the Earth's surface and the atmosphere.

– MISR (Multi-angle Imaging SpectroRadiometer) comprises nine cameras configured for multiple viewing angles in four spectral bands (blue, green, red and near-infrared). The different viewing angles enable the measurement of aerosols, albedo and other important parameters for studying climate change.

– MODIS (Moderate Resolution Imaging Spectroradiometer) is a wide-swath (2,330 km) sensor comprised of 36 spectral bands in the range of 0.4–1.4 μm, providing information on land use, aerosols, cloud properties, ocean color, water vapor, ozone and temperatures of the surface, atmosphere and cloud tops.

– MOPITT (Measurement of Pollution in the Troposphere) measures emitted and reflected radiances from the Earth in the 4.7 μm, 2.3 μm and 2.2 μm infrared spectral bands, using correlation radiometry to provide information on carbon monoxide and methane in daylight.

Figure 3.6. *Aqua satellite (source: NASA)*

Terra has been operating continuously for more than two decades, providing an unprecedented scientific data record. In early 2020, the last orbital maintenance maneuver was conducted, which will alter the satellite's altitude and crossing times during the mid-2020s.

The second EOS flagship satellite mission, Aqua, was launched in 2002 and was named for the mission's focus on observations related to the Earth's water cycle. Its six instruments provide global coverage of the atmosphere, land, ocean and the cryosphere. Aqua complements Terra by flying in a Sun-synchronous polar orbit with an afternoon equatorial crossing time, thereby optimizing the temporal coverage with Terra's morning crossing time. Aqua's instrument suite includes the following.

– AIRS (Atmospheric Infrared Sounder) is a wide-swath (1,650 km) instrument consisting of a grating spectrometer with 2,378 spectral channels within three infrared bands (3.74–4.61 μm, 6.2–8.22 μm and 8.8–15.4 μm), and a visible-to-near-infrared photometer with four spectral bands (0.4–0.94 μm), providing temperature and moisture profiles, surface temperature and emissivity, and total column ozone.

– AMSU (Advanced Microwave Sounding Unit) is a wide-swath (1,650 km) microwave sounder with 15 spectral bands (0.3–1.3 cm), providing atmospheric temperature, cloud properties (cloud drop and cloud ice effective radius, total column liquid water and ice, freezing level height) and integrated water vapor.

– AMSR-E (Advanced Microwave Scanning Radiometer for EOS) is a wide-swath (1,445 km) dual polarization, conically scanning, passive microwave radiometer with 12 spectral bands (six frequencies each with two polarizations, 0.33–4.3 cm), providing total precipitable water, integrated cloud liquid water, surface rain rate and rainfall accumulation amount, sea surface wind speed, sea surface temperature, sea ice concentration, snow water equivalent and soil moisture content.

– CERES (Clouds and the Earth's Radiant Energy System) comprises two broadband instruments each with three spectral bands (shortwave, longwave and total radiance), measuring emitted and reflected radiations from the Earth's surface and the atmosphere.

– HSB (Humidity Sounder for Brazil) is a wide-swath (1,650 km) microwave sounder with four spectral bands (0.16–0.2 cm).

– MODIS (Moderate Resolution Imaging Spectroradiometer) is a wide-swath (2,330 km) sensor comprised of 36 spectral bands in the range of 0.4–1.4 μm, providing information on land use, aerosols, cloud properties, ocean color, water vapor, ozone and temperatures of the surface, atmosphere and cloud tops.

Figure 3.7. *Aura satellite (source: NASA)*

Aqua has been operating continuously since its launch, and four of its six instruments have been providing quality data. HSB generated nearly nine months of data before the instrument failed in February 2003. In 2011, AMSR-E was placed in a reduced data collection mode for a year, and then ceased operations in 2016. Among the many scientific advancements made by Aqua, the predictive skill of operational weather prediction was significantly improved through the assimilation of AIRS temperature retrievals (Zhou et al. 2010).

The third EOS flagship satellite mission, Aura, launched in 2004, was primarily focused on atmospheric composition. The primary objective was to characterize the ozone layer, the processes that control tropospheric pollutants, and the roles of upper tropospheric aerosols, water vapor and ozone in climate change. On board the satellite are four science instruments.

– HIRDLS (High Resolution Dynamics Limb Sounder) is a 21-channel infrared radiometer (6.2–17.76 µm), with a vertical resolution of 1.2 km to an altitude of 50 km. The instrument is designed to measure a variety of atmospheric constituents (see Figure 3.8).

– MLS (Microwave Limb Sounder) is a passive microwave radiometer that was one of the instruments flown on UARS. It makes observations in spectral bands centered near five frequencies: 0.25 cm for temperature and pressure; 0.16 cm for water vapor and nitric acid; 0.12 cm for ozone and carbon monoxide; 0.05 cm for water vapor, nitrous oxide, chlorine monoxide, hypochlorous acid, bromine monoxide, the hydroperoxyl radicals and sulfur dioxide; and 0.01 cm for hydroxide ions.

– OMI (Ozone Monitoring Instrument) is a hyperspectral imager contributed by the Finnish Meteorological Institute (FMI). It was designed to continue the TOMS record of total ozone and other atmospheric chemistry parameters. OMI collects data in 780 spectral bands (270–504 nm) with a swath width of 2,600 km.

– TES (Tropospheric Emission Spectrometer) is a spectrometer measuring in the range of 3.2–15.4 µm with a spectral resolution high enough (0.025 cm^{-1}) to discriminate the vertical distribution of radiatively active molecular species in the troposphere. The instrument was capable of "step and stare" observations going from limb to nadir view.

Aura will likely continue operating through the 2020s. There have been some instrument failures since launch, however. The HIRDLS instrument stopped collecting data in 2008, and the TES instrument was decommissioned in 2018. The observations made from Aura, like the EOS flagships before it, have resulted in thousands of journal articles (Vandemark 2017).

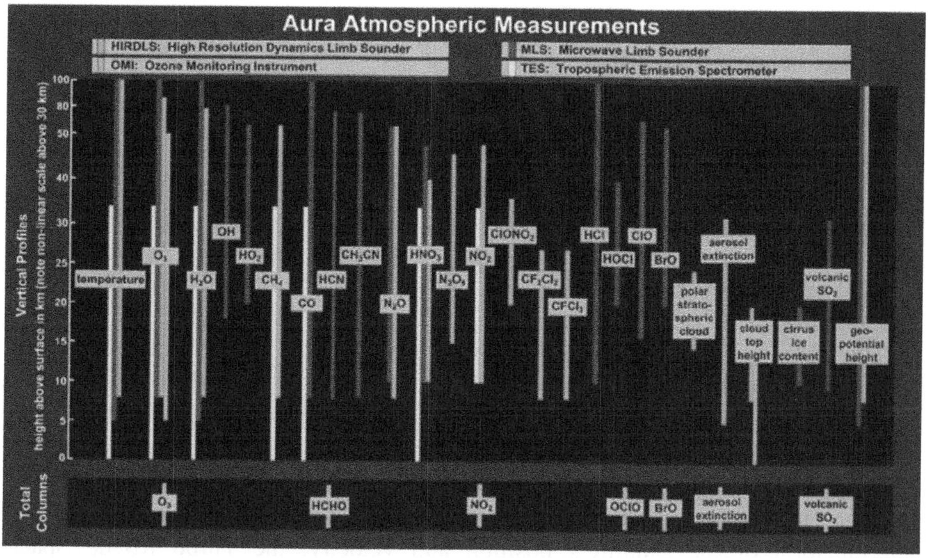

Figure 3.8. *Depiction of Aura measurements and their height range in the atmosphere; column measurements are shown in the panel at the bottom (source: NASA Aura)*

In addition to the three flagships, several other missions were launched under the EOS program between 1999 and 2004.

Landsat-7 was made part of EOS in 1994, and was part of a successful partnership between NASA and the U.S. Geological Survey (USGS), NOAA, and other agencies. It continued a long, successful line of missions to study terrestrial land use, deforestation monitoring, implications of natural disasters and others (United States Geological Survey 2020). The mission, which launched in 1999, contained the Enhanced Thematic Mapper (ETM+) radiometer, with six bands in the range of 0.48–2.2 μm, a thermal infrared band (10.4–12.5 μm) and a panchromatic band (0.52–0.9 μm). The visible bands have a resolution of 30 m, the thermal infrared 60 m, and the panchromatic 15 m, with a swath width of 183 km. In 2003, the Scan Line Corrector failed, reducing the data collection by about 25% per scene. The mission is expected to continue throughout the 2020s, complementing follow-on missions Landsat-8 launched in 2013 and Landsat-9 launched in 2021.

The QuikScat mission launched in 1999 was designed to measure wind vectors over the ice-free sea surface. The satellite carried a single instrument, SeaWinds, a conically scanning scatterometer with two beams operating at a frequency of 2.2 cm and a swath width of 1,800 km. The mission lasted 10 years, and its observations were successfully used in a variety of applications, most notably for hurricane research (Chelton et al. 2004).

ACRIMSAT (Active Cavity Irradiance Monitor Satellite) was launched in 1999 to monitor solar radiation, and was an early example of how small, cost-effective satellites could be used for Earth observation. The mission consisted of three independent active cavity radiometers used for measurement in the range of 0.5–5 μm, and for calibration. The mission lasted for four years, at which time contact with the spacecraft was lost.

The EOS altimeter became the Jason-1 mission, the result of a partnership between the US and France. Launched in 2001, the mission contained five instruments: Poseidon-2, a dual-frequency (2.2 cm, 5.7 cm) radar altimeter; the Jason Microwave Radiometer (JMR), which operated in three spectral bands (1.6, 1.26 and 0.88 cm); the Doppler Orbitography and Radiopositioning Integrated by Satellite (DORIS), a receiver that measures the Doppler shift from ground-based transmitters for determination of the satellite's position and velocity; the Turbo Rogue Space Receiver (TRSR), a GPS receiver; and the Laser Retroflector Array (LRA), which received laser signals from the surface.

SAGE III (Stratospheric Aerosol and Gas Experiment) was launched in 2001 as part of a collaboration between the US and Russia. Measurements of multi-wavelength aerosol extinction, the molecular density of ozone, nitrogen dioxide and water vapor, along with temperature profiles and clouds, were made with an 800-channel spectrometer from 280 to 1,040 nm. A separate photodiode made the aerosol extinction measurements at 1,550 nm. Observations from SAGE III and its successors have been used to study the vertical temperature structure of the stratosphere and mesosphere, as well as trace gases, and have provided reliable, long-term records of aerosols, ozone and water vapor in the upper atmosphere.

SORCE (Solar Radiation and Climate Experiment) was launched in 2003 to measure X-ray, ultraviolet, visible, near-infrared and total solar radiation for climate change studies. Four instruments were on board: the Spectral Irradiance Monitor (SIM), a spectrometer that measured the continuous spectrum 240–2,400 nm; the Total Irradiance Monitor (TIM), containing four cavity sensors; the Solar Stellar Irradiance Comparison Experiment II (SOLSTICE II), a grating spectrometer that measured the continuous spectrum 115–320 nm; and the XUV Photometer System (XPS), which measured eight spectral bands in the range of 1–35 nm, and one Lyman-α band (121.6 nm). The SORCE mission ended in 2020.

ICESat (Ice, Clouds and land Elevation Satellite) was launched in 2003 with its science focus on the characterization of polar ice sheets, aerosols, clouds and land surface canopy. The mission consisted of a single instrument, a laser altimeter that operated at two frequencies: 1.064 µm for surface measurements and 0.532 µm for the atmospheric measurements. The laser footprint was 70 m, with an altimetry precision of 10 cm. The mission produced new insights into ice mass loss over Greenland and both losses and gains over Antarctica (Smith et al. 2020). ICESat ended in 2010.

3.4. The "A-train" (2004–present)

In 2003, following the implementation of the full suite of EOS missions, the ESE began planning the next generation of space-based observing systems to support Earth System science research and applications. Several "exploratory" class missions were identified. These missions, mostly funded under the Earth System Science Pathfinder (ESSP) Program, included the Orbiting Carbon Observatory, which made use of a diffraction grating spectrometer to make indirect measurements of carbon dioxide (the first OCO experienced launch failure and was replaced with OCO-2 in 2014). The CALIPSO (Cloud-Aerosol Lidar and Infrared Pathfinder Satellite Observation) mission is a joint US–France satellite that hosts a two-wavelength (1,064 nm, 532 nm) lidar to measure high-resolution vertical profiles of aerosols and clouds, a nadir viewing imager (620–270 nm), and a three-channel infrared radiometer (8.7 mm, 10.5 mm, 12.0 mm) to enable retrievals of cloud emissivity and particle size. Through a partnership with Germany, the Gravity Recovery and Climate Experiment (GRACE), launched in 2002, made highly accurate observations of Earth's gravitational field through precise measurements of a satellite pair in polar orbit. GRACE provided critical information on changes in underground water storage, soil moisture and ice sheets until the mission ended in 2017. A successor mission, GRACE Follow-On, was launched in 2018. CloudSat, developed jointly with the CSA, hosts a cloud profiling radar that operates at 3.19 mm, providing unprecedented sensitivity for measuring small particles of liquid water and ice in clouds (Stephens et al. 2002).

CloudSat and CALIPSO provide critically needed information on climate change by filling major gaps in the observations of cloud properties and aerosols. Clouds have a significant effect on Earth's radiation balance, and their adequate characterization is critical to our understanding of the planetary water cycle. CloudSat was the first spaceborne millimeter wavelength radar. It measures the vertical structure of clouds and precipitation from the surface to 30 km using radar reflectivity measurements. In fact, it was the first mission to provide detailed statistics on the vertical structure of clouds, accurately determining the water/ice ratio of clouds and the amount that is converted to precipitation. It was also the first mission to detect snowfall from space. The mission contributed significantly to global modeling efforts through improved measurements of the vertical distribution of heating due to clouds. The CALIPSO mission measures backscatter from aerosols and enables the accurate classification of aerosol types. With these coordinated measurements, the A-train enabled new research into the relationship between pollution and clouds.

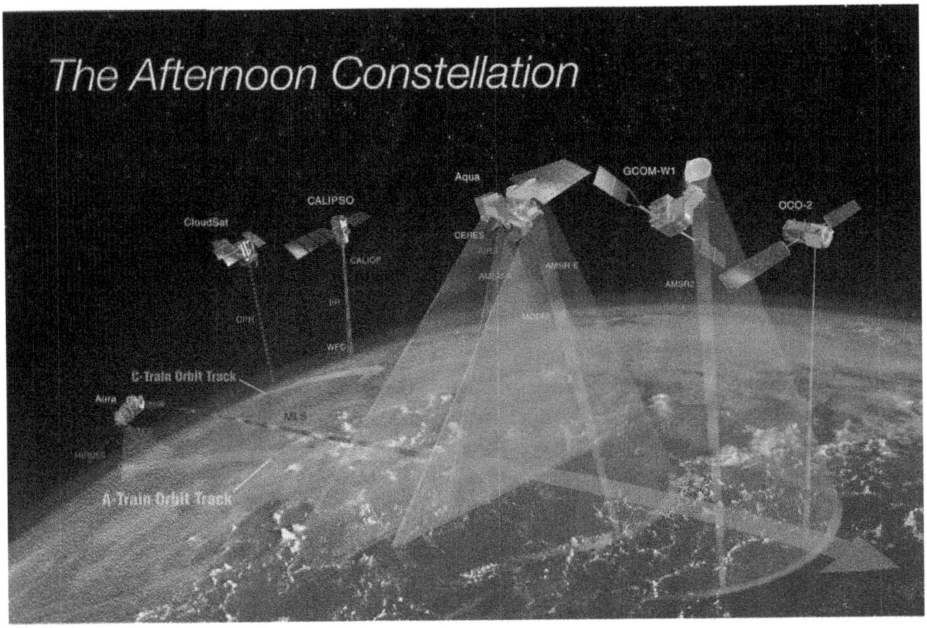

Figure 3.9. *A depiction of the A-train constellation of international satellites. Microwaves (observed by both AMSR instruments, AMSU-A, CPR, MLS) are represented as red-purple to deep purple colors; yellow represents solar wavelengths (OMI, OCO-2); gray represents solar and infrared wavelengths (MODIS, CERES); and red represents other infrared wavelengths (IIR, AIRS, TES, HIRDLS) (source: NASA)*

Through international cooperation between NASA, the Centre National d'Etudes Spatiales (CNES) and JAXA, between 2003 and 2006 the recently launched satellites were placed in a closely following constellation orbital track. This began with the launch of CloudSat and CALIPSO joining with Aqua and Aura, and eventually with PARASOL (the Polarization and Anisotropy of Reflectances for Atmospheric Science coupled with Observations from Lidar) mission launched by the CNES in 2004. The name "A-train" was applied because Aura and Aqua were the endpoints of the constellation, and because it referenced the 1:30 PM local solar time equatorial crossing. The train expanded with the addition of the JAXA Global Change Observation Mission – Water (GCOM-W1) in 2012, and OCO-2 in 2014. Over the years, the A-train has provided a unique Earth observing opportunity because it consists of both active and passive sensors. When combined, these missions provide multiple observations of the Earth in the ultraviolet to microwave wavelength range, and angled viewing geometries important to the assessment of climate change (e.g. L'Ecuyer and Jiang (2010)).

As originally configured, the synergistic measurements of the A-train were made possible by careful coordination of the formation-flying platforms. CloudSat was designed to make routine orbital adjustments to maintain formation with Aqua and CALIPSO such that radar and lidar observations have frequent overlap and fall within the Aqua MODIS footprint. The combined CloudSat and/or CALIPSO observations allowed for unprecedented 3D representations of how clouds and aerosols interact and evolve. In recent years, the satellites comprising the A-train changed. PARASOL was removed from the constellation in 2011, and in 2018 CloudSat and CALIPSO were placed in a separate "C-Train" orbit 17 km beneath the A-train. Figure 3.9 depicts the current state of the A-train1[1]. The observation geometry of each constellation instrument is shown, with active instruments aboard CALIPSO and CloudSat indicated with dashed lines. This illustration color-codes instrument swaths – the area of Earth's surface, or the surface of its atmosphere, over which data are collected – based on observed wavelength ranges.

Figure 3.10. *The SMAP satellite (source: NASA JPL/CalTech)*

1. The figure and the constellation status are updated at: https://atrain.nasa.gov.

3.5. Decadal surveys and technological disruption (2007–present)

With the implementation of the Vision for Space Exploration in 2004, another reorganization of NASA Headquarters occurred, this time placing Earth Science management under the newly created Science Mission Directorate. With growing concern over maintaining the measurement capabilities that were established after the turn of the century, NASA, along with NOAA's National Environmental Satellite Data and Information Service (NESDIS), and the USGS, tasked the U.S. National Research Council with a study ("Decadal Survey") to help define a new suite of measurements that would extend the predictive skill of weather forecasts, as well as longer-term scientific understanding of global change (National Research Council 2007). For NASA, the committee recommended a series of missions, along with corresponding recommendations for research, applications and technology development that would address measurement needs aimed at understanding the Earth System processes and dynamics. Two of the recommended missions have been implemented: the Soil Moisture Active Passive (SMAP) mission launched in 2015, and ICESat-II – the next-generation laser altimeter launched in 2018. The SMAP mission acquires measurements using an L-band radar and an L-band radiometer over the entire global land surface to estimate surface soil moisture. While the radar failed shortly after its 2015 launch, data from radiometer is being used to produce research-quality data. A third Decadal Survey mission, the joint NASA/CNES Surface Water and Ocean Topography (SWOT) mission, slated for launch in 2022, will contain a Ka-band radar interferometer capable of measuring ocean and surface water levels over a 120 km swath, an altimeter, a microwave radiometer and a laser reflector assembly for ground targeting.

In addition to these Decadal Survey missions, the launch of the Global Precipitation Mission (GPM) in 2014 represented a major leap forward in the study of the global water and energy cycle. GPM, a collaboration between NASA and JAXA, contains a dual-frequency (2.2 cm, 0.84 cm) precipitation radar and a microwave imager (0.16–3.0 cm). GPM has built upon the success of TRMM through its ability to monitor and observe both rainfall and snowfall globally (TRMM operated within 35° of latitude from the equator), and with its advanced instrumentation, GPM can generate three-dimensional precipitation profiles and highly accurate measurements of precipitation intensity. GPM operates as a "core observatory", designed to assimilate data from a constellation of partner satellites, and perform continuous calibration from those measurements.

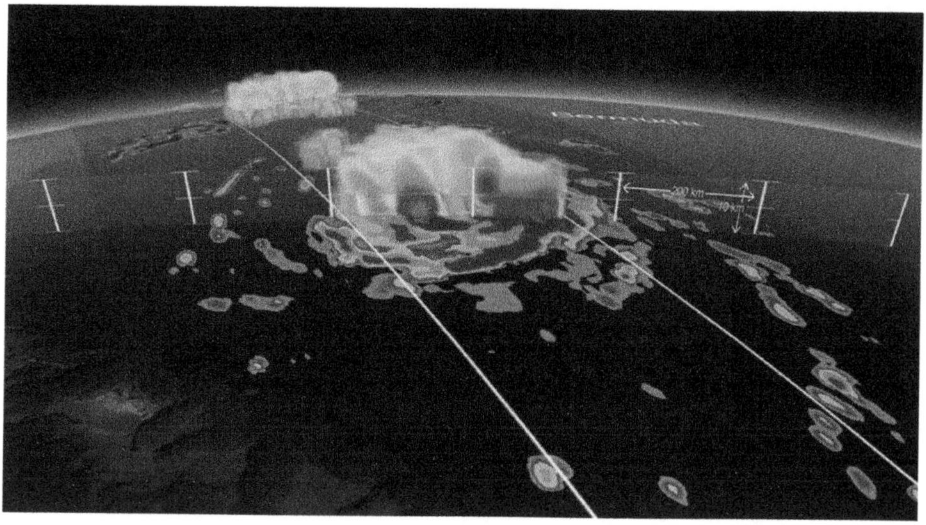

Figure 3.11. *GPM captures the three-dimensional storm structure of a hurricane (source: NASA GSFC Scientific Visualization Studio)*

By 2010, the emergence of disruptive technologies began to influence strategic planning for new Earth Science missions. Advancements in electronics, detectors and advanced subsystems began to significantly reduce the resource requirements for many of the recommended missions, enabling higher acceptance of risk. At the same time, a growing commercial market in CubeSats and other types of small satellites, with the potential for rapid access to space through the various Rideshare programs. By 2013, ESTO's Instrument Incubator Program (IIP) began soliciting for miniaturized science instruments for use on small satellites, resulting in 7 of 17 of its 2016 portfolio targeted at new instrument technology for small satellites. ESTO also established the In-space Validation of Earth Science Technologies (InVEST) program for validation of instrument technologies on CubeSat platforms and, as of 2021, there have been eight projects awarded. A notable breakthrough occurred with the development of RainCube (Peral et al. 2017), a Ka-band precipitation radar hosted on a 6U CubeSat platform. The early success of RainCube in characterizing the spatial structure of clouds and precipitation might indicate that future constellations of similar, low-cost radars could complement, or possibly replace a flagship mission such as GPM, with continued advancements in technology.

Figure 3.12. *RainCube (source: NASA JPL/CalTech)*

Small satellites for NASA have already moved beyond technology demonstrations. A Venture Class program, targeted at funding low-cost, innovative science missions, was established following a recommendation in the 2007 Decadal Survey. As of 2020, four Venture Class awards have been made involving small satellite missions. Two were operational: the Cyclone Global Navigation Satellite System (CYGNSS), which measures ocean surface winds and soil moisture over land using reflected signals from the Global Positioning System (GPS); and the Temporal Experiment for Storms and Tropical Systems – Demonstration (TEMPEST-D), which measures the temporal evolution of clouds at the onset of precipitation using a millimeter-wave radiometer. There is also a rapidly growing commercial market for Earth observation from small satellites, and NASA expects to leverage such capabilities going forward.

The National Academies released a second Earth Science Decadal Survey in 2018 (National Academies of Sciences, Engineering, and Medicine 2018), which largely maintained the overarching science recommendations of the previous study, but with recognition of new advances in small spacecraft technology, artificial intelligence, the use of "sensor webs" (i.e. use of multiple observing platforms for a common observation goal), enhanced numerical models and the use of commercial products to augment NASA's measurement requirements. In 2021, NASA's Earth Science Division responded to the recommendations by establishing the Earth System Observatory (ESO), a collection of missions that will focus on climate change impacts, disaster mitigation, addressing wildfires and improving real-time agricultural processes. The initial focus areas for the observatory include aerosols, clouds/convection/precipitation, mass change, surface biology and geology, and surface deformation and change. A pathfinder mission for the observatory will be a partnership between NASA and the Indian Space Research Organisation (ISRO) to measure changes in the Earth's surface at unprecedented scales. The mission, NISAR (NASA-ISRO Synthetic Aperture Radar), will measure changes in the planet's ice sheets as well as earthquakes, volcanoes and landslides. During the 2020s and beyond, the ESO will continue to implement the recommendations of the Decadal Survey through its competed flight programs.

3.6. References

Chelton, D.B., Schlax, M.G., Freilich, M.H., Milliff, R.F. (2004). Satellite measurements reveal persistent small-scale features in ocean winds. *Science*, 303, 978–983.

Gurney, R.J., Foster, J.L., Parkinson, C.L. (1993). *Atlas of Satellite Observations Related to Global Change.* Cambridge University Press, Cambridge.

L'Ecuyer, T. and Jiang, J. (2010). Touring the atmosphere aboard the A-Train [Online]. Available at: https://atrain.nasa.gov/publications/A-TrainPhysicsToday201007.pdf.

National Academies of Sciences, Engineering, and Medicine (2018). *Thriving on Our Changing Planet: A Decadal Strategy for Earth Observation from Space.* National Academies Press, Washington, DC.

National Research Council (1988). *Earth System Science: A Closer View.* National Academies Press, Washington, DC.

National Research Council (2006). *Assessment of the Benefits of Extending the Tropical Rainfall Measuring Mission.* National Academies Press, Washington, DC.

National Research Council (2007). *Earth Science and Applications from Space: National Imperatives for the Next Decade and Beyond.* National Academies Press, Washington, DC.

NSSDC (2003). NSSDC Earth Science – Nimbus Program [Online]. Available at: https://nssdc.gsfc.nasa.gov/earth/nimbus.html/.

Peral, E., Imken, T., Sauder, J., Statham, S., Tanelli, S., Price, D., Chahat, N., Williams, A. (2017). RainCube, a Ka-band Precipitation Radar in a 6U CubeSat. *Proceedings of the Small Satellite Conference*, Logan, UT, SSC17-III-03.

Schoeberl, M., Douglass, A., Hilsenrath, E., Bhartia, P., Beer, R., Waters, J., Gunson, M., Froidevaux, L., Gille, J., Barnett, J. et al. (2006). Overview of the EOS Aura mission. *IEEE Transactions on Geoscience and Remote Sensing*, 44, 1066–1074.

Smith, B., Fricker, H., Gardner, A., Medley, B., Nilsson, J., Paolo, F., Holschuh, N., Adusumilli, S., Brunt, K., Csatho, B. et al. (2020). Pervasive ice sheet mass loss reflects competing ocean and atmosphere processes. *Science*, 368, 1239–1242.

Staelin, D.H., Barrett, A.H., Waters, J.W., Barath, F.T., Johnston, E.J., Rosenkranz, P.W., Gaut, N.E., Lenoir, W.B. (1973). Microwave spectrometer on the Nimbus 5 satellite: Meteorological and geophysical data. *Science*, 182(4119), 1339–1341.

Stephens, G., Vane, D., Boain, R., Mace, G., Sassen, K., Wang, Z., Illingworth, A., O'Connor, E., Rossow, W., Durden, S. et al. (2002). The Cloud Sat mission and the A-Train, a new dimension of space-based observations of clouds and precipitation. *Bulletin of the American Meteorological Society*, 83(12), 1771–1790.

United States Geological Survey (2020). Landsat science team publications [Online]. Available at: https://www.usgs.gov/core-science-systems/nli/landsat/landsat-science-team-publications.

Vandemark, D. (2017). NASA Earth Science Senior Review Subcommittee Report [Online]. Available at: https://smd-prod.s3.amazonaws.com/science-pink/s3fs-public/atoms/files/2017-NASA-ESSR_V2_FINALcorrected.pdf.

Zhou, Y., Lau, K.-M., Reale, O., Rosenberg, R. (2010). AIRS impact on precipitation analysis and forecast of tropical cyclones in a global data assimilation and forecast system. *Geophysical Research Letters*, 37(10).

4

The Role of the European Space Agency (ESA)

Paul INGMANN
Retired, European Space Agency's European Space Research and Technology Centre (ESA/ESTEC), Noordwijk, The Netherlands

4.1. Missions in Geostationary Orbit (GEO) – ESA's Start in Earth Observation

ESA – The European Space Agency is an intergovernmental organization dedicated to the exploration of space. Established in 1975 as a merger of ESRO and ELDO with their headquarters in Paris. Current (as of 2021) Member States are Austria, Belgium, Czech Republic, Denmark, Estonia, Finland, France, Germany, Greece, Hungary, Ireland, Italy, Luxembourg, the Netherlands, Norway, Poland, Portugal, Romania, Spain, Sweden, Switzerland and the United Kingdom. Slovenia and Latvia are Associate Members. Canada takes part in some projects under a cooperation agreement. In addition, there are cooperation agreements with Bulgaria, Croatia, Cyprus, Malta, Lithuania and Slovakia.

ESA's mandatory activities, for example Space Science, are funded by a financial contribution from all Member States, calculated in accordance with each country's gross national product (GNP). In addition, ESA conducts a number of optional programs. This includes Earth Observation. Each Member State decides which optional program they

For a color version of all the figures in this chapter, see www.iste.co.uk/phulpin/satellites1.zip. More detailed information on the various ESA missions can be found at https://earth.esa.int/web/eoportal/satellite-missions.

Satellites for Atmospheric Sciences 1,
coordinated by Thierry PHULPIN, Didier RENAUT, Hervé ROQUET and Claude CAMY-PEYRET.

wish to participate in and the amount they wish to contribute. Regardless, for both mandatory and optional programs, geo-return[1] has to be met.

The Council is ESA's governing body and provides the basic policy guidelines within which ESA develops the European space program. Each Member State is represented on the Council and has one vote, regardless of its size or financial contribution.

Background information on the first series of European geostationary satellites can be found in Chapter 1.

In March 1983, an inter-governmental conference at ESA decided that the execution of an Operational Program of Meteosat would be entrusted to ESA, which would cover inter alia:

– the development and launching of three Meteosat Operational Program (MOP) spacecraft of an improved design compared with the preoperational satellites;

– the exploitation of the spacecraft observations (from the Preoperational to the Operational Program) from November 1983 until November 1995 on behalf of EUMETSAT.

The success of the Meteosat preoperational program led a large group of European nations to establish firm plans for sustained operations by an independent agency. On May 24, 1983, 16 countries signed the EUMETSAT (European Organisation for the Exploitation of Meteorological Satellites) convention, which came into force in June 1986. As of January 1, 1987, responsibility for the operation of the Meteosat satellites was transferred from ESA to EUMETSAT. EUMETSAT has been the owner of the Meteosat series beginning with Meteosat-4. Since then, ESA continued to build the MOP (Meteosat Operational Program) spacecraft but now on behalf of EUMETSAT. In a 1993 agreement, EUMETSAT was given the full operational mandate. This means that EUMETSAT provides all operational functions and services of its satellites with a dedicated EUMETSAT ground segment. On November 15, 1995, EUMETSAT took over the operations of the Meteosat system and implemented the Meteosat Transition Program (MTP). This program included the provision and launch of a further satellite of the same design (Meteosat-7), the development of a new ground system and routine operations from December 1995 onwards. The so-called First Generation Meteosat consisted of seven satellites summarized in Table 4.1 (pre-operational Meteosat satellites (-1 to -3) and Meteosat operational satellites (-4 to -7)).

1. The ratio between the share of a country in the weighted value of contracts and its share in the contribution paid to the Agency must be balanced by the end of a given period. This ratio is called the industrial return coefficient or geo-return.

Spacecraft	Launch Date	Comments
Meteosat-1	November 23, 1977	Service provision until November 1979 when the imager failed. DCS ended in 1984.
Meteosat-2	June 19, 1981	Service operation from August 12, 1981 until August 11, 1988. Deorbited from GEO in December 1991.
Meteosat-3	June 15, 1988	Meteosat-3 became prime shortly after launch since Meteosat-2 had reached a critical inclination. Retired in 1995.
Meteosat-4 (MOP-1)	March 06, 1989	First image on April 19, 1989. Entered service on June 19, 1989. Deactivated in November 1995.
Meteosat-5 (MOP-2)	March 2, 1991	Primary S/C for the 0deg service over Europe from February 1994 until 1997. Decommissioned in April 2007.
Meteosat-6 (MOP-3)	November 20, 1993	In October 2002, Meteosat-6 moved from 9deg W to 10deg E to accommodate MSG-2. Decommissioned in April 2011.
Meteosat-7 (MOP-4)	September 2, 1997	Part of the MTP (Meteosat Transition Program) towards MSG (Meteosat Second Generation). In June 1998, Meteosat-7 became the primary operational S/C at 0deg location. Decommissioned in March 2017.

Table 4.1. *Meteosat preoperational and operational program (MOP) missions*

In parallel with the pre-operational Meteosat program, preparations of Meteosat Second Generation (MSG) were initiated in the early 1980s in order to avoid any gap in services after MOP. The concept of the MSG satellites was developed through a series of workshops organized by ESA with the European meteorological community, which started in Avignon, France in June 1984. Assessment studies lead to the conclusion that the imager proposed by a group of imagery experts could be expanded to accommodate 12 channels in total, that is, one high-resolution visible, three in the VNIR, and eight in the IR. Finally, ESA initiated the development of a spin-stabilized, geostationary MSG satellite with a 12-channel imager, called the Spinning Enhanced Visible and Infra-Red Imager (SEVIRI). Also accommodated was a science payload called GERB (Geostationary Earth Radiation Budget); see Harries et al. (2005). The first launch of MSG-1 took place on August 28, 2002. MSG-1 became Meteosat-8 on January 29, 2004 when the mission was declared "operational". More details about the MSG system can be found in Schmetz et al. (2002) and a comparison of MFG versus MSG is shown in Table 4.2.

More details about these programs can be found in Chapter 5.

	Meteosat First Generation (MFG)	Meteosat Second Generation (MSG)
Main instrument	VIS and IR imager (MVIRI)	VIS and IR imager (SEVIRI)
Spectral channels	3 (1 VIS and 2 IR)	12 (4 VIS, VNIR and 8 IR)
Sampling distance at SSP	2.5 km (VIS) and 5 km (IR)	1 km (HRV) and 3 km (IR)
Repeat cycle	30 minutes	15 minutes
Stabilization concept	Spin-stabilized – 100 rpm	Ditto
Dimensions (overall height in m, diameter in m)	3.2, 2.1	3.7, 3.2
Design life (years)	5	7

Table 4.2. *Comparison of key properties of MFG and MSG*

4.2. Missions in low Earth orbit (LEO)

Based on the success of meteorological missions in GEO, ESA Member States decided that more missions addressing environmental issues should be prepared. The ERS (European Remote-Sensing Satellite) program was the first ESA program in Earth observation with the overall objective of providing environmental monitoring. ERS-1 focused on the microwave spectrum (i.e. regular monitoring of land–surface and ocean–surface processes for change detection). Coverage of a broad range of disciplines and topics, namely observation of oceans, polar ice, land ecology, geology, forestry, wave phenomena, bathymetry (water depth), atmospheric physics, meteorology, etc. Scientific research was also a target. Besides, the demonstration of new concepts and technologies for space and ground segments and overall high performances plus operational capabilities were aimed at.

4.2.1. *ERS*

The ERS program had a pioneering character in the field of instrument development (active sensors), and the introduction and demonstration of advanced observation technologies – it became a European success story. The ERS preparatory and development program was initiated at ESA in 1981. The satellite carried a suite of instruments, namely:

– an Active Microwave Instrument (AMI) operated in three modes, namely “SAR imaging mode”, “wave mode” and “wind scatterometer mode”;

– a Radar Altimeter (RA), later called RA-1;

– an Along-Track Scanning Radiometer (ATSR) in the IR, later called ATSR-1, with a microwave radiometer to reduce errors in IR and mw measurements due to water vapor and liquid water content;

– a Laser Retro-Reflector (LRR);

– the Precise Range And Range-Rate Equipment (PRARE).

The launch of the ERS-1 spacecraft took place on July 17, 1991 on an Ariane-4 launcher from Kourou, French Guiana.

It was in 1989 when ESA decided to continue the ERS program beyond the lifetime of ERS-1. ERS-2 was the follow-up ESA mission of ERS-1 and was essentially a copy of ERS-1. The differences are a number of enhancements (e.g. ATSR with more channels, i.e. ATSR-2) and a new payload instrument to measure the chemical composition of the atmosphere, named GOME (Global Ozone Monitoring Experiment). Background was that a margin in payload capability permitted to fly an additional instrument. As at that time preparations for Envisat had already started (see below) and an instrument called Sciamachy was foreseen as a payload, ESA Member States came to the conclusion that a slimmed down version of Sciamachy should be added to the payload.

The launch of the ERS-2 spacecraft took place on April 21, 1995 from Kourou on an Ariane-4 vehicle.

In the late 1980s, ESA member states decided that the monitoring and management of the Earth's resources, both renewable and non-renewable and continuation and improvement of the services provided to the worldwide operational meteorological community, should be continued. A concept called POEM (Polar Orbit Earth-observation Missions) was developed as a successor to ERS-1/2 using a Polar Platform concept. However, in 1992, at the ESA Ministerial Council in Granada, Spain, it was decided to split POEM into two dedicated missions called EnviSat (Environmental Satellite), focused on environmental issues with a more research-oriented package of essentially preoperational instruments and MetOp (Meteorology Operational Programme) devoted primarily to operational meteorology and climate monitoring in a morning orbit.

4.2.2. *Envisat*

The development of the payload instruments for Envisat began in 1994. The payload of Envisat consisted of the following:

– MERIS (Medium Resolution Imaging Spectrometer);

– ASAR (Advanced SAR);

– MIPAS (Michelson Interferometer for Passive Atmospheric Sounding);

– RA-2 (Radar Altimeter-2);

– MWR (Microwave Radiometer);

– GOMOS (Global Ozone Monitoring by Occultation of Stars);

– SCIAMACHY (Scanning Imaging Absorption Spectrometer for Atmospheric Cartography);

– AATSR (Advanced Along-Track Scanning Radiometer; see also ATSR (ERS-1,-2);

– DORIS (Doppler Orbitography and Radiopositioning Integrated by Satellite);

– LRR (Laser Retro-Reflector).

MIPAS, GOMOS and SCIAMACHY addressed atmospheric composition issues

– MIPAS: a high-resolution FTS (Fourier-Transform Spectrometer) instrument operating in the IR in limb viewing mode designed to measure pressure and temperature profiles, and profiles of trace gases including nitrogen dioxide (NO_2), nitrous oxide (N_2O), methane (CH_4), nitric acid (HNO_3), ozone (O_3), various chlorofluorocarbons (CFCs) and water (H_2O) in the stratosphere and mesosphere.

– GOMOS: a UV/visible/near-infrared limb viewing grating spectrometer, operating in occultation mode with the objective of high accuracy (altitude-resolved) and global long-term monitoring of stratospheric and mesospheric NO_2, nitrogen trioxide (NO_3), chlorine dioxide (OClO), O_3 and aerosols present between about 20 and 80 km (12–50 mi) altitude.

– SCIAMACHY: a passive spectrometer (of GOME heritage) for the measurement of a variety of trace gases like O_3, oxygen (O_2), bromine monoxide (BrO), OClO, chlorine monoxide (ClO), sulfur dioxide (SO_2), nitric oxide (NO), NO_2, NO_3, carbon monoxide (CO), carbon dioxide (CO_2), CH_4, H2O, N_2O and aerosol in the troposphere and stratosphere using differential optical absorption spectroscopy.

Envisat, the largest civilian Earth observation satellite ever built, was launched on March 1, 2002 and operated until April 8, 2012, when a problem arose. Contact with Envisat was unexpectedly lost, preventing the reception of any data. Finally, on May 9, 2012, ESA declared the end of the Envisat mission.

4.2.3. *MetOp*

The MetOp series consisting of the three satellites MetOp-A, -B and -C, although an independent development, is complementary to: a) the NOAA POES system (in LEO, i.e. the European component of the Initial Joint Polar System (IJPS)); b) the EUMETSAT/ESA MSG (Meteosat Second Generation) system (in GEO); and c) the ESA Envisat system (in LEO) until Envisat failed in 2012.

Instrument	Mission objectives	Instrument provider
AVHRR/3 (Advanced Very High Resolution Radiometer)	Global imagery, global sounding, ocean measurements (SST), clouds and Earth radiation budget, land measurements	NOAA
HIRS/4 (High Resolution Infrared Sounder)	Global sounding of temperature and humidity, atmospheric minor constituents (ozone)	NOAA
AMSU-A (Advanced Microwave Sounding Unit-A – A1 & A2)	Global sounding of temperature, sea ice	NOAA
MHS (Microwave Humidity Sounder)	Global sounding of humidity, clouds and Earth radiation budget, sea ice	EUMETSAT
IASI (Infrared Atmospheric Sounder Interferometer)	Global sounding, ocean measurements (SST), clouds and Earth radiation budget, some atmospheric trace constituents, land measurements	CNES/EUMETSAT
ASCAT (Advanced Scatterometer)	Ocean measurements, surface stress and surface wind	ESA
GOME-2 (Global Ozone Monitoring Experiment-2)	Atmospheric trace gases (ozone content and profile)	ESA/EUMETSAT
GRAS (GNSS Receiver for Atmospheric Sounding)	Atmospheric refractive index measurement in limb-sounding mode	ESA/EUMETSAT

Table 4.3. *Overview of the main MetOp payload complement*

MetOp-A was Europe's first polar-orbiting LEO satellite dedicated to operational meteorology. It is the space segment of EPS, i.e. the EUMETSAT Polar system. MetOp satellites carry 13 instruments provided cooperatively by EUMETSAT, ESA, NOAA and CNES. While AVHRR/3, HIRS and AMSU-A1/A2 are recurrent copies of instruments that had been developed within the U.S. Polar Orbiting Environmental Satellite (POES) Program, three completely new instruments have been developed specifically for MetOp, namely IASI, ASCAT and GRAS (see Table 4.3).

With respect to the ESA or ESA/EUMETSAT developed instruments, the following should be noted:

– ASCAT – the scatterometer is based on AMI-SCAT flown on ERS-1 and -2, although with improved coverage because of double swath operation;

– GOME-2 – the spectrometer is based on GOME-1 flown on ERS-2 but with improvements concerning spatial sampling (40 km x 40 km vs 320 km x 40 km), four slightly different spectral channels, and improvements in polarization measurements and calibration;

– GRAS – this is an entirely new instrument. Its development was motivated by the success of GPS/MET (Kursinski et al. 1997) providing atmospheric temperature and humidity profiles. The measurement principle is based on GNSS radio occultation (see Chapter 9).

4.2.4. *The Earth Explorer and Earth Watch Concept*

In Toulouse in 1995, the ESA Council at the Ministerial Level set the scene for ESA's Earth Observation strategy and issued a "Proposal for a European Policy for Earth Observation from Space", put forward jointly by ESA, the European Commission (EC) and EUMETSAT. Three fundamental objectives were set in Toulouse:

– develop our knowledge of the Earth;

– preserve the Earth and its environment;

– manage life on Earth in a more efficient way.

And the principle of two complimentary types of Earth Observation missions serving user needs was adopted:

– Earth Explorers for research, including the demonstration of new observing techniques;

– Earth Watches, i.e. prototype operational missions serving operational application needs.

This helped in establishing ESA's Living Planet Programme. The most recent update took place in 2015 (ESA 2015).

4.2.4.1. *ESA's Earth Explorer missions*

Earth Explorer (EE) missions were originally split into two kinds of missions, namely Core and Opportunity (Opp) missions: Core missions are larger missions where only a conceptual idea is required to start from. In addition, fast track missions were introduced. In essence, they are similar to Core missions although with a shortened feasibility phase with the overall objective to shorten the time span between proposing an idea and mission implementation. Opportunity missions, however, should be smaller and based on an existing conceptual design, that is, basic feasibility should have been demonstrated. For both kinds of missions "Calls" are issued, that is, a "Call for Ideas" for Core and fast track missions and a "Call for Proposals" for Opportunity missions. To formulate the detailed user requirements, the Agency is supported by Mission or Science Advisory Groups (MAGs or SAGs), usually one per mission. User consultation meetings involving the wider user community are part of the process. A first such meeting took place in 1996, leading to nine concepts for Core missions. From that, the first two Core missions were selected in 1999, namely GOCE (ESA 1999a) and ADM-Aeolus (ESA 1999b). The most recent user consultation meeting took place in 2022 which led to the selection of Harmony as EE10, a mission that measures various aspects of ocean dynamics (see ESA (2020a)). The last atmospheric mission selected was FORUM in 2019 (ESA 2019). As part of the process, the Earth Science Advisory Committee (ESAC, now called Advisory Committee for Earth Observation) provides scientific recommendations giving guidelines for mission implementation.

At present, EarthCARE, Biomass, FORUM and Harmony are Core missions in the process of implementation. The first two Opportunity missions selected were Cryosat and SMOS, followed by Swarm. At present, FLEX is in the process of being implemented. The most recent Call for Ideas for EE11 was issued in May 2020.

Table 4.4 summarizes the status of ESA's Earth Explorer missions that have flown, are in operation or are in the process of being implemented.

Earth Explorer	Name	Mission Objectives	Launch	Status (Mar23)
1 – Core	GOCE	Study of the Earth's **G**ravity Field and Steady-State **O**cean **C**irculation **E**xplorer	March 17, 2009	Science mission ended on October 21, 2013
2 – Opp	CryoSat	Measure extent and thickness of polar ice	-1: launched October 8, 2005	Launch failure
			-2: launched April 8, 2010	**Operational**
3 – Opp	SMOS	Study of **S**oil **M**oisture and **O**cean **S**alinity	November 2, 2009	**Operational**
4 – Opp	Swarm	Mapping the Earth's magnetic field with a constellation of three satellites	November 22, 2013	**Operational**
5 – Core	Aeolus	The Atmospheric Dynamics Mission Aeolus measures wind profiles globally	August 22, 2018	**Operational**
6 – Core	EarthCARE	The **Earth** **C**louds **A**erosols and **R**adiation **E**xplorer measures cloud and aerosol profiles	2024 (TBC)	In preparation
7 – Core	Biomass	Designed to calculate the amount of carbon stored in the world's forests	2024 (TBC)	In preparation
8 – Opp	FLEX	Will monitor steady-state chlorophyll fluorescence in terrestrial vegetation globally	2025 (TBC)	In preparation
9 – Fast track	FORUM	**F**ar-infrared **O**utgoing **R**adiation **U**nderstanding and **M**onitoring will measure the Earth's outgoing radiation in the far-infrared	2027 (TBC)	In preparation
10	Harmony	Stereo thermo-optically enhanced radar for Earth, ocean, ice and land dynamics	2029 (TBC)	In preparation

Table 4.4. *ESA's Earth Explorer missions*

Three of the Earth Explorer missions are of particular importance with respect to the atmosphere, namely Aeolus, EarthCARE and FORUM.

Aeolus

Together with temperature, pressure and humidity, wind is one of the basic variables describing the state of the atmosphere. Wind speed and direction observations are needed in support of weather forecasts and for the prediction of long-term climate change. Improved knowledge of the global wind field is widely recognized as fundamental to advancing the understanding and prediction of weather and climate. The standard method to determine wind profiles is radio sounding from ground-based networks and is therefore limited in coverage, that is, mostly in the Northern Hemisphere extra-tropics.

The prime aim of the Atmospheric Dynamics Mission (ADM) Aeolus is to demonstrate that measurements of vertical wind profiles from space are possible and beneficial. The technique that has been selected is a high-performance Doppler wind lidar based on direct-detection interferometric techniques. Such a system, with a laser operating at 355-nm wavelength, uses both Rayleigh scattering from molecules and Mie scattering from thin cloud and aerosol particles. Measurement of the residual Doppler shift from successive levels in the atmosphere provides vertical wind profiles. More details can be found in Chapter 9.

Figure 4.1. *Aeolus in orbit (artist's view) (© ESA – P. Carril)*

The lidar is embarked on a satellite flying in a Sun-synchronous dawn–dusk orbit, at an altitude of ~320 km providing near-global coverage. Processing of the backscatter signals provides about 3,000 globally distributed wind profiles per day, above thick clouds or down to the surface in clear air along the satellite track. Knowledge over large parts of the Tropics and major oceans is presently quite incomplete leading to major difficulties in studying key processes in the climate system and in improving numerical simulations and predictions. Data are available in near real-time (NRT), that is, here within between 30 mins and 90 mins after sensing, as required by NWP models. Figure 4.1 shows an artist's impression of Aeolus in orbit. Difficulties in the hardware development, for example stability of the laser, lead to significant delays before the mission could be launched.

More details about Aeolus and the need for atmospheric wind profiles can be found in ESA (2011) and Stoffelen et al. (2020). Results can be found in Chapter 2 of Volume 2. As Aeolus is a so-called "demonstrator" mission, that is, a mission demonstrating a new observing technique, and primary use is in operational meteorology, work is underway to implement an operational follow-on mission in cooperation with EUMETSAT.

EarthCARE

The Earth's radiation budget is primarily governed by clouds. An accurate quantification of their impact on radiation is particularly difficult, however, due to their high spatial and temporal variability and complex vertical structures. Satellite-based observations of the three-dimensional structure of clouds and aerosols collocated with top-of-atmosphere radiation measurements are therefore highly desirable in order to directly link and verify the impact of clouds and aerosols on radiation on a global scale.

EarthCARE is a joint ESA-JAXA mission. It embarks two active and two passive instruments. A lidar with a high-spectral resolution receiver operating at 355 nm built in Europe, a W-band radar with Doppler capability built in Japan, a multi-spectral imager (with seven channels in the VIS, near, short-wave and thermal IR) and a three-view broad-band radiometer for solar and thermal flux estimates at the top of the atmosphere. Both passive instruments are provided by ESA. Problems with the laser and the radar have significantly delayed the launch.

The unprecedented complexity of the synergistic exploitation of EarthCARE data products in the ground segment will provide the vertical distribution of ice and liquid cloud profiles, heights and overlaps, vertical motion within clouds and precipitation. Furthermore, aerosol profiles will be measured. From these measurements, heating profiles and solar and thermal flux will be modeled.

Collocated estimates of top-of-atmosphere flux will be provided by the broad-band radiometer's observations, allowing us to verify the modeled fluxes and will thus allow us to link cloud, aerosol and radiation measurements. In addition, EarthCARE cloud profile data will be the first cloud radar and lidar data operationally assimilated in the ECMWF operational forecast model. Data are available in NRT, that is, here within between 30 mins and 90 mins after sensing as required by NWP models. Applications can be found in Chapter 14 of Volume 2. Figure 4.2 shows an artist's impression of EarthCARE in orbit. More details about EarthCARE and its applications can be found in ESA (2004) and Illingworth et al. (2015).

Figure 4.2. *EarthCARE in orbit (artist's view) (© ESA – P. Carril)*

FORUM

Roughly one-third of the Earth's outgoing longwave radiation (OLR) is emitted in the far-infrared (FIR) part of the spectrum, that is, at wavenumbers in the range of 0–500 cm^{-1}. Emission is primarily due to water vapor located in the upper and mid-troposphere. High, cold cirrus clouds have a large effect on the intensity of the OLR here. Until today, no spectrally resolved FIR measurements of the OLR have been made from space. Technology limitations have, until recently, prohibited any progress in this direction.

The far-infrared outgoing radiation understanding and monitoring (FORUM) missions aim to measure the Earth's top-of-atmosphere emission spectrum in the 100–1,600 cm^{-1} (i.e. 6.25–100 μm) spectral region, filling the observational gap in the far-infrared (100–667 cm^{-1}, i.e. from 15 to 100 μm), which has never been observed from space, spectrally resolved, and in its entirety. This measurement will improve the understanding of the climate system by supplying, for the first time, most of the spectral features of the far-infrared contribution to the Earth radiation budget, particularly focusing on the water vapor contribution related to the continuum absorption in the rotational band, the cirrus cloud properties and the ice/snow surface emissivity.

By flying in "loose formation"[2] with MetOp-SG-A, at a time difference less than 1 min and with the ground tracks located within 300 km of distance, FORUM will complement the spectral measurements performed in the 645–2,760 cm^{-1} range by the Infrared Atmospheric Sounding Instrument New Generation (IASI-NG). More details can be found in ESA (2019) and Palchetti et al. (2020).

4.2.4.2. *ESA's Earth Watch missions*

Strictly speaking, Earth Watch (EW) missions can be split into two categories, namely missions in support of operational meteorology, that is, design and develop new mission concepts for EUMETSAT, see Chapter 5, and other missions with an operational nature. The latter, for the time being, is related to Copernicus missions. In addition, dedicated small missions for operational purposes have been introduced in the EW context recently. An example is ALTIUS, a gap-filling atmospheric limb sounder mission with the objective of delivering accurate ozone profiles globally.

EUMETSAT

The roles of ESA and EUMETSAT – according to the co-operation agreements between them, ESA is responsible for the development of the space segment and EUMETSAT for the overall system, including launch services, ground segment and operations for the duration of the mission.

As a general rule, ESA funds a major part of the development costs of the first satellite of a series; EUMETSAT financially contributes through a fixed share to its development costs and fully funds the costs of the recurrent satellites, the launch services and operations of all the satellites, including their ground segment.

2. No tight control of the relative movement of the two spacecraft.

At present, that is, in 2023, two such satellite implementation programs are in place, namely for Meteosat Third Generation (MTG) and for MetOp-Second Generation (MetOp-SG).

MTG is a constellation and consists of a pair of geostationary satellites with complementary payload addressing imaging (MTG-I) and sounding (MTG-S). The first MTG satellite (MTG-I) was launched in December 2022. The second, MTG-S, is scheduled for launch in 2024. The planned MTG series will consist of three pairs.

Furthermore, MetOp-SG is a constellation consisting of a pair of polar-orbiting satellites with complementary payload called MetOp-SG-A and -B. MetOp-SG-A focuses on optical imaging and sounding, while MetOp-SG-B focuses on microwave imaging. The planned MetOp-SG series will consist of three pairs. The first MetOp-SG satellite (MetOp-SG-A) is scheduled for launch in 2025.

More details about MTG and MetOp-SG can be found in Chapter 5.

Copernicus

Copernicus is the European Union's (EU) Earth observation program, looking at the Earth and its environment to benefit all European citizens. It offers information services that draw from satellite Earth Observation and in situ (non-space) data.

The roles of ESA and the European Commission (EC) for Copernicus – the initiative is headed by the EC, acting on behalf of the European Union, in partnership with the EU Member States, EUMETSAT, the European Centre for Medium-Range Weather Forecasts (ECMWF), EU Agencies, Mercator Océan and ESA.

ESA coordinates the development of new mission concepts and the delivery of data. The EC is responsible for the overall initiative, setting requirements and managing the services. ESA is developing a new family of satellites called Sentinels, specifically for the operational needs of the Copernicus program.

Sentinels – overview

The space-related needs of the Copernicus services are met by the ESA Sentinel missions. Table 4.5 provides an overview of the ESA Sentinel missions. While several missions require two satellites in orbit at the same time (satellites-A and -B), that is, S-1, S-2 and S-3, additional copies are available to replace either at end-of-life or after a failure of one of the satellites in orbit.

Sentinel	Mission objectives	Launch	Status
S-1	Polar-orbiting radar imaging mission for land and ocean services	-A: launched April 3, 2014	**Operational**
		-B: launched April 25, 2016	**Operational**
S-2	Polar-orbiting, multispectral high-resolution imaging mission for land monitoring	-A: launched June 23, 2015	**Operational**
		-B: launched March 7, 2017	**Operational**
S-3	Polar-orbiting, multi-instrument mission to measure sea–surface topography, sea– and land–surface temperatures, ocean color and land color	-A: launched February 16, 2016	**Operational**
		-B: launched April 25, 2018	**Operational**
S-4	A payload devoted to atmospheric composition monitoring that will be embarked upon an MTG-S satellite in geostationary orbit	2024 (TBC)	In preparation
S-5	A payload that will monitor the composition of the atmosphere from polar orbit aboard a MetOp SG-A satellite	2025 (TBC)	In preparation
S-5P	Sentinel-5P (P for "precursor") is a mission developed to reduce the data gap between Envisat, in particular the Sciamachy instrument, OMI and the launch of Sentinel-5	October 13, 2017	**Operational**
S-6	Polar-orbiting radar altimeter mission to measure global sea–surface height, primarily for operational oceanography and for climate studies	November 21, 2020	**Operational**

Table 4.5. *The ESA Sentinel Missions. Sentinels-1, -2 and -3 are identical twins, that is, always two of them (-A and -B) have to be in orbit at the same time*

More candidate Sentinel missions are in preparation: six high-priority candidate missions are being studied to address EU policy and gaps in Copernicus user needs, and to expand the current capabilities of the Copernicus space component.

The background of Sentinels S-4, S-5 and S-5P – the Sentinels for atmospheric composition

The former GMES atmospheric service (GAS), now called Copernicus Atmosphere Monitoring Service (CAMS) (see GMES Atmosphere Core Service (2009)), is one out of six existing services (the other five are "marine", "land", "climate change", "security" and "emergency") and provides continuous data and information on atmospheric composition. The service describes the current situation, forecasts the situation a few days ahead and analyzes consistently retrospective data records for recent years.

CAMS supports many applications in a variety of domains including health, environmental monitoring, renewable energies, meteorology and climatology.

The service focuses on five main areas:

– air quality and atmospheric composition;

– ozone layer and ultra-violet radiation;

– emissions and surface fluxes;

– solar radiation;

– climate forcing.

The S5P, S4 and S5 missions are dedicated to monitoring Earth's atmospheric composition and are designed to meet the needs of CAMS. More details about S5P can be found in ESA (2016), about S4 in ESA (2017) and about S5 in ESA (2020b).

The atmospheric Sentinel mission requirements are fully established and aim to safeguard the consistency and complementarity of the three missions (see Ingmann et al. 2012). While the S4 primary mission requirements are driven by the need to meet specific air quality monitoring applications, as secondary objectives the mission will also contribute to other application areas of the CAMS. The S4 mission requirements have been set to ensure that the mission yields high-quality data products for the target chemical species, and that the related product performance will meet the needs of the CAMS. Beyond that, there are European plans to continue this observational capability by Sentinel Next Generation missions after the end of life of S4 and S5.

The concept of observing atmospheric ozone from space by measuring backscattered solar radiation in the UV spectral domain was formulated already in 1957 (Singer and Wentworth 1957). The first spaceborne nadir sounder that provided such measurements was the Backscatter UltraViolet (BUV) radiometer, launched by NASA in 1970. More details can be found in Chapter 3.

Currently, many atmospheric constituents are measured by a variety of sensors feeding a wide range of atmospheric applications. The main steps of this development are outlined in this section, with a focus on nadir sounding spectrometers covering the UV, visible, near-infrared and shortwave infrared (UVNS) spectral ranges (see Table 4.6).

Instrument	Spectral range	Sampling distance	Swath	Coverage	Revisit time	Operational
Dimension	(nm)	(km x km)	(km)			
GOME (on ERS-2, see section 4.2.1)	UVN (240–790)	320 x 40	960	Global	3 days	1995–2011 ℘
Sciamachy (on Envisat, see section 4.2.2)	UVNS (240–2,400)	30 x 215	960	Global	6 days	2002–2012 ℘
GOME-2 (on MetOp-A, MetOp-B, MetOp-C, see section 4.2.3)	UVN (240–790)	40 x 40 80 x 40 80 x 40	960 1,920 1,920	Global	1.5 days	2006–present, 2012–present, 2018–present
OMI (on NASA's Aura satellite)	UV–vis (270–500)	13 x 24	2,600	Global	One day	2004–present

Table 4.6. *Heritage European nadir sounding spectrometers (℘= out of service)*

The main properties of the heritage instruments can be summarized as follows.

– The Global Ozone Monitoring Experiment (GOME), developed by ESA and operated on ERS-2 from 1995 to 2011 (see section 4.2.1) offered a wider product portfolio, covering, besides ozone, various chemical constituents related to ozone chemistry (H_2O, SO_2, NO_2 and several halogen oxides). This was possible due to its ability to spectrally resolve continuous ranges in the UV, the visible and the near-infrared.

– The Scanning Imaging Absorption Spectrometer for Atmospheric Cartography (Sciamachy) instrument on ESA's Envisat mission; see section 4.2.2. The instrument also covered the shortwave infrared, which enabled the retrieval of CO and the potent greenhouse gas CH_4, and offered a combination of nadir and limb-sounding capabilities for ozone profiling.

– The series of GOME-2 instruments, developed by ESA and flown aboard the MetOp-A, -B and -C satellites of the EUMETSAT Polar System (EPS) (see section 4.2.3) have continued this dataset (albeit without shortwave infrared or limb-sounding capability) from 2006 to the present day. Launched in 2006, 2012 and 2018, respectively, all three are still operating and are providing continuous atmospheric data.

The footprint of the sensors mentioned thus far is of the order of tens of kilometers, which is well suited to applications with a focus on stratospheric composition and ozone chemistry. Tropospheric composition and above all air quality applications call for a higher spatial resolution in the planetary boundary layer.

A step in this direction was taken by the Dutch–Finnish Ozone Monitoring Instrument (OMI) imaging spectrometer launched on NASA's Aura satellite in 2004, which offers ozone profile and total column data for NO_2, SO_2, HCHO and CHOCHO at a horizontal resolution of 13 km × 24 km at nadir. Enhancement of the spatial resolution while maintaining a high signal-to-noise ratio was enabled by two-dimensional detector technology. Years after the end of its nominal lifetime, OMI is still operating, albeit with degraded performance.

Sentinel-5P (S5P), -4 (S4) and -5 (S5) build on the heritage from these instruments and set a new performance benchmark for nadir atmospheric composition observation. As compared to OMI, the atmospheric Sentinel sensors offer a substantially enhanced spatial resolution and, at the same time, a significantly enhanced signal-to-noise ratio. These enhancements boost the yield of clear-sky observations, help to resolve the signatures of localized emission sources at the Earth's surface and allow the detection of weaker signatures even in single overpass data.

Instrument	Spectral range	Sampling distance	Swath	Coverage	Revisit time	Operational
Dimension	(nm)	(km x km)	(km)			
TROPOMI (on Sentinel-5P)	UVNS (270–2,385)	3.5 x 7 (UVN) 7 x 7 (SWIR)	2,600	Global	1 day	2017–present
Sentinel-4 (S4) (on MTG-S)	UVN (305–775)	8 x 8	3,900 (N/S)	Europe, North Africa	1 h	*Planned launch: 2024*
Sentinel-5 (S5) (on MetOp-SG-A)	UVNS (270–2,385)	7 x 7	2,670	Global	1 day	*Planned launch: 2025*

Table 4.7. *The Atmospheric Sentinel Missions*

The LEO missions, first S5P and then S5, and the geostationary mission S4 form a constellation with complementary observation capabilities. Sentinel-4 focuses on the short-lived species with a pronounced diurnal variability and addresses the need for air quality applications for measurements of tropospheric composition over Europe with a short revisit time. S5 and S5P satisfy the needs of global applications related to air quality, ozone and climate for daily measurements of a wide range of atmospheric components. S5P is the mission with the best global performance related to spatial resolution, signal-to-noise ratio (SNR) and spectral coverage since it brings together several of the best characteristics of its predecessors, and even goes beyond them, for example, an unprecedented on-ground spatial resolution of 7 × 7 km. Table 4.7 shows the main characteristics of Sentinels S5P, S4 and S5.

Sentinel-5P

Sentinel-5P (S5P) was launched from Plesetsk (Russia) on October 13, 2017. S4 and S5 will follow in 2024 and 2025 respectively. The S5P instrument TROPOMI provides atmospheric composition observations until S4 and S5 become operational. The expected overlap of the mission lifetimes will be crucial for achieving consistent global mission data sets. Figure 4.3 shows an artist impression of S5P in orbit.

Figure 4.3. *Sentinel-5P (S5P) in orbit (artist's view) (© ESA/ATG Medialab)*

In line with the needs of numerical forecasting systems, S5P data are supplied within 3 h after sensing. This NRT service is provided throughout the operational phase, for the high-priority species O_3, SO_2, NO_2, HCHO and CO, as well as cloud/aerosol distributions. In addition to the NRT service, an extended list of target species can be supplied in an "offline" mode, taking advantage of increased accuracies achievable with specific calibration steps (level-1b processing). Trace gas retrievals (level-2 processing) are performed with relaxed timing and data availability constraints. The offline products comprise the high-priority species and CH_4. Offline products are available to end-users within two weeks after sensing.

Sentinel-4

The concept for a European geostationary atmospheric chemistry mission started in the late 1990s with European initiatives to derive and define measurement concepts for tropospheric measurements from geostationary orbit. The Geostationary Scanning Imaging Absorption Spectrometer (GeoSCIA) was suggested at the beginning of this millennium with similar characteristics to those of the S4 ultraviolet–visible–near-infrared instrument.

The geostationary S4 mission will provide accurate, continuous and timely data on an hourly basis of gaseous tropospheric constituents over Europe mainly for air quality applications. The target species of the S4 mission include key air quality parameters with pronounced temporal variability such as O_3, NO_2, SO_2, HCHO, CHOCHO and aerosols. In addition, the LEO missions, Sentinel-5 and Sentinel-5P will deliver data on not only the Sentinel-4 target species but also longer-lived species such as CO, CH_4 and stratospheric O_3, with global daily coverage for climate, air quality and ozone/surface UV applications. The S4 mission will build on the heritage of a series of spectrometer instruments that are currently or have been measuring atmospheric properties since 1995, and are expected to be exploited together with observations from the other atmospheric Sentinel missions.

Two missions, the Geostationary Environment Monitoring Spectrometer (GEMS) on GEO-KOMPSAT-2, Korea, see Kim et al. (2020), launched in 2020 and the future Tropospheric Emissions: Monitoring of Pollution mission (TEMPO, NASA; see Zoogman et al. (2017)), will measure atmospheric composition over Asia and North America, respectively. These geostationary missions will offer similar observation capabilities to those of S4 and a similar product portfolio to those of the atmospheric Sentinels. These geostationary missions, combined with the LEO atmospheric Sentinels, have been labeled the "Atmospheric Composition Virtual Constellation (AC-VC)", by the Committee on Earth Observation Satellites (CEOS). The space agencies involved will pursue coordination of their respective development in order to enhance the interoperability and consistency of products and eventually to improve the relevance of the missions for science and policy.

All S4 heritage missions are passive backscatter spectrometers operating in low Earth orbit, which allows them to perform global coverage measurements. The GEO orbit, although limiting the coverage of the Earth's surface (for S4, this is limited to Europe and North Africa) and having lower sensitivity due to the higher orbit, will for the first time allow the extremely frequent, that is, hourly, delivery of accurate atmospheric data.

S4 will provide performances equivalent to those delivered by the TROPOMI instrument in terms of spatial resolution. The wide spectral range covered by S4 will allow the retrieval of a huge amount of data related to all the key atmospheric constituents in support of the air quality applications of CAMS. Figure 4.4 shows an artist view of the S4 UVN instrument on board the MTG-S satellite.

The timeliness requirements ensure that the S4 data products are available as inputs to NRT applications such as the CAMS forecasting services and the alert services provided by the Volcanic Ash Advisory Centres (VAAC). The hourly revisit time is driven by the need to capture the diurnal variability of tropospheric composition and can be achieved only by flying S4 in a geostationary orbit.

Figure 4.4. *The Sentinel-4 (S4) UVN instrument on the MTG-S satellite (artist's view) (© ESA – P. Carril)*

Sentinel-5

Sentinel-5 (S5) has capabilities similar to Sentinel-5P. S5 and S5P as its precursor offer spectral reflectance measurements in the near- and shortwave infrared, aimed at the observation of CO and CH_4. This concept was pioneered by Sciamachy and is also exploited by the Greenhouse Gases Observing Satellite (GOSAT), the TANSAT mission and the Orbiting Carbon Observatory (OCO-2 and OCO-3). Details about GOSAT, TANSAT and OCO-2 can be found here. In comparison with these greenhouse gas missions, S5 and S5P provide better spatio-temporal sampling, by offering spatially contiguous observations with high resolution across a wide swath, and with global coverage on a daily basis. This enhanced spatial sampling capability is crucial for capturing horizontal gradients of trace gas abundances and hence for estimating emissions. S5 covers an additional shortwave infrared band, not covered by S5P, in order to allow the retrieval of CH_4 with an advanced multiband scheme. Note that this additional band also offers some sensitivity to CO_2, which is covered as a byproduct of the CH_4 algorithm. However, the challenge of providing highly accurate CO_2 observations will be addressed by a Copernicus High Priority Candidate mission at a later date; see CO_2M mission later in this chapter. Figure 4.5 shows an artist view of the S5 UVNS instrument on board the MetOp-SG-A satellite.

The observational capabilities of UVNS and thermal infrared sounders are complementary, in particular in terms of the vertical sensitivity profiles. The synergy exploiting the complementary profile information has been explored by the MOPITT instrument (Measurement of Pollution in the Troposphere), which combined measurements in these spectral domains, and by combined products, for example, from GOME-2 and IASI (Infrared Atmospheric Sounding Interferometer) or from OMI and TES (Tropospheric Emission Spectrometer). The atmospheric Sentinel missions are expected to yield synergetic products of this kind with the Cross-track Infrared Sounder CRiS on Suomi-NPP (S5P), IASI-New Generation on MetOp-SG-A (S5) and the InfraRed Sounder IRS on MTG-S (S4).

The timeliness requirements are that the Sentinel-5 products must be available to the user very soon after sensing, that is, L1b data within 2 h 15' and L2 data within 2 h 45', in order to enable Sentinel-5 to serve a variety of NRT applications. These include global and regional atmospheric composition monitoring and forecasting, pollution event monitoring, volcanic emission monitoring in support of air traffic control and forecasting of near-surface Down-welling irradiance and dose rates related to human skin sensitivity.

Figure 4.5. *The Sentinel-5 (S5) UVNS instrument on the MetOp-SG-A satellite (artist's view) (© ESA – P. Carril)*

LEO orbiting UVN and UVNS nadir sounder missions populate orbits with equator-crossing local times around 09:30 (S5, GOME-2) and 13:30 (S5P, OMI, OMPS). Measurements from morning and early afternoon missions are complementary, as they capture quite different atmospheric conditions and provide some information about diurnal variation: in the morning, the boundary layer is not yet developed and the diluting effect of vertical mixing on local emission signatures is weak. Later in the day, these mixing processes become more pronounced, while various photochemical processes and also the diurnal signatures of emissions contribute to a buildup of abundances of a number of key pollutants.

As S5P cannot meet any of the stringent CO_2 observations with sufficient accuracy, a study was performed to investigate what would be needed to meet those requirements by adding such a capability to S5. It was concluded that those additional capabilities would lead to an S5 instrument concept not compatible with the MetOp-SG constraints, both in terms of resources and development risk. It was therefore decided to discard any specific CO_2-related requirements for the S5/UVNS instrument. This was picked up later and led to the concept of a CO_2M mission.

CO_2M (Copernicus Anthropogenic Carbon Dioxide Monitoring)

The CO_2M mission will consist of two satellites and carry a near-infrared and a shortwave-infrared spectrometer to measure atmospheric carbon dioxide produced by human activity (Janssens-Maenhout et al. 2020). These measurements would reduce current uncertainties in estimates of emissions of carbon dioxide from the combustion of fossil fuel at national and regional scales. This would provide the EU with a unique and independent source of information to assess the effectiveness of policy measures, and to track their impact towards decarbonizing Europe and meeting national emission reduction targets. The tentative launch date of the first satellite is end 2025/early 2026. Figure 4.6 shows an artist impression of CO_2M in orbit.

Figure 4.6. *CO_2M in orbit (artist's view) (© OHB SE)*

In conclusion, the atmospheric sentinels under implementation (or already in orbit) meet the basic needs of CAMS until 2030.

4.3. ESA's Climate Change Initiative (CCI)

The United Nations Framework Convention on Climate Change (UNFCCC) is leading international efforts to combat climate change and limit global warming to below 2 K (ideally 1.5 K) above pre-industrial levels, as set out in the Paris Agreement in 2015. To achieve this objective and to make decisions on climate change mitigation and adaptation, the UNFCCC requires systematic observations of the climate system.

In response to the needs, ESA has undertaken the Climate Change Initiative (CCI) program. The objective of the CCI is to realize the full potential of the long-term global Earth Observation archives that ESA, together with its Member states, has established over the past 30 years, as a significant and timely contribution to the Essential Climate Variable (ECV) databases required by UNFCCC. It ensures that full capital is derived from past, ongoing and planned ESA missions, including ERS, Envisat, the Earth Explorer missions, relevant ESA-managed archives of Third-Party Mission data and the constellation of Sentinel satellites.

4.4. References

ESA (1999a). Gravity field and steady-state ocean circulation mission – GOCE. ESA SP-1233(1).

ESA (1999b). Atmospheric dynamics mission. ESA SP-1233(4).

ESA (2004). EarthCARE – Earth Clouds, Aerosols and Radiation Explorer. ESA SP-1279(1) [Online]. Available at: http://esamultimedia.esa.int/docs/SP_1279_1_EarthCARE.pdf.

ESA (2011). ADM-Aeolus science report. ESA SP-1311 [Online]. Available at: http://esamultimedia.esa.int/docs/EarthObservation/SP-1311ADM-Aeolus_Final.pdf.

ESA (2015). Earth observation science strategy for ESA – A new era for scientific advances and societal benefits. ESA SP-1329/1. European Space Agency, Noordwijk [Online]. Available at: https://esamultimedia.esa.int/multimedia/publications/SP1329_1/offline/download.pdf.

ESA (2016). Sentinel-5 precursor: ESA's atmospheric chemistry and pollution-monitoring mission. ESA SP-1336. European Space Agency, Noordwijk [Online]. Available at: https://esamultimedia.esa.int/multimedia/publications/SP-1332/SP-1332.pdf.

ESA (2017). Sentinel-4: ESA's geostationary atmospheric mission for copernicus operational services. ESA SP-1334. 2nd edition. European Space Agency, Noordwijk [Online]. Available at: https://www.esa.int/About_Us/ESA_Publications/ESA_Publications_Monographs/ESA_SP-1334_Sentinel-4.

ESA (2019). Report for mission selection: FORUM, ESA-EOPSM-FORM-RP-3549, European Space Agency, Noordwijk [Online]. Available at: https://esamultimedia.esa.int/docs/EarthObservation/EE9-FORUM-RfMS-ESA-v1.0-FINAL.pdf.

ESA (2020a). Report for assessment: Earth Explorer 10 Candidate Mission Harmony, ESA-EOPSM-HARM-RP-3784, European Space Agency, Noordwijk [Online]. Available at: https://esamultimedia.esa.int/docs/EarthObservation/EE10_Harmony_Report-for-Assessment-v1.0_13Nov2020.pdf.

ESA (2020b). Sentinel-5: ESA's polar-orbiting atmospheric composition mission in support of copernicus services. ESA SP-1336. European Space Agency, Noordwijk [Online]. Available at: https://esamultimedia.esa.int/docs/EarthObservation/SP_1336_Sentinel-5_web.pdf.

GMES Atmosphere Core Service (2009). GMES Atmospheric Service Implementation Group Space Observation Infrastructure for the GMES Atmosphere Core Service – Summary for the implementation group: Conclusions and recommendations [Online]. Available at: https://www.researchgate.net/publication/260827233_GMES_Atmosphere_Core_Service_-_Implementation_Group_Final_Report.

Harries, J., Russell, J.E., Hanafin, J.A., Brindley, H., Futyan, J., Rufus, J., Kellock, S., Matthews, G., Wrigley, R., Last, A. et al. (2005). The geostationary earth radiation budget project. *Bull. Amer. Meteorol. Soc.*, 86(7), 945–960 [Online]. Available at: http://journals.ametsoc.org/doi/pdf/10.1175/BAMS-86-7-945.

Illingworth, A.J., Barker, H.W., Beljaars, A., Ceccaldi, M., Chepfer, H., Clerbaux, N., Cole, J., Delanoë, J., Domenech, C., Donovan, D.P. et al. (2015). The EarthCARE satellite: The next step forward in global measurements of clouds, aerosols, precipitation, and radiation. *Bull. Am. Meteorol. Soc.*, 96(8), 1311–1332 [Online]. https://doi.org/10.1175/BAMS-D-12-00227.1.

Ingmann, P., Veihelmann, B., Langen, J., Lamarre, D., Stark, H., Bazalgette Courrèges-Lacoste, G. (2012). Requirements for the GMES atmosphere service and ESA's implementation concept: Sentinels-4/-5 and -5p. *Rem. Sens. Environ.*, 120, 58–69 [Online]. https://doi.org/10.1016/j.rse.2012.01.023.

Janssens-Maenhout, G., Pinty, B., Dowell, M., Zunker, H., Andersson, E., Balsamo, G., Bézy, J.-L., Brunhes, T., Bösch, H., Bojkov, B. et al. (2020). Toward an operational anthropogenic CO_2 emissions monitoring and verification support capacity. *Bull. Amer. Meteorol. Soc.*, 101(8), E1439–E1451 [Online]. https://doi.org/10.1175/BAMS-D-19-0017.1.

Kim, J., Jeong, U., Ahn, M.-H., Kim, J.H., Park, R.J., Lee, H., Song, C.H., Choi, Y.-S., Lee, K.-H., Yoo, J.-M. et al. (2020). New era of air quality monitoring from space: Geostationary environment monitoring spectrometer (GEMS). *Bull. Am. Meteorol. Soc.*, 101(1), E1–E22 [Online]. https://doi.org/10.1175/BAMS-D-18-0013.1.

Kursinski, E., Hajj, G., Schofield, J., Linfield, R.P., Hardy, K.R. (1997). Observing earth's atmosphere with radio occultation measurements using GPS. *J. Geophys. Res.*, 102(D19), 23429–23465 [Online]. Available at: https://doi.org/10.1029/97JD01569.

Palchetti, L., Brindley, H., Bantges, R., Buehler, S.A., Camy-Peyret, C., Carli, B., Cortesi, U., Del Bianco, S., Di Natale, G., Dinelli, B.M. et al. (2020). FORUM: Unique far-infrared satellite observations to better understand how earth radiates energy to space. *Bull. Am. Meteorol. Soc.*, 101(12), E2030–E2046 [Online]. Available at: https://doi.org/10.1175/BAMS-D-19-0322.1.

Schmetz, J., Pili, P., Tjemkes, S., Just, D., Kerkmann, J., Rota, S., Ratier, A. (2002). An introduction to METEOSAT second Generation (MSG). *Bull. Am. Meteorol. Soc.*, 83(7), 977–992 [Online]. Available at: https://doi.org/10.1175/1520-0477(2002)083.

Singer, S.F. and Wentworth, R.C. (1957). A method for the determination of the vertical ozone distribution from a satellite. *J. Geophys. Res.*, 62, 299–308.

Stoffelen, A., Benedetti, A., Borde, R., Dabas, A., Flamant, P., Forsythe, M., Hardesty, M., Isaksen, L., Källén, E., Körnich, H. et al. (2020). Wind profile satellite observation requirements and capabilities. *Bull. Am. Meteorol. Soc.*, 101(11), E2005–E2021 [Online]. Available at: https://doi.org/10.1175/BAMS-D-18-0202.1.

Zoogman, P., Liu, X., Suleiman, R.M., Pennington, W.F., Flittner, D.E., Al-Saadi, J.A., Hilton, B.B., Nicks, D.K., Newchurch, M.J., Carr, J.L. (2017). Tropospheric emissions: Monitoring of pollution (TEMPO). *J. Quant. Spectrosc. Radiat. Transf.*, 186, 17–39 [Online]. Available at: https://doi.org/10.1016/j.jqsrt.2016.05.008.

5

The Role of EUMETSAT (Europe)

François MONTAGNER
Retired, European Organisation for the Exploitation of Meteorological Satellites (EUMETSAT), Darmstadt, Germany

5.1. Introduction: What does EUMETSAT do?

5.1.1. *Public service value of weather satellites*

Weather satellites are a key asset in the value chain that produces relevant, reliable and timely weather forecasts. It is estimated (Hallegatte et al. 2012) that the economic value of these weather forecasts for Europe is at least 15 billion euros each year (and probably much more, around 65 billion) and that observations from satellites such as MetOp contribute approximately 25% of this. To this, we should add the inestimable social value of lives protected from severe climatic events. By comparing this value with the contribution provided by the Member States, around 600 million euros per year[1], it is clear that EUMETSAT provides great economic and social value.

5.1.2. *EUMETSAT, a key player in Europe*

The global organization of weather observations and forecasts, under the aegis of the World Meteorological Organization, is a rare example of success in international

For a color version of all the figures in this chapter, see www.iste.co.uk/phulpin/satellites1.zip.

1. Annual report 2019–2020, see: https://www.eumetsat.int.

Satellites for Atmospheric Sciences 1,
coordinated by Thierry PHULPIN, Didier RENAUT, Hervé ROQUET and Claude CAMY-PEYRET.

cooperation for the benefit of all. Owning and maintaining a fleet of dedicated satellites is therefore an asset for international politics. EUMETSAT is a cooperative stakeholder embodying Europe in all discussions related to the coordination and evolution of these fleets within the CGMS, as well as the dissemination of data. When the European Union and the ESA created the Copernicus program, dedicated to satellite observation of the environment (see Chapter 4), EUMETSAT contributed thanks to its operational know-how and its communication and technical cooperation networks.

5.1.3. *Climate and environment*

The daily observation of the terrestrial environment by a series of satellites with extremely stable instruments makes it possible, in addition to providing information to weather forecasting systems, the constitution of time series of data, improving knowledge on aspects of the climate and its recent evolution. EUMETSAT maintains and enhances its archived observations, performing post-processing to improve the quality of its archived data, and supporting climate and meteorological agencies in their reanalysis activities, as well as the Copernicus Climate Change Service (C3S). An important aspect of this reprocessing is the intercalibration between missions. This justifies numerous exchanges and cooperation with other operators. IASI** quickly established itself as a standard in this field.

As the technologies of numerical models for weather forecasting and climate modeling have been converging and become more robust, new observations have entered the scope of EUMETSAT; including, for instance, ocean altimetry measurements, which with Sentinel-6, joined the EUMETSAT satellite fleet in 2020.

5.2. The organization

5.2.1. *First steps*

Since the 1960s, Europe has seen great interest in satellites for weather forecasting, identifying two main types of mission: low Earth orbit for atmospheric sounding and high-resolution imagery, and geostationary Earth orbit for repetitive synoptic imagery. ESRO began construction of the METEOSAT geostationary satellites in 1971, and ESA was the prime contractor and operator until 1995 (see Chapters 1 and 4). Meteorological exploitation of the polar orbit, in cooperation with

**See Appendix 3.

the US, was envisaged for the POEM polar platform at the end of the 1980s. At the Granada conference (1992), the ESA member countries decided to split the POEM platform into two programs: ENVISAT (ENVIronment SATellite) and METOP (METeorological OPerational), making EUMETSAT responsible for the European Polar System (EPS), centered on the MetOp satellites.

EUMETSAT differentiated itself from ESA as follows:

– operational agency versus research and development agency;

– weather agency versus generalist space agency;

during discussions and negotiations concluded in 1983 by the founding treaty of EUMETSAT, which entered into force on June 6, 1986 (Murphy and Mohr 2011).

The respective roles of the two Agencies remain closely linked (see Chapter 4):

– Upstream, ESA is developing key technologies for future missions;

– EUMETSAT develops the specifications of future systems with its users;

– ESA develops satellites and instruments;

– EUMETSAT is responsible for the complete system, including the ground segment and launch services; ESA is responsible for putting the satellites into orbit.

As far as financial contributions are concerned, ESA in principle covers the development costs of the satellite and the instruments, up to the prototype that is ready to launch; EUMETSAT covers those of recurring satellites, the ground segment, launches and operations. The role of ESA is supported by national agencies such as CNES (see Chapter 6) and the German Aerospace Center (DLR). The role of EUMETSAT is supported by the National Meteorological and Hydrological Services of its Member States.

One of the main roles of the EUMETSAT Ground Segment is to provide Near Real-Time (NRT) processing and dissemination of data from the satellites. The operational nature is expressed by commitments of availability, punctuality and quality of these products.

The operational deployment of EUMETSAT was carried out in parallel with that of the Meteosat Transition Program (MTP) series Meteosat-5 to Meteosat-7: the EUMETSAT control center in Darmstadt (Germany) was commissioned on December 1, 1995 and took over the operations of these satellites.

Figure 5.1. *EUMETSAT control center (source: EUMETSAT)*

5.2.2. *Stability and growth*

Despite public expenditure reduction policies in most Member States, affecting their Meteorological Services, EUMETSAT's annual budget has remained stable, at around 500 million euros (MEUR), contributed from Member States each year. Since the delegation of European Union Copernicus program tasks to EUMETSAT, which began in 2014, EUMETSAT's resources have also included the financing of these tasks, ensuring significant growth of the organization.

5.2.3. *Government*

The constitutive EUMETSAT treaty, ratified on June 18, 1986, welcomed Member States and provided a cooperating State status[2].

To date, the following 30 States are members of EUMETSAT: Austria, Belgium, Bulgaria, Croatia, Denmark, Spain, Estonia, Finland, France, Germany, Greece, Hungary, Ireland, Iceland, Italy, Latvia, Lithuania, Luxembourg, Norway, the

2. This status is currently not used; it served to facilitate the accession of candidate countries, for example, those of the former Eastern Bloc.

Netherlands, Poland, Portugal, Czech Republic, Romania, the United Kingdom, Slovakia, Slovenia, Sweden, Switzerland and Turkey.

Each Member State contributes in principle to the Agency's programs in proportion to its Gross Industrial Product (GDP). EUMETSAT has two main institutions: the Council, representing the Member States, with its working groups and the Secretariat based in Darmstadt (Germany). Funding is organized according to the programs and is therefore subject to multi-year commitments. The geostationary program, the polar program and the general budget are compulsory and require the participation of all Member States. Other programs, such as Sentinel-6, are optional. EUMETSAT has its own procurement rules, aimed at maximum efficiency, and its own data policy.

5.2.4. *European pooling: EUMETSAT, ECMWF and EUMETNET*

The role of EUMETSAT in ensuring the pooling of resources common to meteorological services at the European level follows the same logic as the European Centre for Medium-range Weather Forecasts (ECMWF) and EUMETNET agencies. The ECMWF pools computing resources and scientific know-how for weather forecasting that are too costly for most Member States – even though, from the outset, major national services (such as the Met Office, Météo France) have already developed their own resources, and the new generations of supercomputers are accessible to a growing number of Member States, or regional associations, to implement regional forecasting models[3].

EUMETNET consolidates surface observation methods (e.g. drift buoys, sensors on board commercial aircraft), best practices in processing observations (e.g. sounding, radar), as well as know-how in numerical modeling and training, available in each Member State. EUMETNET has very few resources of its own, unlike EUMETSAT and ECMWF[4].

The EUMETSAT–ECMWF axis is crucial to ensure the best efficiency of satellite data in numerical prediction, mainly through assimilation, and therefore the best return on such collective investments as the fleet of meteorological satellites and the Integrated Forecasting System (IFS). Hence, these two agencies have their specific roles, their coordination and cooperation further strengthened under the Copernicus program (see below).

---

3. For more details on ECMWF, see: https://www.ecmwf.int.

4. For more details on EUMETNET, see: https://www.eumetnet.eu.

It should be noted that a number of national space agencies of Member States provide their own contribution to the common effort to develop space resources for meteorology (CNES, DLR, etc., see below), even though none has developed an operational meteorological mission in parallel with ESA-EUMETSAT.

5.2.5. *Global pooling by the World Meteorological Organization (WMO)*

In the early stages of global numerical prediction, the WMO established in principle the exchange of observational data between all of its Member States. The global nature of satellite observations makes them significant contributions to this global exchange. EUMETSAT has been a key player since its inception, distributing a large part of its observations, identified as "essential data", on the WMO networks. In addition to these global exchanges, observations from EUMETSAT satellites are the subject of many bilateral exchanges. Indeed, the exchange of high-value data constitutes a non-financial element, facilitating cooperation between nations.

Coordination between meteorological satellite operators for highest user satisfaction is ensured worldwide by the Coordination Group for Meteorological Satellites (CGMS), of which EUMETSAT is a member (see Chapter 7). It seeks to ensure compatibility and complementarity between weather missions, concerning spatial, spectral and temporal coverage – for example, time slot for the passage of satellites in polar orbit, definition of the spectral channels of imagers and sounders and schedule of geostationary shots.

5.3. Geostationary weather satellites: from synoptic to regional zoom

5.3.1. *Meteosat first generation*

The system architecture of the Meteosat First Generation (MFG) satellites was designed to meet the needs of meteorological agencies in the 1970s (see Chapter 1). The philosophy of using satellite imagery was thus guided by initial observations of the Earth from space, centered on the real-time interpretation by experts of the images acquired and transmitted continuously by the satellites. The position of a satellite in geostationary orbit allows for continuous coverage of almost half of the Earth, and therefore of large geographical regions such as Europe and Africa (Meteosat at longitude 0), Russia (Meteor series), East Asia and Pacific (China, Korea, Japan), Pacific and Western Americas (GOES-W) and East North America and Atlantic (GOES-E). Acquisition of images in the visible and infrared spectral bands (see Table 5.1) accounts for the properties of cloud systems and their evolution, air masses and their movement and surface properties: temperature and albedo. From the outset, Meteosat satellites provided images of excellent quality and radiometric stability. Geometric rectification processing by the "ground segment", by compensating for the apparent movement of the Earth in the images caused by the displacements of the satellite around a theoretical reference position, and by correcting the small irregularities of the scanning by the satellite, created rectified imaging products. These made it possible to build video loops that quickly became essential for forecasters. Initially essentially qualitative, the interpretation gradually took advantage of the excellent accuracy of the measurements to offer products such as atmospheric motion vectors, cloud temperature and surface albedo.

Frequency: one image every 30 minutes			
Channel	Wavelength (μm)	Spatial resolution (km at sub-satellite point)	Signal-to-noise ratio
VIS	0.5–0.9	2.5	3 @ 1% albedo
WV	5.7–7.1	5	NEDT 1 K @ 250 K
IR	10.5–12.5	5	NEDT 0.5 K @ 300 K

Table 5.1. *Main features of the Meteosat Visible and Infrared Imager (MVIRI**). NEDT: Noise Equivalent Differential Temperature*

Figure 5.2. *Image of the Earth's disk taken by Meteosat at noon (source: EUMETSAT)*

In addition to images of the complete terrestrial disk, the imager offers flexibility in defining the area to be observed. In a fraction of the nominal scanning time, the corresponding fraction of the complete disk can be formed. In practice, the Rapid Scanning Service (RSS), introduced in 2001, provides an image centered on Europe every 10 minutes (see an example in Figure 5.4). Its main purpose is the observation of rapidly changing phenomena, such as the development of thunderstorms over Europe in summer.

Figure 5.3. *Meteosat satellite (source: Wikimedia)*

Figure 5.4. *Sample image provided by the RSS (source: EUMETSAT)*

The Meteosat system was originally self-sufficient, insofar as it was the weather satellite itself which served as a relay to distribute these products in the form of digital and analog streams (WEFAX) to user weather services equipped with relatively small antennae.

A total of seven Meteosat satellites have been built to order by ESA, including a first series of three, followed by another series of three Meteosat Operational Program (MOP) satellites and a final one called the Meteosat Transition Program (MTP). The dates of these missions can be found in Chapter 4.

5.3.2. *Meteosat second generation*

Figure 5.5. *MSG satellite (source: Thalès)*

The Meteosat second-generation satellites offer increased performance compared to the first, according to all criteria, in particular:

– doubling of imaging frequency, from one image every 30 minutes to one every 15 minutes;

– spatial resolution of 3 km above the satellite;

– 11-channel visible and infrared imaging and HRV high-resolution panchromatic imaging (1 km).

The rapid imagery operational concept is inherited from that of Meteosat: a reduced amplitude scan produces an image every 5 minutes. A degree of freedom also exists in the definition of the area covered by the HRV channel.

Although the concept of the MSG satellites remains close to that of the first Meteosats, the design of MSG took a relatively long time, in particular because of the parallel study of the design of satellites, stabilized by rotation and on three axes (see section 5.3.4 and see also Chapter 4). In addition to the main SEVIRI** instrument, MSG embeds the experimental Global Earth Radiation Budget (GERB**) instrument.

Frequency: one image every 15 minutes			
Channel	**Wavelength (μm)**	**Spatial resolution (km at sub-satellite point)**	**Signal-to-noise ratio**
HRV	0.6–0.9	1	4.3 @ 1% albedo
0.635 μm	0.56–0.71	3	10.1 @ 1% albedo
0.81 μm	0.74–0.88	3	7.28 @ 1% albedo
1.64 μm	1.50–1.78	3	3 @ 1% albedo
3.92 μm	3.48–4.36	3	NEDT 0.35 K @ 300 K
6.25 μm	5.35–7.15	3	0.75 K @ 250 K
7.35 μm	6.85–7.85	3	0.75 K @ 250 K
8.70 μm	8.30–9.10	3	0.28 K @ 300 K
9.66 μm	9.38–9.94	3	1.50 K @ 255 K
10.8 μm	9.8–11.8	3	0.25 K @ 300 K
12.0 μm	11.0–13.0	3	0.37 K @ 300 K
13.4 μm	12.4–14.4	3	1.80 K @ 270 K

Table 5.2. *Key Features of SEVIRI imaging*

**See Appendix 3.

The first MSG satellite developed by ESA (see Chapter 4) was launched on August 28, 2002 and was called Meteosat-8. Table 5.3 summarizes the operation dates of the MSG missions.

The MSG program also includes scientific, technical and operational contributions from Member States in the form of Satellite Application Facilities (SAF). The purpose of these development and processing centers is to facilitate and amplify the use of satellite data by offering scientific products complementary to those produced in Darmstadt or processing software. The SAFs began their operational existence in 2002 and have since developed in line with EUMETSAT's geostationary and polar programs.

Satellite	Launch	Commissioning as primary satellite	Current state
Meteosat-8	August 28, 2002	January 19, 2004	Meteosat-8 was decommissioned on July 1, 2022
Meteosat-9	December 22, 2005	April 11, 2007	Meteosat-9 has provided IODC service since June 1, 2022
Meteosat-10	July 5, 2012	January 21, 2013	RSS service
Meteosat-11	July 15, 2015	February 20, 2018	Primary satellite

Table 5.3. *Meteosat second-generation operation dates*

Unexpectedly, MSG also gave EUMETSAT the opportunity to modify the dissemination of its data. As the transponder of Meteosat-8, intended to ensure the direct broadcast from the satellite, was not working correctly, this triggered EUMETSAT to redeploy the distribution of its data on the EUMETCast network, based on commercial satellite services of digital television (Digital Video Broadcast (DVB)). Thanks to its flexibility and the low cost of access for the user, EUMETCast quickly scaled up and became the main distribution channel for all EUMETSAT's NRT data, that is, those of its own satellites and those it redistributes.

5.3.3. *Agility of geostationary missions*

As soon as more than one Meteosat satellite in good working order is in geostationary orbit, and additional ground stations are available to take over their control, then there are possibilities to extend their coverage. Priority was given to expanding operational capability by defining a nominal satellite and keeping a backup Meteosat close to nominal – a practice that has been continuously adopted since Meteosat-9; the redundant Meteosat can provide an RSS image service. Spatial coverage has also been extended within the framework of operational or scientific cooperation. Thus, ESA moved Meteosat-3 to longitude 50° W in 1991, then 75° W to help NOAA cope with the difficulties it was having with the GOES-E satellite. EUMETSAT contributed to the INDOEX experimental campaign by moving Meteosat-5 to longitude 63° E in early 1998. This contribution evolved into an operational service, the Indian Ocean Data Coverage (IODC), which is still ongoing, provided by Meteosat-9 at the time of writing.

5.3.4. *Stabilization by rotation or on three axes: system aspects*

The first- and second-generation Meteosat satellites are stabilized by permanent rotation on an axis parallel to that of the Earth ("spin stabilized"). This design has clear advantages in terms of thermal and mechanical stability. The satellite is designed around the imaging instrument, housed in an axial tube. The horizontal axis of the scan is ensured by the spinning of the satellite, the vertical axis by the progressive tilting of a mirror. A reduced number of sensors (per channel and per line in the case of SEVIRI) receive all the radiation observed, coming from the Earth, from space or from the calibration reference, which guarantees high radiometric stability. The price to pay for those advantages is mainly limited information throughput as all data are sent to Earth via a low-gain omnidirectional antenna. Moreover, the example of GERB on MSG illustrates the relative difficulty of integrating one or more secondary instruments on the rotating platform. The design of MSG began with a long phase of comparison between a rotating platform and a stabilized triaxial platform, where the three axes of the satellite are maintained in a fixed orientation with respect to the Earth. Having to meet very stringent pointing stability requirements – a 1-km pixel occupying an angle of 1.6 millidegrees vertically from the satellite – three-axes stabilization has become the choice for the third generation of Meteosat, whose first launch occurred in December 2022. Spread over two Imager (MTG-I) and Sounder (MTG-S) platforms, the payload now includes four instruments with a very high acquisition rate and the digital throughput and latency requirements of the information require the use of a high-gain antenna, easier to fit on a satellite stabilized on three axes. The agile imager and sounder design make the spin-stabilization less relevant.

5.3.5. *Meteosat Third Generation*

The Meteosat Third Generation (MTG) program, in development since 2008 and scheduled to enter into operation in 2023, aims to ensure a 20-year succession to the Meteosat second-generation satellites, all of which are currently in operational service. This succession is a continuation, from the point of view of the users, of the imaging and rapid imaging missions; but it includes significant design breakthroughs, intended to meet the new needs of meteorological services while limiting the increase in the cost of the mission.

– The nominal MTG mission will be carried out by two MTG-I "imager" satellites and one MTG-S "sounder" satellite. This nominal configuration is expected to be reached three years after the first launch.

– Each MTG satellite is based on a three-axis stabilized platform, breaking with previous generations (see above).

The MTG-I mission carries an entirely new agile imager, the Flexible Combined Imager (FCI**), capable of scanning the Earth's disk in 10 minutes as well as rapidly scanning selected regions. A second instrument, the Lightning Imager (LI**), is dedicated to the detection of lightning across the entire terrestrial disk with very short response times: there must be less than 2 minutes from the detection of a flash to the availability of the product to users.

– Atmospheric sounding by MTG-S is provided by two instruments: the hyperspectral Infra-Red Sounder (IRS**) and the Sentinel-4** Ultraviolet, Visible and Near-infrared (UVN) instrument.

Figure 5.6. *MTG-I and MTG-S satellites (artist's view, source: Thales)*

The introduction of IR sounding from geostationary orbit meets the needs of numerical prediction, both global and regional; near real-time monitoring of the atmosphere structure in three dimensions should, among other things, allow for

**See Appendix 3.

better prediction of severe events. It is made possible by successful experiments in low orbit (AIRS**, IASI, CrIS**, etc.); the Chinese agency CMA did the same on their new generation of FY-4 geostationary satellites, launched in December 2016 and June 2021.

The UVN sounder responds to the growing need for monitoring atmospheric pollution by making it possible to characterize the diurnal evolution of gaseous or aerosol pollution for a large number of species, and to locate its sources. This monitoring of atmospheric pollution is a mission shared between the Copernicus program of the European Union and air quality and/or meteorological agencies. Hence, the UVN instrument and the processing of its data are how the Copernicus program contributes to the MTG mission. Chapter 4 provides more details on the space missions of the Copernicus program.

5.4. MetOp satellites, the first source for numerical weather forecasting

The creation of the EPS, determined in 1992, followed the model of cooperation between ESA and EUMETSAT described above. There has also been an important dimension of international cooperation: EPS was intended to be complementary to the NOAA Polar Orbiting Environmental Satellites (POES) (see Chapter 2), together forming the Initial Joint Polar System (IJPS). EPS took charge of some of the observations previously provided by the POES satellites, those of the so-called "morning" heliosynchronous orbit (time of passage of the descending track at around 9:30 a.m. local). NOAA provided Europe with essential instruments for atmospheric imaging and sounding, already used on POES satellites: Advanced High-Resolution Radiometer (AVHRR**), Advanced Microwave Sounder Unit A (AMSU-A**) and High-resolution Infra-Red Sounder (HIRS**).

The receiving stations and real-time data processing systems of EUMETSAT and NOAA have been interconnected and coordinated to ensure that each satellite receives a "pass" in each orbit, that is, the possibility of transmitting its data to a station, either Svalbard or Fairbanks. This made it possible to satisfy a crucial need of numerical forecasting systems which were in development in Europe and the USA: data is available less than 2 hours after acquisition.

Supported by Europe in its operational tasks, NOAA hoped to be able to concentrate on the new polar satellite system (NPOESS) developed in cooperation with NASA and the Department of Defense (see Chapter 2).

**See Appendix 3.

The EPS represented a major challenge in terms of complexity and cost, with the development of new instruments, the delivery of three satellites to ensure at least 15 years of continuous observations, and the distribution in Europe and beyond of near real-time data. MetOp satellites were developed by ESA; more details on their design can be found in Chapter 4.

The first satellite, named MetOp-A when it was put into service, was launched on October 19, 2006. It was declared operational on May 15, 2007. MetOp-A provided more than 14 years of service before EUMETSAT ceased its operations on November 30, 2021. MetOp-B (launched on September 17, 2012) and MetOp-C (launched on July 11, 2018) joined MetOp-A in orbit rather than succeeding it; the configuration of two or three MetOp satellites in orbit, together with adding processing capabilities to the ground segment, allowed for improvement of the geographical coverage of observations, as well as improving certain products such as atmospheric motion vectors (see Chapter 2 of Volume 2).

5.4.1. *Synergy of observations*

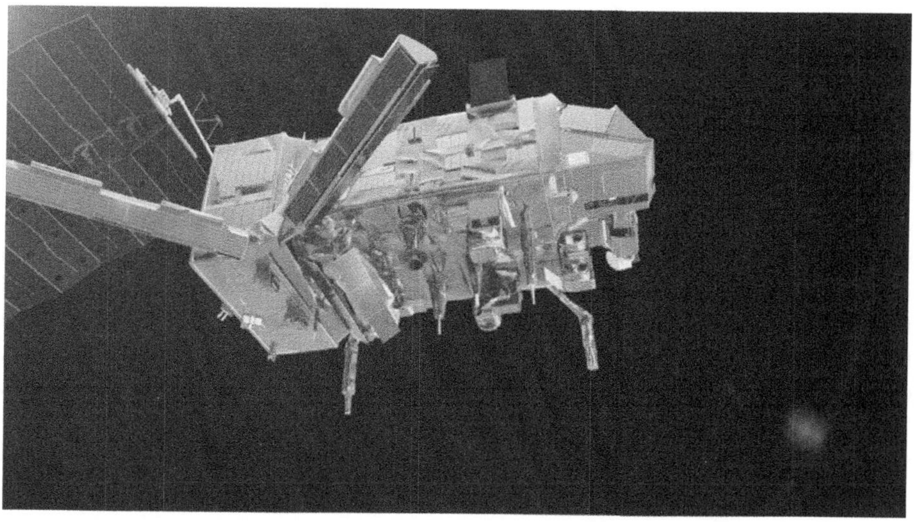

Figure 5.7. *Artistic view of the satellite MetOp-A (source: ESA)*

The design of MetOp satellites, like that of the POES satellites, or that of the ERS and ENVISAT satellites, favors synergy between observations from several instruments. In particular, for vertical sounding, that is, the measurement of the vertical distribution of atmospheric temperature and humidity, MetOp embeds an

HIRS infrared sounder and two microwave sounders, AMSU-A and MHS**. This combination, supported by the AVHRR imager, was already known as the "Advanced TIROS Operational Vertical Sounder" (ATOVS**) on the most recent POES satellites. The new IASI infrared sounder significantly improves temperature sounding while providing information on the chemistry of the atmosphere (see Chapter 6).

The ability to detect clouds by AVHRR is crucial for HIRS and IASI infrared sounders. All of these instruments, as well as microwave sounders, are synchronized to simultaneously observe the same region.

For numerical prediction, the generalization of variational observation assimilation schemes considerably extends the domain of these synergies. For example, the measurement of winds by the Advanced SCATterometer (ASCAT**) works in synergy with sounding, by constraining the energy fluxes at the ocean–atmosphere interface. Radio occultation sounding provides information on the stratosphere and ionosphere, complementing the tropospheric sounding of ATOVS and IASI.

5.4.2. *Continuity and innovation*

MetOp's mission was largely based on NOAA's ATOVS-AVHRR legacy. Europe contributed with the new microwave radiometer, the Microwave Humidity Sounder (MHS), heir to the European instrument AMSU-B flying on POES satellites. The GOME-2** instrument dedicated to atmospheric chemistry bears heritage from the GOME (ERS-2 mission) and SCIAMACHY** (ENVISAT) instruments (see Chapter 4). As for the ASCAT scatterometer radar, it was derived from the Advanced Microwave Instrument (AMI**) radar from ESA's ERS-1 and ERS-2 missions (see Chapter 4). In all of these cases, the performance of MetOp's onboard instruments were significantly improved compared to their predecessors. This enabled the first MetOp satellite to provide quality data in near real-time very soon after its start-up; European numerical prediction centers began the operational assimilation of the AMSU-A sounder in April 2007.

The MetOp mission also carried two new instruments, which proved to have a major impact on numerical prediction and on other Earth Observation applications: the GNSS radio occultation sounder, the GNSS Radio-occultation Atmosphere Sounder (GRAS**) and the IASI infrared hyperspectral sounder (see Chapter 6).

**See Appendix 3.

Details on the impact of EPS on knowledge of the atmosphere can be found in Parts 1 and 2 of Volume 2.

5.4.3. *The second generation of the European Polar System*

EUMETSAT and ESA have started preparatory activities for a second generation of MetOp and European Polar System satellites in order to take over from EPS at its end of life, initially scheduled 15 years after the launch of MetOp-A, that is, from 2022. As with the successive generations of geostationary satellites, this second generation must ensure the continuity of the services of the first and also meet the new requirements of the Member States, through a general improvement in performance and the introduction of new sensors.

The mission will be shared between two separate platforms flying in the same orbital plane, MetOp-SG-A and MetOp-SG-B. Three pairs have been built, each with a lifespan of 7.5 years. The A satellites provide temperature, humidity and atmospheric composition sounding, imaging and radio occultation sounding. The B satellites carry a scatterometer radar, two microwave imagers – a novelty compared to MetOp – a radio occultation sounder, and support the ARGOS data collection system.

Like EPS, EPS-SG is the result of international cooperation involving ESA (according to the organic scheme described above), NOAA, CNES and new partners: the German Space Agency (DLR) and the European Copernicus program. Unlike EPS, the satellites are entirely of European origin: the DLR supports the design and development of the new-generation imager, MetImage**, replacing AVHRR; vertical sounding is provided by IASI-NG**, developed by CNES, and by a new microwave sounder, developed by ESA. The cooperation with NOAA focuses on the inter-operation of European and American receiving stations to make NRT products available to users 70 minutes after acquisition. The Copernicus program provides the Sentinel-5** instrument for ultraviolet, visible and near infrared sounding to monitor atmospheric composition and air quality – continuation and expansion of the GOME-2 mission on MetOp. The launch of the first MetOp-SG-A satellite is scheduled for 2025, the first MetOp-SG-B satellite a few months afterwards.

EPS-SG will produce a throughput of products for end-users of around 170 Megabits/second. To process and route such a volume of data, a new design is

**See Appendix 3.

required for data processing and the ground segment must make extensive use of multi-mission elements (see below).

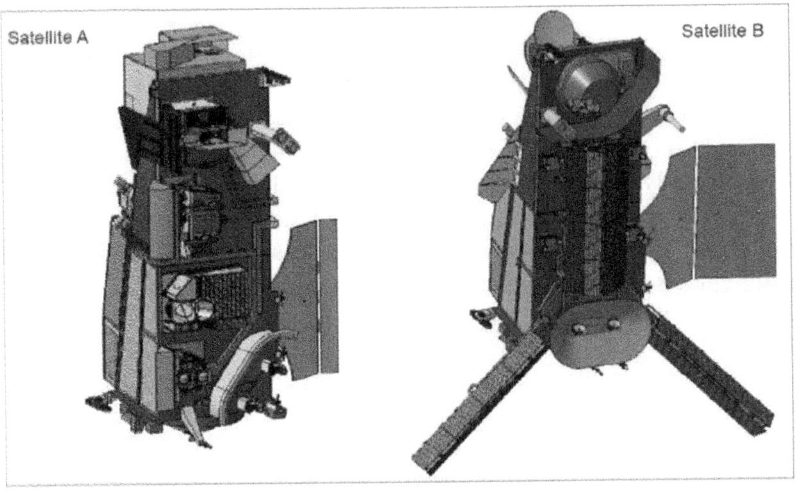

Figure 5.8. *Configuration of satellites MetOp-SG-A and MetOp-SG-B (source: ESA)*

5.4.4. *Scale economies*

Since its inception, EUMETSAT has sought to optimize the efficiency of its investments through scale economies. In terms of satellite manufacturing, EUMETSAT places an order for the number of satellites needed to complete the program for mass production. Satellites intended for future operations are kept in storage and only removed from storage in preparation for their launch. In the case of MetOp-C, launched on November 17, 2018, the storage period was approximately eight years. Such a long storage time is not without its risks, such as the contamination of the sensors, the aging under gravity of mechanisms designed to operate without it, etc. The Meteosat-11 satellite is an exception: ordered from industry several years after the first three MSGs, it was launched on July 15, 2015, only three years after Meteosat-10, and stored in orbit until the end of 2017.

EUMETSAT also seeks scale economies in the ground segment. Initially, the ground segment of each mission was almost totally autonomous. By developing multi-mission elements and standardizing internal services such as archiving or network management around them, EUMETSAT can spread development costs over several programs and manage them independently.

5.4.5. *Cooperation regarding the polar orbit*

As we saw above, the EPS system was created to provide observation services from the so-called "morning" orbit (9:30 a.m. local time), in addition to the NOAA polar system focusing on the so-called "afternoon" orbit (1:30 p.m. local time). Through the commitments of these organizations with respect to their Member States and the WMO, this scheme guarantees meteorological agencies around the world the availability of global observations spaced 6 hours apart on average at almost all points. Within the CGMS (see Chapter 7), discussions between agencies have resulted in the Chinese Meteorological Agency (CMA) exploiting the early morning orbit (5:40 a.m. local time). For users such as large numerical prediction centers assimilating measurements from multiple missions, the regularity of observations is thus improved.

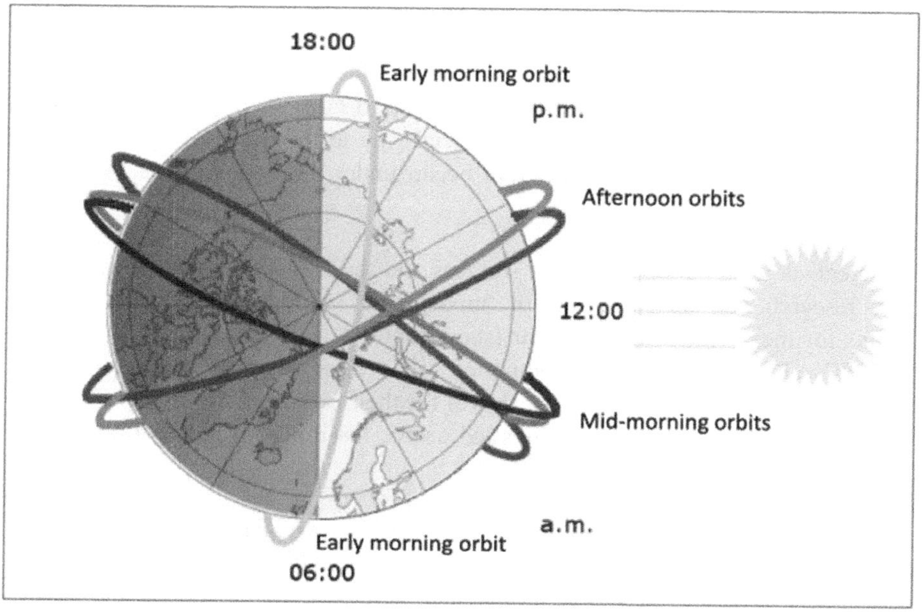

Figure 5.9. *Distribution of polar orbit missions according to their Equatorial Crossing Time (source: J. Lafeuille)*

5.5. Weather perspective and innovation

Seeking to best meet the needs and demands of its users, EUMETSAT has supported or taken charge of space missions of an experimental nature or indirectly related to meteorology.

Space altimetry, in particular the Topex-Poseidon and Jason-1 missions dedicated to observing the oceans, provides observations of great interest to meteorologists: wave height and wind speed. In cooperation with NASA, CNES and NOAA, EUMETSAT has been working to ensure that this data is distributed in near real-time to meteorological services. Beginning after the commissioning of Jason-2, launched on June 20, 2008, this service continued with Jason-3 launched on January 17, 2016.

A significant evolution of numerical forecasting models, coupling atmospheric and oceanic circulation models will accentuate the impact of space altimetry on the weather forecasts. For this reason in particular, the Jason-CS altimetry program decided by Europe to give operational succession to Jason includes the participation of EUMETSAT. Also responding in a crucial way to the needs of the Copernicus ocean monitoring service, this program began under the double name Sentinel-6-Jason-CS.

Scientists from national weather services and the ECMWF are also involved in ESA research and development for atmospheric observation missions such as Aeolus for measuring winds by lidar (see Chapter 4 and Chapter 2 of Volume 2). EUMETSAT has frequently supported the dissemination of experimental measurements via EUMETCast (see above) for their evaluation or temporary use by these meteorological services and worked on the feasibility of possible operational extensions.

The Global Navigation Satellite Systems (GNSS) navigation satellites, the best known of which is the American GPS, due to the stability of their signal and their availability, have proven to be an excellent reference for observing the Earth's atmosphere by radio occultation since 1995 (see Chapter 4 of Volume 2). After a few experimental missions (Genesis, Oersted, CHAMP, COSMIC), the GRAS instruments on board MetOp have brought major improvements from an operational point of view. The success of these measurements in numerical prediction assimilation systems motivated not only their continuation with EPS-SG but also their inclusion in Sentinel-6, in addition to the main altimetry mission.

5.6. Climate

EUMETSAT has continuously accumulated long and homogeneous time series of satellite observations in its archives since its inception. Interest in these series for the study of recent climate change has grown proportionally, encouraged by the reanalyses, such as those carried out by ECMWF since the 2000s. To meet this demand of increasing scientific and societal importance EUMETSAT has taken an active role by (see Part 3 of Volume 2):

– supporting re-calibrations and re-processing of past observations;

– integrating needs directly related to the climate in the specifications of its missions, starting with EPS;

– facilitating access to archived data via IT portals, including WEkEO (see section 5.7.2).

5.7. EUMETSAT and Copernicus

5.7.1. *A convenient partnership*

The reasons for the partnership between the European Copernicus program for operational Earth observation and EUMETSAT are numerous and arise mainly from:

– the interest in Copernicus satellite data to EUMETSAT members, and reciprocally that of EUMETSAT data to Copernicus services;

– the relevance of EUMETSAT resources and operational know-how for the implementation of Copernicus.

The European policies that the Copernicus program aims to serve, and in particular the monitoring of the ocean environment and the composition of the atmosphere and the climate, are often implemented by the same national and international services who are customers of EUMETSAT. There are many synergies between Copernicus data and EUMETSAT data for these services. For example, we saw above (section 5.5) the contribution of space altimetry, essential for monitoring the oceans and the climate, to meteorology. On the contrary, the data from the IASI and GOME-2 instruments are essential contributions to the knowledge of the gases and aerosols present in the atmosphere, from which the CAMS service benefits.

For the space component of Copernicus, made up of the Sentinel missions developed by ESA (see Chapter 4), their operational deployment, and their data processing and distribution services, constituting the majority of NRT services, EUMETSAT provides operational know-how, appropriate infrastructure (e.g. the EUMETCast data dissemination network) and links to its own MTG and MetOp-SG satellites.

Within the framework of the Copernicus program, activities have therefore been delegated by the European Union to EUMETSAT, in particular:

– Sentinel-3 satellite operations and data processing for near real-time (NRT) oceanographic and atmospheric applications;

– the accommodation of the UVN Sentinel-4 sounder on the MTG-S satellites and the processing of NRT data;

– the accommodation of the UVN Sentinel-5 sounder on the MetOp-SG A satellites and the processing of NRT data;

– the operations of the Sentinel-6 satellites including the processing of their data.

Development and preparation activities, in conjunction with ESA, began even before the start of the operational phase of Copernicus in 2014. Since then, the Sentinel-3A and Sentinel-3B satellites have been launched (respectively on February 16, 2016 and April 25, 2018), as well as Sentinel-6 Michael Freilich on November 21, 2020, and provide all operational services planned in real time and in deferred time.

5.7.2. *EUMETSAT and the Copernicus services*

Just as the Meteorological and Hydrological Services of its Member States and ECMWF are the preferred recipients of EUMETSAT data, EUMETSAT also provides a priority service to Copernicus services for Sentinel data – otherwise open and free for all users. Especially:

– Copernicus Marine Environment Monitoring Service (CMEMS) for the Sentinel-3 and Sentinel-6 missions;

– Copernicus Atmosphere Monitoring Service (CAMS) for the atmospheric products of Sentinel-3, and the preparation of Sentinel-4 and Sentinel-5;

– Copernicus Climate Change Monitoring (C3S) for all missions.

For all Copernicus services participating in the coordination of the quality and development of products and operational deadlines, etc. These privileged relations have led these services and EUMETSAT to jointly propose and develop the WEkEO platform for access to Copernicus data.

5.7.3. *Continuity and expansion: the challenge of CO_2*

It is expected that the second phase of the Copernicus program, starting in 2022, will include the continuation of current activities, in particular the start-up of Sentinel-4 and Sentinel-5 operations (see Chapter 4), and also new missions. Among these, the development by ESA, ECMWF and EUMETSAT of a European capacity for monitoring and verification of anthropogenic carbon dioxide is built around the space mission Copernicus Anthropogenic Carbon Dioxide Monitoring (CO2M), described in Chapter 4.

The scientific principles for the detection of CO_2 by satellite are described in Chapter 13 of Volume 2. For the stakeholders, the main challenge is to develop the satellite and the initial elements of data processing within a few years (the first launch is planned for 2025). There is a precursor, the CNES Microcarb mission (see Chapter 6), with which EUMETSAT is associated for near real-time data processing.

The task of monitoring and verifying anthropogenic CO_2 is part of the international commitments made in the Paris climate agreements. The role of ESA, ECMWF and EUMETSAT with regard to CO2M gives them a new political importance, well beyond their programmatic, scientific, technical and operational responsibilities.

5.8. References

Hallegatte, S., Eyre, J., McNally, T., Potthast, R. (2012). The case for the EUMETSAT Polar System (EPS)/MetOp second-generation programme: Cost benefit analysis [Online]. Available at: https://www-cdn.eumetsat.int/files/2020-04/pdf_report_eps-sg_cost-benefit.pdf.

Murphy, D. and Mohr, T. (2011). EUMETSAT 25 years. EUMETSAT annual reports [Online]. Available at: https://www.eumetsat.int/all-annual-reports.

6

The Role of the National Center for Space Studies (CNES, France)

Carole DENIEL[1] and Pierre TABARY[2]
[1] *CNES, Paris, France*
[2] *Météo-France, Toulouse, France*

6.1. The CNES and its scientific missions

The National Center for Space Studies (*Centre national d'études spatiales*, CNES) is a French public establishment whose role is to propose to the public authorities the space policy in France and to implement it, among other things, for Earth observation. In this area, CNES works closely with the French scientific community, and space mission projects stem primarily from the needs identified during the scientific prospective seminars it organizes approximately every four years. After collecting, analyzing, and then having the scientific community's proposals evaluated by groups of scientific and technical experts, these are put into perspective with the strengths and weaknesses of the French laboratories and with the technical expertise of CNES and the industry. A roadmap is then established and makes it possible to put together, on the one hand, a policy for research and technology (R&T) in order to prepare the technological building blocks and, on the other hand, mission definition studies in which the requirement specifications of the observable are compared with the technological possibilities available.

For a color version of all the figures in this chapter, see www.iste.co.uk/phulpin/satellites1.zip.

Satellites for Atmospheric Sciences 1,
coordinated by Thierry PHULPIN, Didier RENAUT, Hervé ROQUET and Claude CAMY-PEYRET.

This project incubator is subject to harsh budgetary and programmatic realities. Collaborations with other space agencies, whether sought by CNES, proposed by other agencies, or guided by national policy, are often the key to the continuation of the project and the development of the mission.

To develop its global scientific strategy, CNES relies on the recommendations of recognized scientists, through a Scientific Programs Committee supported by thematic groups: TOSCA for Earth and environmental sciences (with its four subgroups: Earth, Ocean, Continental Surfaces and Atmosphere) and CERES (Committee for the Evaluation of Space Research) for the study and exploration of the Universe. These committees also guide CNES in the selection and monitoring of support provided in the laboratories (thesis grants, post-doc grants, support for research actions, etc.).

From the 1960s and 1970s, CNES took an interest in meteorological satellites by technically studying the future European Meteosat satellite and its visible-infrared imager MVIRI**. This historical aspect is discussed in Chapter 1. In this chapter, we will focus on recent, current and future satellites and instruments developed alone or in partnership by CNES.

Space agency programs are changing rapidly. The reader should also note that this chapter reflects the status of CNES programs in April 2021.

6.2. Greenhouse gases and composition of the atmosphere

Many uncertainties remain regarding the balance of greenhouse gases (atmospheric concentration, sources and sinks), the main causes of global warming. In addition, understanding phenomena related to the chemical composition of the atmosphere is a major challenge for our societies because reactive chemical species directly or indirectly affect air quality and climate. The observation of the atmosphere from space, due to its spatio-temporal coverage, is one of the fundamental pillars for significant progress in these fields, in combination with in situ measurements and in close association with numerical modeling of the Earth System.

Space observation – complemented by measurements taken by balloons, planes, or ground stations – allows for the measurement of an increasing number of atmospheric variables. However, for greenhouse gases, measuring concentrations at the required level of precision still represents a challenge from a technical and

**See Appendix 3.

measurement physics perspective. Indeed, as the objective is to provide information on surface fluxes, it is necessary to access low horizontal concentration gradients and to overcome auxiliary parameters such as aerosols, clouds and dynamics of air masses (see Chapter 13 of Volume 2).

To meet these challenges, several instrument or satellite projects have been born in CNES over the past 20 years in order to provide the scientific community with global, reliable, precise and continuous observations of atmospheric constituents, most of the time thanks to international cooperation with major space agencies or within Europe itself with the European Commission, ESA or EUMETSAT.

6.2.1. *Merlin, a political French–German will*

This is the case of the iconic French–German space mission, the MEthane Remote sensing LIdar missioN (MERLIN), dedicated to measuring atmospheric methane (Figure 6.1). It was founded in 2009 by both countries, then led by Nicolas Sarkozy and Angela Merkel, to carry out a joint space mission devoted to the climate, the launch of which would take place in 2014. To respond to this political initiative, the CNES and the German space agency DLR[1] quickly analyzed the various possibilities. The priorities of the French scientific prospective seminar of 2009 had clearly identified the need to measure greenhouse gases, but instead betting on the passive technique (using a spectrometer) to measure CO_2 with the Microcarb project. The active technique with a DIfferential Absorption Lidar (DIAL) in IPDA[2] version (see Chapter 9) did not yet seem ready to be spatialized and was comparatively more expensive. But the DLR quickly convinced the CNES that it had the appropriate DIAL technology at 1.64 μm, of which an airborne model on a helicopter, CHARM-F, had already proved the feasibility. For its part, the CNES had a Myriade platform* that was relatively inexpensive and potentially compatible at the calendar level. This is how the first cooperation agreement was signed at the end of 2010 and phase 0*, bringing together engineers and scientists from both countries, was initiated, leading to phase A* one year later (Bousquet et al. 2018). However, it quickly became apparent that the spatialization of the German DIAL was scarcely compatible in terms of power and mass with the possibilities of the Myriade platforms. It was therefore quite opportune that the new Myriade Evolution platform project was founded in France, supported by the main future investment plan put in place by the government. Thus, the CNES was able to

---

*See Glossary.

1. DLR: Deutsches Zentrum für Luft- und Raumfahrt.

2. IPDA: Integrated Path Differential Absorption.

offer the DLR a more comfortable platform to house the lidar which, at the same time, continued to be more and more greedy.

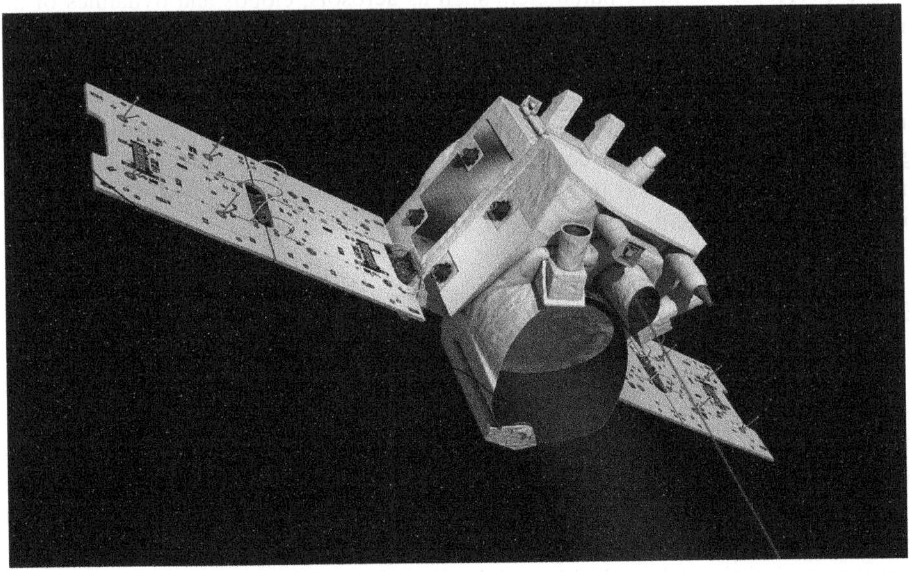

Figure 6.1. *Artist's impression of the Merlin satellite showing its emitted and backscattered lidar beams (© CNES/illustration David Ducros, 2016)*

At present, the project is still in development and the launch of Merlin has now been postponed beyond 2025 given the complexity of this lidar technology. This gives the teams time to develop an innovative and efficient ground segment. Indeed, Merlin should be the first satellite to carry a DIAL in space!

French DIAL CO_2 projects have also been researched and then proposed to ESA but have not yet been deemed technologically mature with regard to the two critical elements, laser power and stability, on the one hand, and the sensitivity of the detector, on the other hand.

6.2.2. *Microcarb, a strategic and continuous project...*

However, the Microcarb project (Deniel et al. 2017), dedicated to the measurement of atmospheric CO_2, had not said its last word! Founded in 2001 from an idea proposed to ESA and deemed technically immature at the time, it was the subject of numerous R&T actions at CNES around the technique known as static

Fourier transform, which gave rise to a CNES patent. During phase 0 of the project, this technique actually proved to be unsuitable, although very compact and precise. The study therefore concluded at the end of 2009 that Microcarb was feasible (Figure 6.2), but by using a grating spectrometer, like the American mission OCO-2, in the wavelength range of 0.7–2.1 μm.

Figure 6.2. *Artist's impression of the satellite Microcarb (© CNES/Oliver Sattler, 2021)*

Then, phase A began, which lasted nearly five years, during which the two industrial consortia led by Airbus Defence and Space France and Thales Alenia Space France developed expertise by competing in innovation and ingenuity. At the end of 2015, the year of the COP21 climate conference in Paris, the need to measure CO_2 from space and ensure its long-term monitoring became obvious at the political level and in the eyes of manufacturers. The French government therefore decided to finance the development of the Microcarb project for a launch before the end of the decade. On the European side, at the same time, ESA did not retain, against all expectations, the Carbonsat mission, which also aimed to measure atmospheric CO_2 and was the favorite to become the 8th mission of the Earth Explorer program!

Today, the Microcarb project, which aims to gain more precise access to the global carbon balance, essentially dominated by biogenic fluxes, has overcome technical difficulties and its launch is expected in 2023. The mission benefits from cooperation with the United Kingdom, in particular for the integration of the satellite, the contribution of two universities and a partnership with EUMETSAT for the operational processing of data.

On the European side, the Commission has declared that the highest priority of the future Sentinel satellites of the Copernicus program goes to the CO2M (Copernicus Anthropogenic Carbon Dioxide Monitoring) mission developed by ESA (see Chapter 4). Thanks to its wide swath (more than 200 km) and its narrow pixel (4 km^2), CO2M must acquire images of the concentration of CO_2, which should make it possible to detect plumes from "hot spots" by 2026, thus distinguishing anthropogenic CO_2 emissions from biogenic emissions, and thus contributing to the building of the carbon balance for 2028 provided for by the Paris Treaty.

6.2.3. *TRAQ, Geotrope, Mageaq, promising projects but no future developments...*

Although the static Fourier transform spectrometry technique was not adapted to the precision required for Microcarb and the measurement of CO_2, it did seem to be suitable for the measurement of gaseous constituents involved in air quality. This is how CNES used this concept for the SIFTI** instrument (Figure 6.3), which was part of the payload of the TRAQ (TRopospheric composition and Air Quality) mission, proposed to the ESA Earth Explorer program in collaboration with the Dutch space agency but was unfortunately discontinued.

For the purposes of air quality monitoring, CNES also carried out a definition study on an infrared spectrometry instrument designed to fly in geostationary orbit in order to observe the diurnal variations of reactive species over Europe. This was the objective of the Geotrope and then Mageaq projects proposed by French laboratories to the ESA Earth Explorer program, again without success. Note, however, that spectrometers operating in the infrared (Infrared Sounder, IRS**) and in the ultraviolet-visible-near infrared (Sentinel-4**) are in development within the framework of the respective programs of EUMETSAT and ESA/Copernicus, to be boarded on the Meteosat Third Generation (MTG) geostationary satellite sounder version, whose launch is scheduled for the end of 2023 (see Chapters 4 and 5). These instruments will make it possible to measure important constituents for

**See Appendix 3.

monitoring air quality (O_3, CO, NO_2, SO_2, etc.) over Europe and Africa with a temporal sampling well adapted to high diurnal variability. Together with a spectrometer similar to Sentinel-4, GEMS** (Geostationary Environment Monitoring Spectrometer) on board South Korea's GEO-Kompsat-2 geostationary satellite, launched in February 2020 over Asia, and TEMPO (Tropospheric Emissions: Monitoring of POllution) planned by NASA in 2022 over North America, the whole globe should be well covered.

Figure 6.3. *Development of the SIFTI instrument at the CNES optics laboratory (© CNES/LE DOARÉ Pascal, 2008)*

6.3. IASI and IASI-NG, for meteorology, atmospheric composition and climate

Three examples of the IASI** hyperspectral infrared sounder fly or have flown on board the MetOp-A (2006–2021), MetOp-B (since 2012) and MetOp-C (since 2018) satellites of the European program EPS[3] (see Chapters 4 and 5). This innovative instrument, based on the Fourier transform interferometry, was developed in the 1990s and 2000s by CNES, in partnership with EUMETSAT and the industrial company Thales. Due to its spectral range (from 3.6 to 15.5 μm) and its spectral resolution (more than 8,000 channels), IASI is able, in cloudless skies, to probe the temperature and humidity of the troposphere with an accuracy of 1K and 10%, a vertical resolution of 1 to 2 km and a horizontal resolution of 25 km (Cayla 2001). MetOp's 14 Sun-synchronous orbits allow each copy of IASI to cover the entire globe twice a day.

**See Appendix 3.

3. EPS: EUMETSAT Polar System.

Initially, IASI's ability to probe temperature and humidity addressed meteorologists' need for numerical weather prediction. This objective has been fully achieved: the characteristics of IASI, best used by means of the assimilation of observed radiances in numerical weather forecasting models, have largely met the expectations of meteorologists, as shown in Box 6.1.

This box gives an estimate of the contribution of the different observation systems to the reduction of the 24-hour forecast error of the global Météo-France Arpege model (Courtier et al. 1991), throughout the year 2020.

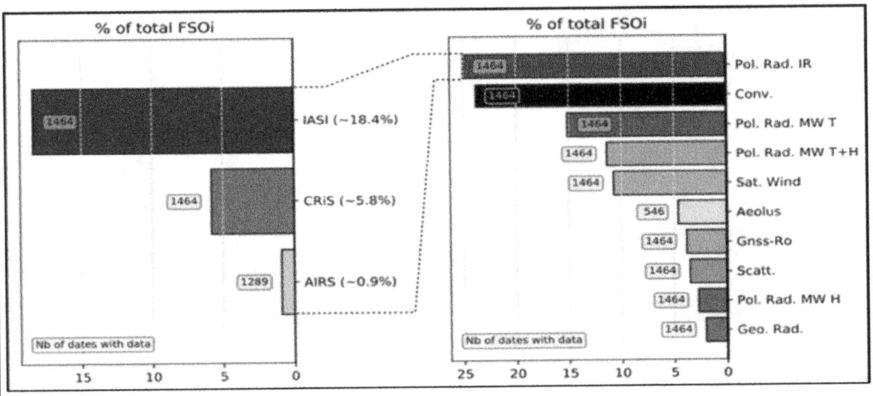

Figure 6.4. *Contribution (in percentage) of different observing systems to the reduction of the 24-hour forecast error of the Arpege global model (©Météo-France, Alex Doerenbecher and Dominique Raspaud)*

The estimate in Figure 6.4 is obtained by the FSOI method (Forecast Sensitivity to Observation derived Impact) (Lorenc and Mariott 2014) throughout 2020. All non-satellite data is grouped under the acronym Conv. The other data are satellite data: microwave data (Pol Rad MW) for temperature T and humidity H, Aeolus data, scatterometer data (Scatt), etc. The data from the infrared sounders in polar orbit (Pol Rad IR) represent 25% of the total reduction in the forecast error and, among them, the data from the three IASI sounders 18.4% alone. Note that for IASI, the assimilation concerns the "temperature" channels and the "humidity" channels.

We observe that the greatest contribution to the quality of the forecasts is by the infrared sounders in polar orbit, and in particular that of the three IASI sounders. It should, however, be noted that the quantitative result of such analyses may vary depending on the model used, the number of instruments considered, the lead time and the geographical area of the forecast.

Box 6.1. *Role of different observation systems in the quality of numerical weather prediction*

In addition, IASI observations made it possible to restore vertically integrated information on aerosols (Capelle et al. 2014) and, in some cases, their three-dimensional distribution. Cuesta et al. (2020) thus document the temporal evolution of the 3D distribution of desert aerosols in connection with atmospheric dynamics at different scales. In addition, IASI, sensitive to clouds in the upper troposphere, has contributed to the establishment of climatologies of upper tropospheric ice clouds and, consequently, to a better understanding of their formation processes by linking, for example, the size of anvil clouds to convection intensity (Stubenrauch et al. 2017).

With regard to the monitoring of atmospheric composition and climate, the IASI instrument also had significant scientific objectives. Faced with the increasing acquisition of important results (Hilton et al. 2012; Clerbaux et al. 2015), with the detection of more than 30 unprecedented chemical species from space and many essential climatic variables, IASI quickly became an essential instrument, not only for the meteorological services but also for the atmospheric science community in the broad sense with the monitoring of episodes of pollution, large fires, volcanic eruptions and lately, the impact of the Covid-19 health crisis on air quality (see Part 2 of Volume 2). Therefore, the continuity and improvement of IASI products became crucial in the eyes of the French research community, as well as for the national meteorological services in Europe.

It was therefore fairly easy for CNES to be convinced – following the prospective seminars in 2009 and then in 2014 – of the value of continuing its partnership with EUMETSAT for the development of a new generation of innovative instruments intended to succeed IASI, named IASI-NG** (NG for new generation), which will be carried on European MetOp-SG satellites from 2023 (Crevoisier et al. 2014). Thanks to the ingenuity of CNES and the manufacturer Airbus Defence and Space France, a new instrumental concept based on a Mertz interferometer[4] with field compensation will make it possible to take over from the IASI measurements with an improvement of a factor of 2 in the signal-to-noise ratio and in the spectral resolution (Figure 6.5). By obtaining more accurate infrared spectra, the measurements of IASI-NG will be even more precise and more vertically resolved (especially in the lower layers of the atmosphere) than those of IASI and new chemical species could be identified. In addition, with the continuous series of stable measurements spanning almost 40 years, beginning with IASI, IASI-NG will undoubtedly become a crucial tool for the long-term monitoring of climate trends (Figure 6.6).

**See Appendix 3.

4. Mertz interferometer: Fourier transform spectrometer (see Chapter 9) with field compensation, used for hyperspectral infrared sounding.

Figure 6.5. *Artist's impression of the IASI-NG sounder on the MetOp-SG satellite (© CNES/ill./REGY Michel, 2021)*

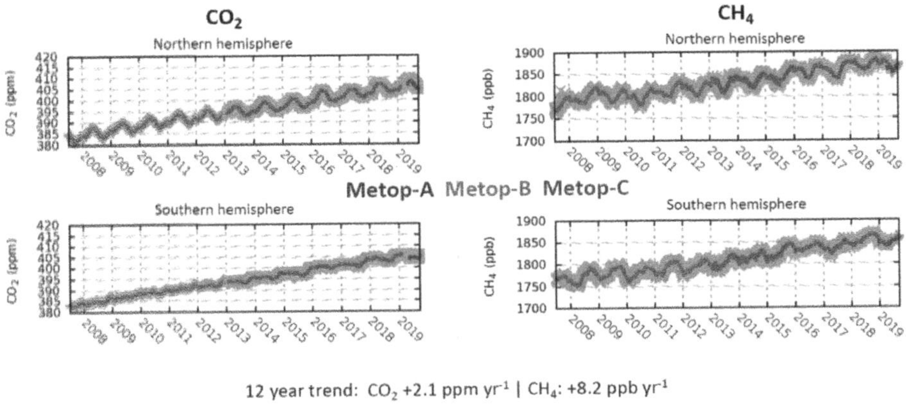

Figure 6.6. *Increase observed by IASI since 2007 in the concentration of greenhouse gases CO_2 (left) and CH_4 (right) in the middle troposphere of the Northern Hemisphere (top) and the Southern Hemisphere (bottom) (figure kindly provided by Cyril Crevoisier, LMD[5]/IPSL[6])*

5. LMD: *Laboratoire de météorologie dynamique* (France).

6. IPSL: *Institut Pierre-Simon Laplace* (France).

6.4. Physical properties of the atmosphere

Whether for meteorology or for the study of the climate, scientific questions and their application and operational issues relating to clouds, precipitation, water vapor, aerosols and wind mobilize a very large community today. This community has been developed and structured over the years, with the support of CNES, around several space missions carried out within the framework of partnerships with other space agencies or within the framework of European programs, such as with EUMETSAT.

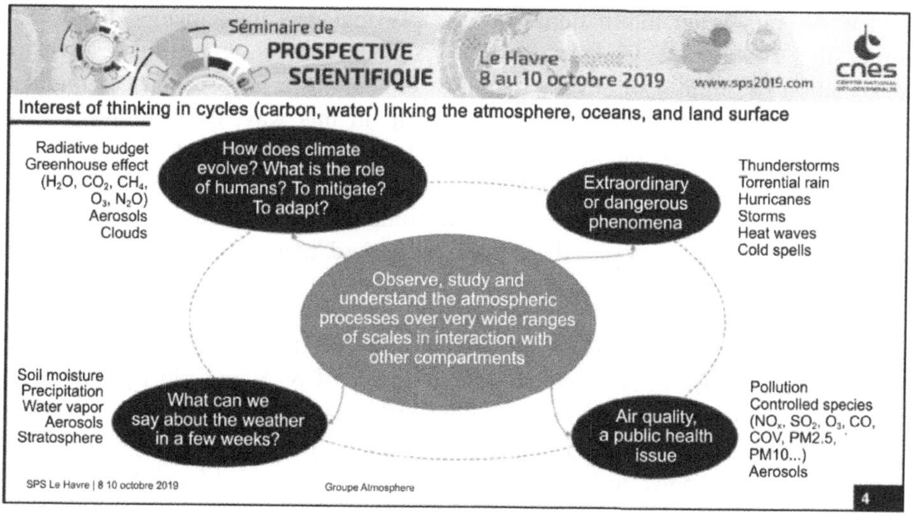

Figure 6.7. *Presentation made by Alain Dabas, president of the Tosca-Atmosphere group, during the 2019 prospective seminar in Le Havre (figure kindly provided by A. Dabas, CNRM[7] and translated here for ease of reading)*

The main questions on the subject (weather forecasting, hazardous atmospheric phenomena, climate change, air quality) are summarized in Figure 6.7. They include, as always in Earth observation, issues relating to the observation, understanding, modeling and then forecasting or projection, in support of the protection of people and property, public policies for mitigation and adaptation to climate change and socio-economic development.

7. CNRM: *Centre national de recherches météorologiques* (National Center for Meteorological Research) (France).

It should be remembered here that, for a given variable, it is generally necessary to access several of its characteristics in order to deduce the processes at work in the atmosphere. For example, understanding the coupling between aerosols and clouds requires the characterization of both physical (altitude, layer thickness) and microphysical and radiative parameters (size and shape of particles, optical depth, emissivity, etc.), for which the levels of maturity of space techniques are different.

6.4.1. *Aerosols and clouds: PARASOL, CALIPSO and the A-Train*

The A-Train is an international observatory that was implemented by NASA during the 2000s (see Chapter 3). The A-Train, as its name suggests, is a constellation (a "train") of several satellites circulating a few seconds or minutes apart in the same Sun-synchronous polar orbit. In this way, the sightings of the sensors carried by these satellites towards the same area of the Earth's surface and the atmosphere are almost simultaneous, which makes it possible to compare or combine their observations. The A-Train instruments are dedicated in particular to the study of clouds, aerosols, their interactions and their effects on the climate. CNES contributed to this in two ways.

First, through the PARASOL mission (2004–2013) carrying the POLDER** polarimeter, designed in close collaboration between CNES and LOA[8], in order to measure the polarized and multidirectional reflectances of the Earth–atmosphere system in the visible and near infrared (Tanré et al. 2011). Indeed, if the radiation from the sun is not polarized it becomes so when it is scattered by certain particles such as aerosols or ice crystals. The PARASOL mission therefore aimed to better characterize the radiative properties of clouds and aerosols in order to better understand their impact on the climate. For aerosols, its observations have notably contributed to specifying the quantity, optical thickness and size distribution of aerosols above the ocean (Figure 6.8); with regard to clouds, it made it possible to determine their thermodynamic phase and their altitude, as well as to estimate the flux reflected in the solar domain. Ultimately, PARASOL will have contributed to a better understanding of how clouds and aerosols influence our climate.

The single scattering albedo makes it possible to characterize the type of particles according to their absorption. For the wavelength considered (865 nm), it is mainly soot-rich particles that absorb solar radiation. Figure 6.8 shows that aerosols from anthropogenic biomass fires transported over the southeast Atlantic are particularly absorbing (SSA of around 0.80). It can be noted that the dust plumes of

8. LOA: Atmospheric Optics Laboratory (France).

**See Appendix 3.

Saharan origin are associated with high optical depths but have low levels of absorption for the wavelength considered (SSA close to 1).

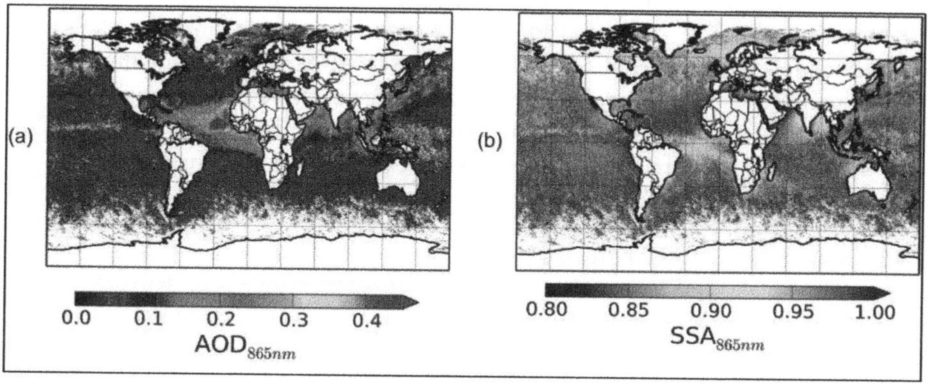

Figure 6.8. *Global distribution of the optical depth AOD of the aerosols (a) and their absorption measured by the single scattering albedo SSA (b), in both cases at 865 nm. Annual averages for the year 2006. Results obtained with the algorithm described in Waquet et al. (2016) applied to POLDER/PARASOL data acquired in clear skies over the ocean (source: Waquet et al. (2016), © 2016 American Geophysical Union)*

CNES's second contribution to the A-Train was through the CALIPSO mission, conducted jointly with NASA. Launched in 2006, the satellite carries a dual-wavelength lidar, the main instrument provided by NASA, as well as an infrared imager provided by CNES and a visible camera (Winker et al. 2010). The platform was provided and is operated by CNES. CALIPSO has been flying in tandem with the Cloudsat satellite carrying a cloud radar since 2006, despite a few separations and the reconfiguration in 2018 of their common orbit, now called C-Train (see Chapter 3). A new extension of the CALIPSO mission is planned until 2023.

The duration of the CALIPSO mission is quite exceptional, particularly with regard to the three-year lifespan initially envisaged. In its 15 years in orbit, 8.8 billion laser pulses have been emitted by the CALIPSO lidar! The scientific results are immense: CALIPSO and its companion Cloudsat will have revolutionized our knowledge of the three-dimensional distribution and characteristics of clouds and aerosols on a global scale. Thanks to its ability to measure vertical profiles, CALIPSO is systematically used to validate the restitutions made from passive instruments (IASI, for example) and is considered a climate watch, duly cited in the IPCC reports. In 2020, the rate of publications citing CALIPSO was 365 articles per year, or one per day! French scientists feature prominently in these publications, with a number of highly innovative and internationally recognized products and tools. For example, we can cite the Dardar product (Delanoë and Hogan 2008), which combines radar and lidar data to restore the microphysical properties of clouds, or the GOCCP[9] lidar product, which makes it possible to make the link with climate models (Chepfer et al. 2010). The CNES–NASA partnership with regard to the A-Train has also been very powerful in boosting scientific collaborations on both sides of the Atlantic (with, in particular, the IPSL, LMD, LATMOS[10], LOA laboratories, etc.), with numerous cross-student exchanges, joint projects, etc.

6.4.2. *Next: 3MI and EarthCare*

The programmatic continuation of the monitoring of aerosols and clouds is, on the one hand, at the European level, and on the other hand, at the American level, which is longer term. The 3MI** instrument (Figure 6.9), which will be on board EUMETSAT's MetOp-SG European polar-orbiting satellites from 2023, alongside other instruments including IASI-NG, is the direct heir of the POLDER instrument (Fougnie et al. 2019). It has been developed by ESA and should thus give rise to nearly 20 years of continuous observations with a polarimeter, which opens up great prospects for understanding phenomena, climate monitoring and taking aerosols into account in models, particularly for air quality. CNES strongly supports the preparation of 3MI, in particular at LOA, which concentrates a lot of expertise and tools on this type of instrument. CNES also brings its expertise to EUMETSAT, in particular for the preparation of the processing chains and the in-flight calibration of the instrument.

On the lidar side, the sequel to CALIPSO in the short term (from 2022) will be provided by the ESA-JAXA (European and Japanese space agencies) EarthCare mission (see Chapter 4), which will carry an ultraviolet lidar, a cloud radar, an imaging spectrometer and a broadband radiometer. A good part of the French scientific know-how acquired with CALIPSO should be brought to EarthCare, but with challenges to overcome: thus, the fact that the wavelength of the EarthCare lidar (355 nm) is different from those of the CALIPSO lidar (532 and 1,064 nm) will pose difficulties when connecting the corresponding measurement series for climate research. As a member state of ESA, France contributes to EarthCare, and the CNES strongly supports the French laboratories involved in the preparation of this mission (approximately 10). It is interesting to note that this mission is being prepared together with the German scientific community, which has set up a dedicated office at the national level to support the activities.

---

9. GOCCP: GCM (General Circulation Model) Oriented Cloud CALIPSO Product.

10. LATMOS: *Laboratoire atmosphères, milieux, observations spatiales* (France).

Figure 6.9. *The instrument 3MI (courtesy Leonardo S.p.A.)*

6.4.3. *A study in the longer term: ACCP*

The longer-term future in this area is currently being defined as part of a NASA pre-formulation study called ACCP (Aerosols Clouds Convection Precipitation). This study was launched following the Decadal Survey prospective report produced in early 2018 by the American Academy of Sciences, which highlighted five priority observables for the next decade, including the aerosols-clouds-convection-precipitation set. The future ACCP international observatory aims for nothing less than to ensure continuity by 2030 with performance improvements to both the A-Train and the NASA-JAXA satellite GPM[11] Core Observatory (see Chapter 3). Strengthened by the legacy of the A-Train and the GPM constellation with Megha-Tropiques, CNES, encouraged by its scientific committees, immediately expressed a strong interest in participating in the study to become, if possible, a NASA partner within this future observatory. Several proposals have been made by CNES and a number of partner laboratories; they gave rise to preliminary studies in France about the contribution to a three-wavelength lidar and the supply of two microwave radiometers that would fly in tandem. This second proposal was accepted by NASA (within a very constrained budgetary, calendar and technical framework) and the study of its implementation in the future ACCP observatory is currently in progress.

11. GPM: Global Precipitation Mission.

6.4.4. *Megha-Tropiques and rainfall*

Figure 6.10. *SAPHIR microwave sounder being tested (© CNES/Girard Sébastien, 2010)*

Another international observatory established under the aegis of NASA and in which CNES participates is the GPM constellation. It is a "virtual" international constellation made up of various microwave space instruments in orbit and dedicated to measuring precipitation on a global scale (see Chapter 15 of Volume 2). CNES contributes to this through the Megha-Tropiques mission, developed and operated in cooperation with the Indian Space Research Organisation (ISRO). Megha-Tropiques has been in orbit since 2011, carrying several instruments including a 183 GHz microwave sounder called SAPHIR** and a radiometer called SCARAB**, designed to measure the radiation budget (Desbois et al. 2007). Combined with the very original orbit of Megha-Tropiques, a very inclined orbit allowing for revisits up to five times a day in the intertropical zone (see Chapter 8), these two instruments, whose quality and stability of measurements have been proven, have over the years become unavoidable for calibrating other instruments on low Earth orbit or geostationary satellites. SAPHIR (Figure 6.10) has also proven to be a key instrument for forecasting high impact weather events in the Tropics in clear skies (Chambon et al. 2015), and also in cloudy atmospheres (Duruisseau et al. 2019), thanks to the work carried out by Météo-France in terms of data assimilation. Its data is now assimilated into the prediction models of nearly 10 meteorological services around the world. Finally, combined with other microwave instruments and

**See Appendix 3.

infrared data from geostationary satellites, SAPHIR has also contributed to the emergence of a product for estimating rainfall in the tropical region at a resolution of 1 degree–1 day called TAPEER[12], whose performance is superior to all other estimators, especially in Africa (Gosset et al. 2018).

6.5. Additional facilities and means of observation

Scientific questions relating to the Earth's atmosphere can only be properly addressed by combining space, ground, balloon and airborne observations, and numerical models with human expertise. For several decades, CNES has therefore supported a certain number of facilities and ground equipment considered strategic for the proper preparation and operation of space missions. We can thus cite the support for the SAFIRE French fleet of research aircraft (Lamorthe et al. 2016), together with Météo-France and the CNRS[13]. SAFIRE aircraft have taken part in countless preparation and validation campaigns for space missions, both French and European. Among these campaigns, we can cite the annual multi-instrumented Magic campaigns which, since 2018, have aimed to better understand the vertical exchanges of greenhouse gases CO_2 and CH_4 in the atmosphere, as part of the preparation for the Merlin, Microcarb and IASI-NG missions (Crevoisier and Bès 2018). CNES also supports a certain number of instruments on board SAFIRE aircraft, in particular the PMA airborne measurement platform, operated by LAMP[14] and dedicated to in situ microphysical measurement, the new generation Leandre airborne lidar, maintained by LATMOS, and the Osiris airborne polarimeter operated by LOA for the preparation of 3MI.

CNES also contributes financially to keeping in operational conditions of a certain number of networks and instrumented sites being grouped together at national level within the framework of the Actris-FR research infrastructure. These include the instrumented sites SIRTA[15] and COP-DD[16], and international networks NDACC[17], TCCON[18] and Aeronet, crucial for the validation of satellite products.

12. TAPEER: Tropical Amount of Precipitation with an Estimate of ERrors.

13. CNRS: The French National Centre for Scientific Research.

14. LAMP: Laboratoire de météorologie physique, France.

15. SIRTA: Instrumental research site using atmospheric remote sensing located in Palaiseau, France.

16. COP-DD: Cézeaux-Opme-Puy-de-Dôme site situated in Clermont-Ferrand, France.

17. NDACC: Network for the Detection of Atmospheric Composition Change.

18. TCCON: Total Carbon Column Observing Network.

In addition, CNES maintains, develops and deploys a wide range of balloons (boundary layer balloons, weather balloons, stratospheric balloons) serving the scientific community, particularly in connection with space missions. We can mention here the Stratéole-2 campaigns (Haase et al. 2018) dedicated from 2019 to 2025 to the study of the troposphere–stratosphere interface in the intertropical zone using a flotilla of stratospheric pressurized balloons[19] with a flight duration of several months (Figures 6.11 and 6.12). Stratéole-2 campaigns are also used to validate data from the pioneering Aeolus mission launched in 2018 by ESA (see Chapter 4). Aeolus measures wind profiles on a global scale and its data is assimilated into several operational weather prediction models, including that of Météo-France; in addition, Aeolus cloud-aerosol products could be used to bridge CALIPSO and EarthCare in climate monitoring studies.

Figure 6.11. *Preparation of a CNES stratospheric pressurized balloon during the Stratéole-2 campaign in 2019 (© CNES)*

Finally, CNES has strongly supported the establishment of data and service hubs in France, as part of the national research infrastructure Data Terra, in particular the hub Aeris[20], dedicated to the atmosphere component. Aeris brings together four data and service centers located in Lille, Paris, Toulouse and Lannion and aims to

19. Pressurized balloon: closed gas balloon, which makes it possible to reconcile flight duration (several months), large payload capacity (up to 30 kg) and sufficient altitude (up to 20 km).

20. https://www.aeris-data.fr/.

provide the user community with easy and unified access to space or in situ atmospheric data, as well as the associated visualization and processing tools.

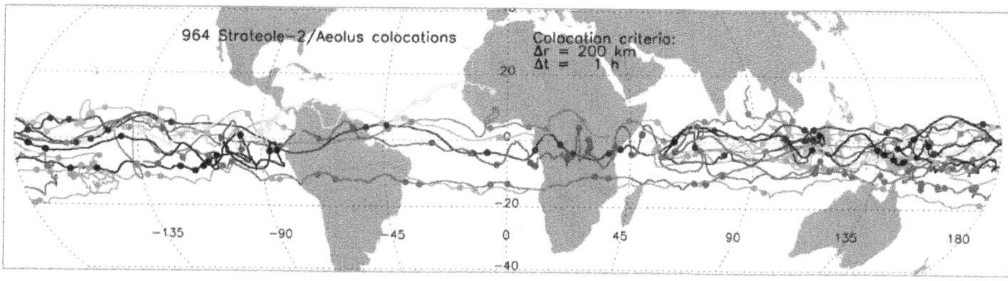

Figure 6.12. *Trajectories of the balloons of the first Stratéole-2 campaign (November 2019–February 2020). Round dots indicate co-locations with the Aeolus satellite (© LMD/CNRS)*

6.6. The role of numerical models

The definition, preparation and exploitation of satellite missions in terms of the physics and chemistry of the atmosphere also benefit from the exceptional development of numerical models. For France, this is the case for the Mocage and Chimere air quality models, the Arpege and Arome weather prediction models and the CNRM and IPSL climate models. These models have now reached spatial resolutions and capacities for representing phenomena and variables that open up new perspectives for satellite data, whether for their validation or for their assimilation. All of the space missions in the field are thus linked to one or more models, either because the model was used to specify the instrument or the observation system, for example, through so-called OSSE experiments (Observing Systems Simulation Experiments, see Chapter 4 of Volume 2), because the model assimilates the observations of the mission, or because these observations lead to the validation or invalidation of certain components of the model: climatic trends, representation of hydrometeors, radiation budget, etc.

Thus, the C3IEL mission project (Cluster for Cloud Evolution Climate and Lightning) currently in phase A (feasibility study), led by CNES in cooperation with the Israeli space agency ISA, aims to restore the three-dimensional envelopes of clouds at decametric resolution, as well as their rate of vertical expansion. For this, a train of coordinated nanosats[21] is planned, embarking cameras in the visible range, whose simultaneous images on cloudy scenes will be combined in a stereoscopic

21. Nanosat: a nanosatellite (or nanosat) is a small satellite, composed of one or six units of 10x10x10 cm^3, which can have a mass of 1–10 kg.

way. The characteristics of the cameras and the sampling strategy are studied through high-resolution numerical simulations of the Meso-NH model (Lafore et al. 1997). In return, the data from the mission once in orbit – the launch is envisaged in 2025 – will be very valuable for validating the future generation of weather prediction models at hectometric resolution that Météo-France must develop by 2030.

6.7. References

Bousquet, P., Pierangelo, C., Bacour, C., Marshall, J., Peylin, P., Ayar, P.V., Ehret, G., Bréon, F.-M., Chevallier, F., Crevoisier, C. et al. (2018). Error budget of the methane remote lidar mission and its impact on the uncertainties of the global methane budget. *J. Geophys. Res. Atmos.*, 123(20), 11766–11785. https://doi.org/10.1029/2018JD028907.

Capelle, V., Chédin, A., Siméon, M., Tsamalis, C., Pierangelo, C., Pondrom, M., Crevoisier, C., Crepeau, L., Scott, N.A. (2014). Evaluation of IASI-derived dust aerosol characteristics over the tropical belt. *Atmos. Chem. Phys.*, 14, 9343–9362. https://doi.org/10.5194/acp-14-9343-2014.

Cayla, F.-R. (2001). L'interféromètre IASI : un nouveau sondeur satellitaire haute résolution. *La Météorologie*, 32, 23–39. doi: 10.4267/2042/36150.

Chambon, P., Meunier, L.-F., Guillaume, F., Piriou, J.-M., Roca, R., Mahfouf, J.-F. (2015). Investigating the impact of the water-vapour sounding observations from SAPHIR on board Megha-Tropiques for the ARPEGE global model. *Q.J.R. Meteorol. Soc.*, 141, 1769–1779. https://doi.org/10.1002/qj.2478.

Chepfer, H., Bony, S., Winker, D., Cesana, G., Dufresne, J.-L., Minnis, P., Stubenrauch, C.J., Zeng, S. (2010). The GCM oriented CALIPSO cloud product (CALIPSO-GOCCP). *J. Geophys. Res.*, 115, D00H16. doi:10.1029/2009JD012251.

Clerbaux, C., Hadji-Lazaro, J., Turquety, S., George, M., Boynard, A., Pommier, M., Safieddine, S., Coheur, P.-F., Hurtmans, D., Clarisse, L. et al. (2015). Tracking pollutants from space: Eight years of IASI satellite observation. *Comptes rendus géoscience*, 347(3), 134–144. https://doi.org/10.1016/j.crte.2015.06.001.

Courtier, P., Freydier, C., Geleyn, J.-F., Rabier, F., Rochas, M. (1991). The ARPEGE project at METEO-FRANCE. *Proc. ECMWF Workshop*. Numerical methods in atmospheric modelling, 9–13 September 1991, 2, 193–231. ECMWF, Shinfield Park, Reading.

Crevoisier, C. and Bès, C. (2018). Magic, une campagne inédite de mesures des gaz à effet de serre. *La Météorologie*, 102, 5–6. https://doi.org/10.4267/2042/68202.

Crevoisier, C., Clerbaux, C., Guidard, V., Pequignot, É., Pasternak, F. (2014). IASI-NG : un concentré d'innovations technologiques pour l'étude de l'atmosphère terrestre. *La Météorologie*, 86, 3–5.

Cuesta, J., Flamant, C., Gaetani, M., Knippertz, P., Fink, A.H., Chazette, P., Eremenko, M., Dufour, G., Di Biagio, C., Formenti, P. (2020). Three-dimensional pathways of dust over the Sahara during summer 2011 as revealed by new infrared atmospheric sounding interferometer observations. *Q.J.R. Meteorol. Soc.*, 146, 2731–2755. https://doi.org/10.1002/qj.3814.

Delanoë, J. and Hogan, R.J. (2008). A variational scheme for retrieving ice cloud properties from combined radar, lidar, and infrared radiometer. *J. Geophys. Res.*, 113, D07204. doi:10.1029/2007JD009000.

Deniel, C., Millet, B., Buisson, F., Pierangelo, C., Jouglet, D., Bréon, F.-M., Bousquet, P., Chevallier, F., Crevoisier, C., Ehret, G. (2017). MERLIN and MICROCARB: Preparation of 2 space missions for CO_2 and CH_4. *19th EGU General Assembly, EGU2017, Proc., Vienna, Austria*, 23–28 April, 5276.

Desbois, M., Capderou, M., Eymard, L., Roca, R., Viltard, N., Viollier, M., Karouche, N. (2007). Megha-Tropiques : un satellite hydrométéorologique franco-indien. *La Météorologie*, 57, 19–27.

Duruisseau, F., Chambon, P., Wattrelot, E., Barreyat, M., Mahfouf, J.-F. (2019). Assimilating cloudy and rainy microwave observations from SAPHIR on board Megha-Tropiques within the ARPEGE global model. *Q.J.R. Meteorol. Soc.*, 145, 620–641. https://doi.org/10.1002/qj.3456.

Fougnie, B., Marbach, T., Lacan, A., Lang, R., Schlüssel, P., Poli, G., Munro, R., Couto, A.B. (2019). The multi-viewing multi-channel multi-polarisation imager – Overview of the 3MI polarimetric mission for aerosol and cloud characterization. *J. Quant. Spectrosc. Radiat. Transfer*, 219, 23–32.

Gosset, M., Alcoba, M., Roca, R., Cloché, S., Urbani, G. (2018). Evaluation of TAPEER daily estimates and other GPM-era products against dense gauge networks in West Africa, analysing ground reference uncertainty. *Q.J.R. Meteorol. Soc.*, 144(Suppl. 1), 255–269. https://doi.org/10.1002/qj.3335.

Haase, J.S., Alexander, M.J., Hertzog, A., Kalnajs, L., Deshler, T., Davis, S.M., Plougonven, R., Cocquerez, P., Venel, S. (2018). Around the world in 84 days. *Eos*, 99. https://doi.org/10.1029/2018EO091907.

Hilton, F., Armante, R.A., August, T., Barnet, C., Bouchard, A., Camy-Peyret, C., Capelle, V., Clarisse, L., Clerbaux, C., Coheur, P. et al. (2012). Hyperspectral Earth observation from IASI: Five years of accomplishments. *Bull. Am. Meteor. Soc.*, 93, 347–370.

Lafore, J., Stein, J., Asencio, N., Bougeault, P., Ducrocq, V., Duron, J., Fisher, C., Hereil, P., Mascart, P., Masson, V. et al. (1997). The Meso-NH atmospheric simulation system. Part I: Adiabatic formulation and control simulations. *Ann. Geophys.*, 16, 90–109.

Lamorthe, C., Borbon, A., Schwarzenboeck, A., Canonici, J.-C. (2016). Safire : des avions au service de la recherche en environnement. *La Météorologie*, 93, 30–41. doi: 10.4267/2042/59939.

Lorenc, A.C. and Marriott, R.T. (2014). Forecast sensitivity to observations in the met office global numerical weather prediction system. *Q.J.R. Meteorol. Soc.*, 140, 209–224. doi: 10.1002/qj.2122.

Stubenrauch, C.J., Feofilov, A.G., Protopapadaki, E.-S., Armante, R. (2017). Cloud climatologies from the InfraRed Sounders AIRS and IASI: Strengths and applications. *Atmospheric Chemistry and Physics*, 17(13), 13625–13644. https://doi.org/10.5194/acp-17-13625-2017.

Tanré, D., Bréon, F.-M., Deuzé, J.-L., Dubovik, O., Ducos, F., François, P., Goloub, P., Herman, M., Lifermann, A., Waquet F. (2011). Remote sensing of aerosols by using polarized, directional and spectral measurements within the A-Train: The PARASOL mission. *Atmos. Meas. Tech.*, 4, 1383–1395. doi:10.5194/amt-4-1383-1395-2011.

Waquet, F., Péré, J.-C., Peers, F., Goloub, P., Ducos, F., Thieuleux, F., Tanré, D. (2016). Global detection of absorbing aerosols over the ocean in the red and near-infrared spectral region. *J. Geophys. Res. Atmos.*, 121, 10902–10918, doi:10.1002/2016JD025163.

Winker, D.M., Pelon, J., Coakley Jr., J.A., Ackerman, S.A., Charlson, R.J., Colarco, P.R., Flamant, P., Fu, Q., Hoff, R.M., Kittaka, C. et al. (2010). The CALIPSO Mission: A global 3D view of aerosols and clouds. *Bull. Am. Meteor. Soc.*, 91, 1211–1230, https://doi.org/10.1175/2010BAMS3009.1.

7

A Coordinated International Effort

Jérôme LAFEUILLE

Retired, World Meteorology Office (WMO), Geneva, Switzerland

7.1. The challenges of international coordination

From its inception, the use of space for the purposes of observing the Earth and its atmosphere has proven to be the setting for remarkably active and mutually beneficial international collaboration. Five years before the adoption of the Outer Space Treaty[1] affirming the supranational character of space, the General Assembly of the United Nations (UN) had already laid the foundations for global cooperation aimed at using space for the benefit of humanity. This is the purpose of the resolutions adopted in 1961[2] and 1962[3], which notably invited the World Meteorological Organization (WMO) to develop a scientific and technical program to improve knowledge of the climate system and its evolution and to develop weather forecasting capabilities "with emphasis on the use of meteorological satellites".

For a color version of all the figures in this chapter, see www.iste.co.uk/phulpin/satellites1.zip.

1. Treaty on Principles Governing the Activities of States in the Exploration and Use of Outer Space, including the Moon and Other Celestial Bodies, Res. 2222 (XXI) of December 19, 1966.

2. Res. 1721 (XVI), December 20, 1961.

3. Res. 1802 (XVII), December 14, 1962.

Satellites for Atmospheric Sciences 1,
coordinated by Thierry PHULPIN, Didier RENAUT, Hervé ROQUET and Claude CAMY-PEYRET.

A space observation system whose footprint extends beyond national borders is naturally of interest to the international community, especially if it is designed for observing the atmosphere and climate. Thus, it is from the community of users represented by the WMO that coordination between satellite operators has been shaped to meet the needs of weather forecasting. Later, the emergence of climate change reinforced the requirement for global coordination in extending the field of observed variables.

Another essential driver of international coordination has been the need to optimize the significant resources required for the development and implementation of a space program. This has led space agencies to share the burden of providing the different components of a mission such as the instrument payload, the platform, the launcher or the ground segment. For the operation of non-geostationary satellites, the need to have ground stations distributed around the globe has also been a trigger for international collaboration.

With the observation of the Earth and the atmosphere becoming a privileged field of scientific and technical cooperation, international coordination has been established at several levels between satellite operators:

– long-term planning of missions with the objective of coordinating the choice of orbits, orbital positions and variables to be measured, so as to ensure the continuity of the observations recognized as necessary by the international community;

– interoperability* of measurements, based on the harmonization of instrument characteristics and processing algorithms, the intercalibration* of instruments and the development of a quality assurance framework;

– cooperation in the implementation of the programs, in particular for the supply of instruments (such as AVHRR** provided by the United States for the European MetOp satellites and MHS** provided by EUMETSAT for the American NOAA satellites), the sharing of ground facilities and even mutual aid in the event of in-orbit failure of geostationary satellites;

– allocation of radio frequencies and standardization of telecommunications, storage and data exchange formats.

---

*See Glossary.

**See Appendix 3.

7.2. Multilateral coordination instances

7.2.1. *Overview*

Different organizations have been set up over the years to coordinate various aspects of this coordination. In 1967, the WMO established the World Weather Watch (WWW) of which the space-based part of the Global Observing System was an important sub-system, although still at the experimental stage. Meteorological satellite operators then formed with the WMO in 1973 the Coordination of Geostationary Meteorological Satellites, which later became the Coordination Group for Meteorological Satellites (CGMS). With a more targeted objective, the Space Frequency Coordination Group (SFCG) was set up in 1980 to represent to the International Telecommunication Union the needs of space programs for Earth observation, meteorology and astronomy, for the allocation of radio frequencies. In 1982, as a joint initiative of NASA and the European Space Agency, the Consultative Committee for Space Data Systems (CCSDS) was created. In 1984, in response to the summit of the seven most industrialized nations (G7), it was decided a Committee on Earth Observation Satellites (CEOS) would be established whose initial mission was to coordinate the use of satellite missions for the high-resolution observation of the Earth's surface. The activities of the main coordination bodies are detailed below.

7.2.2. *The World Weather Watch and its space component*

7.2.2.1. *The beginnings of the WMO Space-based Observing System*

Following the first UN resolution on international cooperation in the use of outer space, it was proposed to create the WWW whose observation component would include a space-based sub-system. In the first WWW implementation plan adopted in 1967, this space-based sub-system included only two satellites in polar orbit, one American and the other Russian, and provided for the establishment of a coordination of space programs of countries or groups of countries. In the 1970s, missions in geostationary orbit were prepared by the United States, the European Space Agency, Japan and the Soviet Union for the First GARP Global Experiment (FGGE), planned for 1978–1979 under the WMO Global Atmospheric Research Program (GARP). The WMO member countries then decided to coordinate their efforts by establishing the CGMS in 1973, which made it possible, in time for the FGGE, to set up a belt of five geostationary satellites, including three satellites deployed by the United States over the Western Atlantic, the Eastern Pacific and the Indian Ocean, one by Europe at 0° and one by Japan at 140° E.

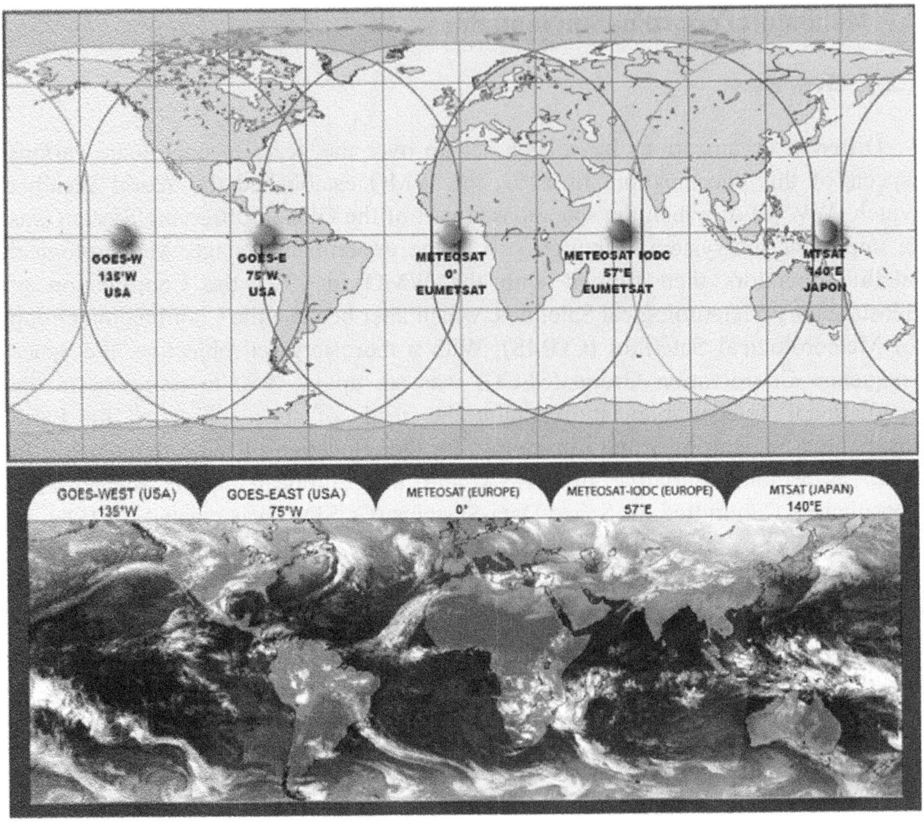

Figure 7.1. *The belt of geostationary meteorological satellites is the backbone of the global space-based observing system. Top: visibility areas of the five basic satellites for a maximum zenith angle of 85°. Bottom: fusion of the five images obtained at the same time, after calibration adjustment and application of a color palette. Note that these are infrared images because in visible light, the night-time half of the scene would be black (source: Météo-France, Centre de Météorologie Spatiale)*

7.2.2.2. *The WMO Space Program*

Because of the increasing importance of space-based observation for meteorology and climate, WMO decided in 2003 to establish its satellite activities as a "WMO Space Program". This program plays a leading role in international cooperation between WMO member countries that operate meteorological satellites and the programs that use these data: public weather services, aeronautical, marine or agricultural meteorology, the Tropical Cyclone Programme, climatology, atmospheric and climate research, oceanography, hydrology, etc.

Thus, the WMO Space Program:

– identifies, analyzes and prioritizes the user requirements of the various programs, building in particular on simulation experiments conducted by numerical weather prediction centers to quantify the respective impact of the various existing or future instruments;

– leads the development of a common long-term "vision" of the space-based observing system, in consultation between operators and user groups;

– highlights and encourages "best practices" in product development, instrument calibration, data access or preparation for the use of new systems, for example;

– advocates the unrestricted and free exchange of essential data;

– undertakes awareness-raising and information actions in anticipation of the commissioning of new series of satellites, facilitating the transfer of information between space agencies and future users[4];

– maintains online resources such as OSCAR/Space[5], allowing users and agencies alike to have reference information on all the satellite programs of interest to WMO programs;

– organizes and implements with the CGMS an extensive remote training plan for users through the network of Virtual Laboratories for Education and Training in Satellite Meteorology[6].

7.2.2.3. *The vision of the observing system by 2040*

Through critical dialog between scientific users, space agencies, industry and societal expectations, the development of a common long-term vision of the space-based system by WMO facilitates consistency between the programs of the various stakeholders. This involves anticipating the needs of global numerical modeling, the future performance of emerging instrumental technologies, in expanded spectral domains, the upheavals in data management, and also the evolution specific to space technology which makes it possible to consider a variety of orbital scenarios in addition to the classic geostationary and heliosynchronous polar configurations: highly elliptical orbits, low-inclination orbits, constellations* of nanosatellites, "formation

4. WMO Satellite User Readiness Navigator (SATURN), see: https://www.meted.ucar.edu/training_module.php?id=1181.

5. Observing System Capability Analysis and Review (OSCAR), see: https://www.wmo-sat.info/oscar/.

6. WMO-CGMS Virtual Laboratory for Education and Training in Satellite Meteorology (VLab), see: https://www.wmo-sat.info/vlab/.

*See Glossary.

flying" satellites. Table 7.1 summarizes the key elements of the space-based observing system prefigured in the WMO vision for 2040 (WMO 2019) for missions whose instrumental techniques and orbital configurations are explicitly defined.

The WMO also indicates other operational and experimental missions for which the concepts of instruments or orbits remain open. This vision is part of a global vision of the observing system in which space-based observation, observation from the land or ocean surface and in situ observation in altitude play complementary roles.

Core components of the space-based observing system (orbital configurations and payloads determined to meet the most essential needs)	
Geostationary core constellation (at least five satellites providing complete Earth coverage)	Multispectral visible and infrared imagery with fast repeat cycle Infrared hyperspectral sounders Lightning mappers Ultraviolet, visible and near-infrared sounders
Sun-synchronous polar orbit core constellation in three regularly distributed orbital planes (equator crossing: early morning, morning and afternoon, for example, 5 h 30, 9 h 30, 13 h 30)	Infrared hyperspectral sounders Microwave sounders Visible and infrared imagery, including day/night band (0.5–0.9 μm) Microwave imagery Scatterometers
Constellation in Sun-synchronous polar orbit at other equator crossing times	Sounders and imagers (especially for precipitation)
Other low Earth orbit missions (various inclinations and altitudes depending on the missions)	Wide-swath radar altimeters and high-altitude, inclined, high-precision orbit altimeters Dual-angle viewing infrared imagers Surface microwave imagery Low-frequency microwave imagery Microwave cross-track stratospheric/mesospheric sounders Ultraviolet/visible/infrared, nadir and limb sounders Precipitation radars, cloud radars Microwave sounders and imagers in inclined orbit Absolutely calibrated broadband radiometers and total/spectral solar irradiance radiometers GNSS radio occultation Narrow-band or hyperspectral imagers High-resolution multispectral visible/infrared imagers Synthetic aperture radar imagers and altimeters

		Gravimetry missions
Other low Earth orbit missions (measurement reference standards)		Visible, near-infrared and infrared reference standards Absolute measurement in microwave
Other missions (space weather)	Lagrange point L1 between the Earth and the Sun	Solar wind and in situ plasma, energetic particles and magnetic field measurement Solar coronagraph and radio spectrograph
	Geostationary orbit	In situ plasma probes Energetic particle spectrometers and magnetometers X-ray spectrograph
	Low Earth orbit	In situ plasma probes, energetic particle spectrometers and magnetometers

Table 7.1 *Summary of the core components of the integrated space-based observing system according to the WMO vision for 2040, with regard to the essential operational missions defined in terms of orbits and instruments. The vision includes other operational or experimental missions that may use different instrument or orbit concepts*

7.2.3. CGMS

7.2.3.1. *Objectives and achievements*

The Coordination Group for Meteorological Satellites (CGMS)[7], established in 1973, aims to ensure the technical coordination of satellite constellations in the field of meteorology and climate monitoring. It emphasizes missions of an operational nature with a perspective of long-term continuity, contributing to WMO's own or co-sponsored programs. The CGMS initially focused only on geostationary satellites but expanded its mandate in 1992 to coordinate all meteorological satellite programs, including low Earth orbit satellites. EUMETSAT provides the secretariat. Through an annual plenary session and regular technical dialogue between operators, the CGMS quickly proved to be an effective structure for ensuring the consistency and robustness of the operational atmospheric observing system. As a first step, the CGMS defined and adopted common standards for the broadcasting of images by fax (Weather Facsimile, "WEFAX") and for the international satellite system for data collection from mobile stations such as drifting buoys. It was within the CGMS that new standards and formats for data transmission were then adopted. More generally, the exchange of experience between operators has enabled the

7. https://www.cgms-info.org.

CGMS to define common strategies for the dissemination of data (in particular through redistribution systems such as EUMETCAST) or for their archiving.

Figure 7.2. *CGMS members 20 years apart. Top: 23rd session in Darmstadt (Germany) in May 1995. Bottom: 43rd session in Boulder (United States) in June 2015. The number of participating organizations has doubled with the addition of space agencies involved in the research and development alongside operational agencies*

The members of the CGMS have entered into mutual support agreements within the framework of a global contingency plan according to which – in the event of failure in orbit of an operational geostationary satellite of one of the members, or in the event of a delay in the deployment of a new generation of satellites – a backup satellite could be relocated and made available by another member to avoid any interruption to the geostationary watch. Such mutual backup scenarios were implemented in the 1980s with GOES-4 backing up Meteosat, in the 1990s with Meteosat-3 backing up GOES, and then in 2003 with GOES-9 backing up MTSAT.

Within the CGMS, operators exchange their experience and compare the performance of their algorithms for the development of products such as the horizontal wind field based on the displacement of atmospheric structures, the quantitative estimation of precipitation or the detection of volcanic ash.

Since the early 2000s, the CGMS has paid increased attention to the needs of climate system observation. The membership of the CGMS has been expanded to include research and development space agencies whose missions make a major contribution to the study of climate. The Intergovernmental Oceanographic Commission of UNESCO (IOC) has also become a member.

To strengthen the link with the main user communities, in addition to the participation of the WMO and the IOC, the CGMS relies on international scientific groups specialized in atmospheric sounding (ITWG[8]), satellite wind observation (IWWG[9]), rainfall estimation (IPWG[10]) and radio occultation (IROWG[11]). Finally, the CGMS makes an essential contribution to the Virtual Laboratory for education and training in satellite meteorology with the WMO.

7.2.3.2. *The reference observation system*

A major achievement of the CGMS is the adoption of a "CGMS Reference for Sustainable Contributions to the WMO Global Observing System" (CGMS 2018), which summarizes the means that CGMS members commit to making available in a sustainable manner to meet the WMO's needs in line with its long-term vision for the global observing system. The reference system includes satellites in Sun-synchronous or non-Sun-synchronous low Earth orbit, in geostationary Earth orbit, in highly elliptical orbit or at the Lagrange point L1. It lists the types of observations: infrared or microwave sounding; broadband microwave, infrared and visible imaging; high spectral resolution imaging; radio occultation; active

8. International TOVS Working Group.

9. International Winds Working Group.

10. International Precipitation Working Group.

11. International Radio Occultation Working Group.

measurements of altimetry, diffusiometry or radar imagery; observations of the sun, magnetic field and energetic particle fluxes for space meteorology. The reference system also defines the principles to be followed concerning the provision of data, the calibration of instruments, the validation and the interoperability of derived products. The CGMS members periodically evaluate the current and planned configuration of the space-based system with respect to the reference observing system and, if necessary, take risk reduction measures or urgent contingency measures to sustainably meet the needs of meteorology and climate.

7.2.4. *CEOS*

7.2.4.1. *Objectives and achievements*

The Committee on Earth Observation Satellites (CEOS) was established in 1984 in response to a recommendation from a group of the seven most industrialized countries in the world (G7). It was first interested in optical Earth observation missions such as Landsat and SPOT, and then in synthetic aperture radars. The specifications of the first Earth observation missions were then very different from those of meteorological satellites: high spatial resolution in one case and real-time requirement in the other case. This led to different ground-segment designs: while meteorological data had to be transmitted in real time, high-resolution so-called "Earth observation" data – more voluminous – had to be downloaded to ground stations and distributed around the globe to third party countries. One of the very first objectives of the CEOS was therefore to facilitate international cooperation to set up a network of ground receiving stations for the high-resolution data of Earth observation satellites, and to facilitate access to this data.

Gradually broadening its scope of activity, the CEOS took up the task of ensuring the international coordination of civil Earth observation programs from space, promoting the exchange of data and informing high-level decision-making processes for the benefit of society as a whole. It promotes cooperation and the exchange of information between its member agencies in the planning of missions, in the development of products, formats, services, applications and implementation policies, in order to encourage complementarity and compatibility of current or future systems.

The CEOS aims to respond as a priority to Earth observation needs for monitoring climate change, disaster risk reduction and sustainable development. It intends to serve as a focal point in this area in terms of the relevant international bodies and in particular with respect to the Group for Earth Observation (see Box 7.1). In particular, initiatives have been undertaken under the aegis of the CEOS to monitor the carbon cycle, forest cover, global agricultural resources, water resources and coastal waters and risks related to natural disasters.

7.2.4.2. *Working groups and virtual constellations*

A CEOS working group develops calibration and data validation procedures and coordinates the establishment of instrumented control sites providing ground truth. This work also led to the definition of a Quality Assurance Framework for Earth Observation (QA4EO). Other CEOS groups focus on promoting satellite applications to user communities, user education, and information systems and services for data discovery, accessibility and long-term preservation. Finally, activities related to the study and monitoring of the climate system are carried out by a joint CEOS–CGMS group within the framework of the common CEOS–CGMS–WMO strategy for climate monitoring from space.

The CEOS refers to the concept of "virtual constellations", whose objective is to create synergy, in given thematic areas, by coordinating infrastructures on the ground or in orbit, by mobilizing skills and by stimulating collaboration on algorithms for processing, presentation or access to data, to better meet the requirements of international users. The first seven virtual constellations deal with the following themes: chemical composition of the atmosphere; land surface imagery; ocean color radiometry; ocean surface topography; vector wind at the surface of the oceans; precipitation and sea surface temperature.

The group on Earth observations

Following the World Summit on Sustainable Development in 2002 in Johannesburg (South Africa), the concept of a global initiative on Earth observation was gradually forged during the "Earth Observation Summits" held in 2003, 2004 and 2005 to lead to the establishment of the Group on Earth Observation (GEO), which met for the first time in 2005 in Geneva. Its secretariat is based at WMO Headquarters. The same year a 10-year plan, 2005–2015, was adopted for the implementation of the Global Earth Observation System of Systems (GEOSS).

The creation of the GEO stems from the observation that in the face of unprecedented social, economic and environmental challenges, humanity must rely on decisions informed by an extensive and sustainable Earth observation system. This requires an interdisciplinary and international effort mobilizing a wide range of stakeholders including governmental and intergovernmental organizations as well as non-governmental organizations, public and private ones and the world of research and business. The GEO brings together more than a hundred governments and other organizations in the same partnership aimed at integrating existing observation networks and at the free and open sharing of data according to common standards.

Box 7.1. *The group on Earth observations*

7.3. The benefits of coordination

7.3.1. *Mission continuity*

The first purpose of this international effort is to ensure, as far as possible, the continuity of satellite observations defined as essential by the international community.

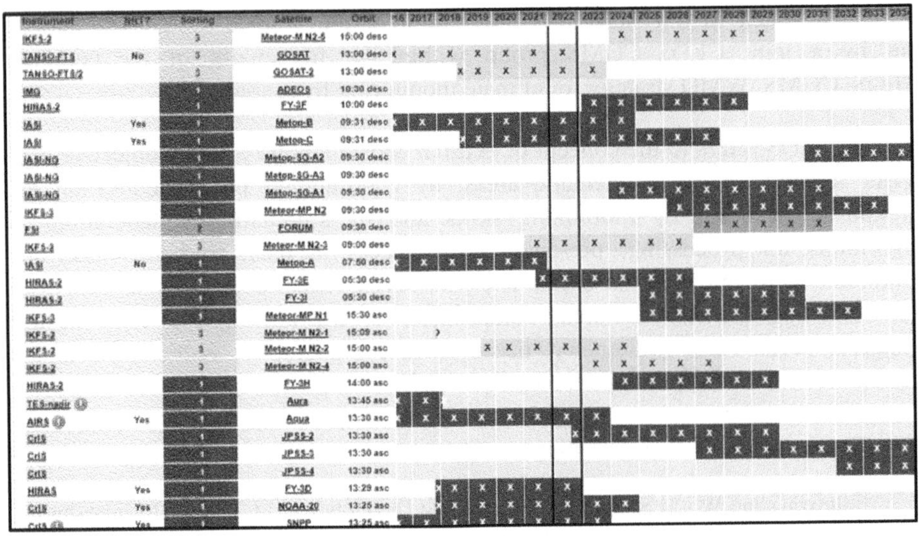

Figure 7.3. *Actual or planned availability of satellite missions. The color codes correspond to the estimated performance classes of the instruments. This type of representation is the starting point for a risk analysis carried out on a recurring basis to ensure the continuity of missions. The example above is for temperature and humidity measurement by high spectral resolution infrared sounding in low Earth orbit for the period 2016–2034 (source: WMO OSCAR/Space online database)*

User communities analyze and prioritize their future requirements, while space agencies strive to factor these requirements into their planning. By promoting the complementarity of their respective contributions, the agencies work together to ensure the spatial and temporal coverage required, with diversified instruments, using a robust system without unnecessary redundancy.

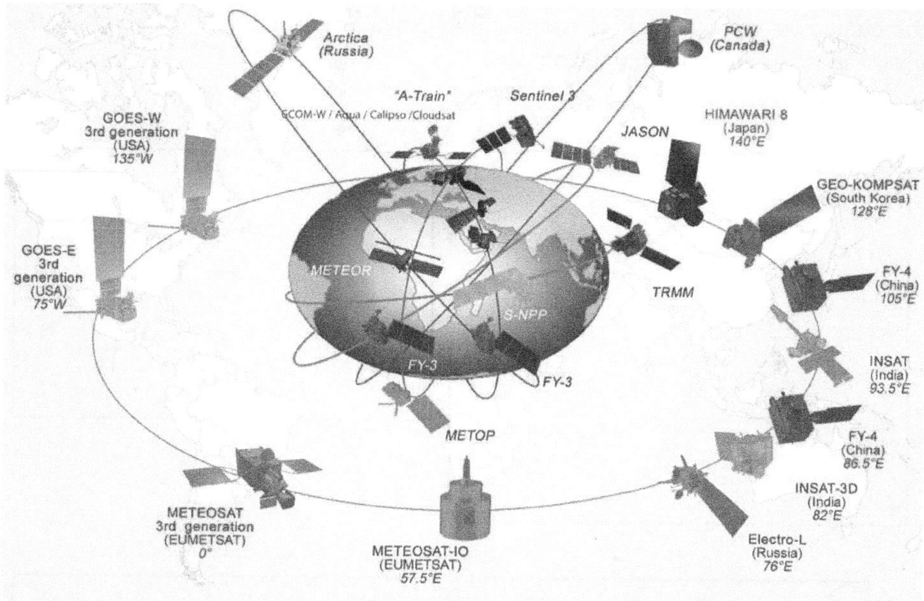

Figure 7.4. *Constellations of operational and additional meteorological satellites. Schematic, non-exhaustive representation of the configuration planned for 2021 (source: WMO, Space Program)*

7.3.2. *Intercalibration of instruments in orbit*

7.3.2.1. *Traceability requirement*

For the observation of climate change, it is essential to have accurate and homogeneous series of measurements over very long periods. Given that these observations are based on different generations of instruments on board different satellites and that these instruments evolve over time, it is necessary to establish metrological traceability for each of them in relation to common references and if possible, during pre-launch tests and again after launch, and throughout the operational phase.

For meteorological applications, the need to merge data from various satellite sources into coherent products implies taking into account the differences in the radiometric calibration of the instruments, their spectral and geometric characteristics, and also the differences resulting from the spatio-temporal sampling, of the location of the observed scenes, the sighting and illumination angles. Finally, for numerical prediction, it is useful to statistically characterize the error of the observations assimilated into the models.

7.3.2.2. *GSICS*

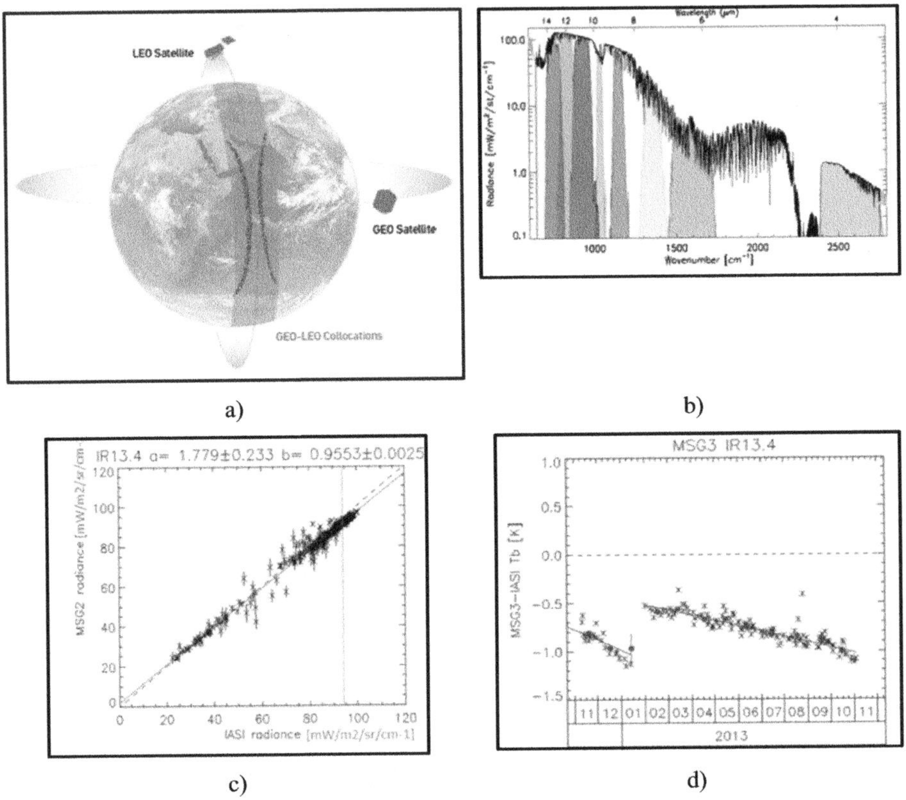

Figure 7.5. *Steps of GEO–LEO calibration according to GSICS: (a) intercalibration of an instrument in geostationary Earth orbit (GEO) and a reference instrument in low Earth orbit (LEO). The measurements acquired simultaneously are compared to the points (shown by two arcs in red) where the viewing angle is the same, and close to the nadir; (b) for the radiances to be comparable, account is taken of the difference in the spectral response functions between the channels of the instrument to be calibrated and the reference hyperspectral instrument; (c) a regression is applied to the radiances thus corrected; (d) the results of the regression make it possible to calculate the evolution of the bias expressed here in brightness temperature (source: T. Hewison, EUMETSAT)*

The aim of the Global Space-based Inter-Calibration System (GSICS) is to develop and implement a set of procedures and software tools that will allow satellite operators to compare their instruments with reference instruments, to analyze and remedy any anomalies and to determine the corrections to calibration to be applied to level 1 data.

GSICS aims to establish an unbroken chain of calibrations to traceable benchmarks against the International System of Units. The methodologies used, which comply with international standards, are subject to critical peer review so that they can be applied in a variety of observation conditions. Evaluation results and corrections are published.

The reference instruments are chosen according to their performance in terms of precision and stability, their spectral range, and also the prospect of continuity of operation of these instruments in orbit. In the field of infrared radiometry, the interferometers of the IASI** and CrIS** series have been retained as references. Instruments can also be developed specifically to serve as a reference, such as the CLARREO** instrument concept proposed by NASA, a spectrometer including phase change cells that would allow for self-calibration against absolute standards. For radiometric measurements in the visible spectrum, observation of the lunar surface, whose geometry and illumination are precisely modeled, can serve as a transfer standard.

By linking the instruments to common references, the quality of the entire observation system is improved. GSICS in-orbit intercalibration activities complement the initiatives of the CEOS concerning ground-based instrumented reference sites, and those of the Global Climate Observing System (GCOS), which set up the network of atmospheric reference sounding stations above the Earth's surface, GRUAN.

7.3.3. *The climate observation strategy*

The urgency to observe, understand and predict climate change has posed a challenge to satellite operators and their coordinating bodies CGMS and CEOS. The needs of climate observation, as formulated by GCOS, exceed, in terms of variables to be measured, accuracy and absolute uncertainty, the needs of operational meteorology, even though the latter are more demanding in terms of time (frequency of observation and time to provide information in near real-time).

**See Appendix 3.

Hence, the CGMS and the CEOS have defined with the WMO a Common Strategy Towards an Architecture for Climate Monitoring from Space (Dowell et al., 2013). A fundamental affirmation of this strategy is that observing the climate system from space relies on a chain of activities that goes far beyond the deployment of instruments in orbit. Upstream, it is a matter of identifying the essential climate variables, determining their resolution, accuracy and stability requirements and planning the missions capable of providing the required data. Downstream, it is necessary to control and document the traceability of observations, the processing intended to calculate the essential climate variables, the validation procedures, the long-term preservation and the provision of data, products and related information (see Part 3 of Volume 2).

7.3.4. *Use of the radio frequency spectrum*

The use of the radio spectrum is another area where international consultation plays a crucial role. The problem is twofold: on the one hand, it is a question of allocating to each operator and to each mission the telecommunication frequencies that will allow the exploitation of the satellite (spacecraft command and control, transmission of the data to the acquisition stations, possible direct distribution to the user) without risk of interference with other satellites; on the other hand, it is a question of allocating the appropriate frequencies to the active or passive microwave telemetry possibly implemented. The issue of protecting microwave frequency bands arises in particular for passive measurements because the frequencies to be used cannot be chosen arbitrarily, they must correspond to the natural emission and absorption frequencies that are specific to the targeted molecules (typically H_2O, O_2, CO_2 or trace gases), or at the atmospheric "windows" between these bands, and are therefore imposed by nature. In addition, the low intensity of natural emissions means that observations are easily interfered with by signals from active systems of terrestrial origin, such as radars, or telecommunication systems.

For the allocation of radio frequencies within the International Telecommunication Union, the operation of meteorological satellites is in competition with fixed or mobile telecommunications services and radio broadcasting and navigation services, which increasingly occupy the radio spectrum. Faced with these economically important activities, the atmospheric emission frequencies are a natural heritage that is difficult but essential to safeguard for the continuity of the microwave space-based observations, necessary for monitoring atmosphere and climate.

7.3.5. *Access to data*

The observations are extracted from the raw data of the satellite instruments through a processing chain. Different data levels are usually considered, as summarized in Table 7.2. The coordination effort initially focused on the transmission of level 0 and level 1 data until the issue of data access was addressed for all data levels, including the products developed.

<table>
<tr><th>Level</th><th colspan="2">Description</th></tr>
<tr><td>0</td><td colspan="2">Instrument data and ancillary data reconstructed from raw satellite data after removing communication artifacts</td></tr>
<tr><td rowspan="3">1</td><td rowspan="3">Data extracted from one instrument, at native resolution, with geolocation and calibration information</td><td>1a (LEO) or 1.0 (GEO): calibration and geolocation information attached but not applied</td></tr>
<tr><td>1b (LEO) or 1.5 (GEO): calibration and geolocation applied</td></tr>
<tr><td>1c, 1d, etc.: optional processing, depending on the nature of the instrument</td></tr>
<tr><td>2</td><td colspan="2">Geophysical product derived from data of one instrument, in its native resolution. The products are developed using algorithms making limited use of external information. Quality information is added</td></tr>
<tr><td>3</td><td colspan="2">Geophysical product derived from data of one instrument, mapped on uniform spatial or temporal grids, possibly using multiple passes (LEO) or multiple scans (GEO). Missing data can be interpolated. Irreversible process due to resampling</td></tr>
<tr><td>4</td><td colspan="2">Composite products resulting from the integration of data produced by several sensors and/or several satellites, possibly including model analysis. It is difficult to recognize the specific contribution of a given satellite instrument</td></tr>
<tr><td>Notes</td><td colspan="2">Data processing at levels 0 and 1, as well as their archiving, are the responsibility of the satellite operator. If the satellite has a direct readout function of raw data, pre-processing software is normally made available to users</td></tr>
</table>

Table 7.2. *Generic description of data levels, to be adapted to each instrument in low Earth orbit (LEO) or in geostationary Earth orbit (GEO) (source: Guide to Instruments and Methods of Observation (WMO-No. 8, Vol. IV))*

Since TIROS-8 in 1963, the first generations of meteorological satellites were equipped with direct readout systems broadcasting data in real time throughout their area of visibility, which led to the emergence of a global community of users. It was necessary to internationally coordinate transmission formats and protocols.

Following the analogue Automatic Picture Transmission (APT) systems for low Earth orbit satellites and facsimile (WEFAX) systems for geostationary satellites, the CGMS has adopted the digital standards LRPT/HRPT and LRIT/HRIT, respectively, for low- and high-resolution direct readout data from low Earth orbit and geostationary satellites, based on the standards of the Consultative Committee for Space Data Systems (CCSDS).

For satellites in polar orbit seeking a compromise between the real-time direct readout of local data, on the one hand, and the delayed reception of global data recorded on board, on the other hand, the WMO has initiated the collaborative network DBNet of local receiving stations, which provides near real-time data acquired by certain instruments over a large part of the globe.

Between local data and global data, the DBNet network

Access to data from polar-orbiting satellites is a dilemma, as it cannot reconcile the advantage of global coverage with that of real-time access. Direct readout allows real-time reception but is limited to "local data", that is, an acquisition zone within approximately 2,500 km around the receiving station. Conversely, on board recording makes it possible to acquire global data, but with a delay due to the storage of the data in orbit until the satellite passes close to a primary acquisition station, which recovers the data.

The primary stations are located at high latitude to be visible from as many orbits as possible. By having two primary stations, one near the North Pole and the other near the South Pole, the latency time of the global data is halved because the data is only stored on board for half an orbit. To further reduce this delay, it would require a network of primary stations at intermediate latitudes distributed over various longitudes, as envisaged with the "SafetyNet" network of the American NPOESS system, a costly concept involving the complex management of data flows both on board and on the ground.

A trade-off between latency and global coverage is to interconnect enough local receiving stations so that the aggregation of their acquisition areas covers most of the globe. This is the principle of the DBNet collaborative data collection network, which was implemented under the aegis of WMO according to a concept initially developed by EUMETSAT. The initial objective of DBNet was to collect near real-time data from atmospheric sounders over 80% of the globe.

Box 7.2. *The DBNet concept*

In the 1990s, progress in communication technologies resulted in that the use of dedicated transponders on board geostationary meteorological satellites proved to be significantly less efficient than the use of commercial telecommunications services, based on the standards of Digital Video Broadcasting by Satellite (DVB-S). Indeed, when the dissemination function is decoupled from the observation satellite, it opens the possibility of integrating data from different sources and can be adjusted over time according to changing needs. This is the case of the EUMETCAST system deployed by EUMETSAT, and the GeoNetCast service promoted by the Group on Earth Observations. International coordination has led to the pooling of certain services and the adoption of common standards for catalogs and portals for accessing this data.

From 2015, with the expansion of cloud technology, access to online data is gaining importance, especially for climate and environmental applications. Even some real-time applications use this approach thanks to advances in data server and cloud infrastructures. Here again, international coordination is important to converge towards standardized portals, catalogs and metadata.

7.3.6. *Bilateral cooperation*

Before the CEOS formalized its so-called virtual constellations, several international satellite constellations had been deployed under bilateral or multilateral cooperation agreements between space agencies. An example is the joint polar system of EUMETSAT and NOAA, composed of at least two satellites in Sun-synchronous polar orbit, one operated by EUMETSAT in a morning orbit (MetOp, descending node at 09:30) and the other by NOAA in an afternoon orbit (JPSS, ascending node around 13:30), with some identical payload elements on both satellites.

Another particularly interesting example is the Afternoon constellation, known as the A-train, deployed in 2004 by NASA in partnership with CNES and JAXA, which includes several satellites flying in formation a few minutes apart on a Sun-synchronous afternoon orbit (ascending node at 13:30). The A-train initially included Aqua, Aura, PARASOL, CloudSat and CALIPSO. After PARASOL was retired, CloudSat and CALIPSO were put into a different orbit to form the C-train, with the A-train giving way to OCO-2 and GCOM-W1. The quasi-simultaneity of the measurements made at the same stage by different instruments, active or passive, in different spectral domains enables synergy in the processing and interpretation of observations concerning clouds, aerosols and chemical constituents of the atmosphere.

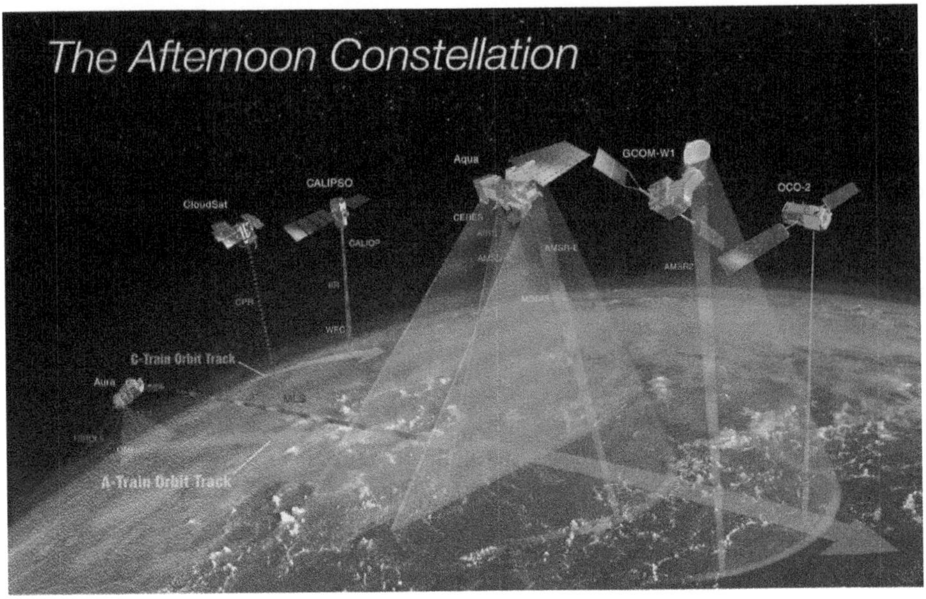

Figure 7.6. *Schematic representation of the A-train and the C-train, and the field of view of the various instruments (source: NASA*[12]*)*

7.4. An extended community of space operators

7.4.1. *A growing number of national operational agencies*

For several decades, operational space programs were only within reach of the greatest world powers, the only ones able to mobilize sufficient resources from a financial, scientific, technical and industrial point of view. Thus, when the World Weather Watch (WWW) was set up, only the US and the USSR were able to operate meteorological satellites. With the advent of geostationary satellites, when the CGMS was created, these pioneers were joined by Japan and Europe (represented by the European Space Agency and then EUMETSAT), followed shortly after by India. In the 1990s, the People's Republic of China became a member of the CGMS and finally, the Republic of South Korea in the early 2000s.

Several so-called research and development national space agencies joined the CGMS in the 2000s, including from the United States (NASA), Russia (ROSCOSMOS), France (CNES), Japan (JAXA), China (CNSA) and India (ISRO). Other national space agencies, members of the CEOS, also contribute to the

12. https://atrain.gsfc.nasa.gov.

observation of the Earth and its atmosphere, such as the agencies in Germany (DLR), Italy (ASI) and Brazil (INPE).

7.4.2. *The emergence of the private sector*

In the US, the SPACE Act (Spurring Private Aerospace Competitiveness and Entrepreneurship Act) adopted in 2015 formalizes the promotion of private operators in the space sector. Its application to Earth observation was specified in the Commercial Remote Sensing Act of 2015 by which the NOAA is required to grant observation missions to the private sector. A similar development has been driven since 2018 within WMO with the launch of a Public–Private Partnership program. The aim is to combine the respective scientific and technical capacities of the private and public sectors of the members so that the countries can better face the threats posed by extreme phenomena linked to weather, climate, water and the environment. Thus, the private space sector is moving from the traditional role of *system supplier* to that of *partner* of the governmental space agency, or even becoming a *data supplier* to the latter.

The agility of the private sector in addition to national agencies

SPIRE Global is a private company founded in 2012 which deploys a constellation that can carry miniaturized sensors capable of providing data for meteorology and other applications. It will eventually include more than a thousand nanosatellites. The agility of the system is due to the very short lead times between design and operation, the possibility of reconfiguring the acquisition software in orbit and the very structure of the company, which is not comparable to a national or international agency. This approach is particularly suitable for atmospheric sounding observations by GPS radio occultation that implement very light receivers whose performance is essentially linked to clock accuracy and the number of units deployed.

The constellation set up by SPIRE will eventually include 1,775 Lemur-2 type nanosatellites including, for meteorology, a radio occultation GPS receiver. With around 100 instruments in operation, SPIRE is already making a significant contribution to the WMO Global Observing System.

Box 7.3. *The example of a private stakeholder: SPIRE Global*

Due to their diversity, private operators are not involved in the traditional consultation and coordination mechanisms. This may result in a lack of transparency in the observation and processing chain, for intellectual property reasons, and limited access rights or data redistribution, for commercial reasons. Furthermore, concepts of missions deploying many satellites pose unique problems with regard to space debris and interference. The rise of commercial operators is, however, an opportunity. By adhering to quality assurance and traceability standards, these operators can guarantee the interoperability of the data they produce and make a valuable contribution to the global observation effort.

7.5. Conclusion

International coordination, which has been one of the keys to the success of space observation for meteorology and climate, is essential to guarantee data continuity, accessibility and interoperability. The high-level planning established within the framework of the WMO or other bodies is based on "best effort" commitments, that is, voluntary commitments based on mutual trust, knowing that each agency follows its own decision-making process, depending on national constraints and priorities. Despite this limitation, by developing a "Common Vision" and cooperating within WMO, CGMS and CEOS, countries have been able to converge their programs and contribute to a global system that over the years has proven to be remarkably robust and consistent, with constantly improved performance, even though its future implementation still presents major technical and financial challenges.

The growing number of space operators and the new role of private operators are leading to a paradigm shift in international coordination. This coordination first materialized, historically, through a belt of geostationary satellites meeting precise specifications. It then evolved towards a more open approach comprising a core set of essential missions supplemented by a variable number of other missions according to the capacities and specific priorities of the various public operators, within the framework of international agreements. With the rise of private operators, inter-governmental coordination is to be supplemented by consultation extended to all stakeholders. Furthermore, coordination between governmental or intergovernmental operators is essential to regulate the use of outer space and to guarantee a response to the priority needs for the security of goods and people and for missions of public benefit, such as the study and monitoring of climate change.

7.6. References

Blondin, C. and Jarraud, M. (2018). Le rôle toujours croissant de l'Organisation météorologique mondiale. *La Météorologie*, 100, 132–138.

Boukabara, S.A., Eyre, J., Anthes, R.A., Holmlund, K., St. Germain, K., Hoffman, R.N. (2021). The earth-observing satellite constellation, a complex, interconnected global system with extensive applications: A review from a meteorological perspective. *IEEE Geoscience and Remote Sensing Magazine*, 9(3), 26–42.

CEOS (2021). 2021–2023 CEOS Work Plan [Online]. Available at: https://ceos.org/document_management /Publications/CEOS_Work-Plans/.

CGMS (2018). CGMS Baseline – Sustained contributions to the global observing system [Online]. Available at: https://www.cgms-info.org/.

Dowell, M., Lecomte, P., Husband, R., Schulz, J., Mohr, T., Tahara, Y., Eckman, R., Lindstrom, E., Wooldridge, C., Hilding, S. et al. (2013). Strategy towards an architecture for climate monitoring from space [Online]. Available at: www.ceos.org and www.wmo.int/sat and http://www.cgms-info.org.

Goldberg, M., Ohring, G., Butler, J., Cao, C., Datla, R.U., Doelling, D., Gärtner, V., Hewison, T., Iacovazzi, B., Kim, D. et al. (2011). The global space-based intercalibration system. *Bulletin of the American Meteorological Society,* 92.

Mohr, T. (2010). The Global Satellite Observing System: A success story. *WMO Bulletin*, 59(1), 7–11.

Rainer, J.-M. (2003). Le système mondial d'observation. *La Météorologie*, 40, 28–32.

WMO (2018). Guide to instruments and methods of observation. No. 8, WMO, Geneva.

WMO (2019). Vision for the WMO integrated global observing system in 2040. Space-based component, no. 1243, WMO, Geneva.

PART 2

The Physical Basis

8

Satellite Orbits for Atmospheric Observation

Michel CAPDEROU
Laboratoire de Météorologie Dynamique (LMD/IPSL), Ecole polytechnique, Palaiseau, France

8.1. Introduction

Knowing the distribution of clouds and the state of the atmosphere for an entire region, or better, for the whole planet, is the dream of any meteorologist. With satellites, the dream has become reality, and even today a reality so commonplace that users forget the technological prowess implemented for such a result.

Nimbus-1, the first satellite for meteorological purposes, was launched in 1964 (Sputnik was launched in 1957) for NOAA, the United States meteorological agency. From then on, the images obtained by the satellites appeared essential for the improvement of meteorological forecasts. With an altitude of around 1,000 km and a fairly wide field of view, these first satellites offered scientists one image per day of the entire Earth, a daily view of the cloud situation at a given time. This daily snapshot, as innovative as it was, showed its limits in the frequency of observation. The idea then came to use a "fixed" satellite (i.e. it remains stationary above a point on the equator) that would permanently observe a given region 24 hours a day. For this, it is necessary to use an orbit that is completely different from that of Nimbus and its successors: the geostationary Earth orbit.

For a color version of all the figures in this chapter, see www.iste.co.uk/phulpin/satellites1.zip.

Satellites for Atmospheric Sciences 1,
coordinated by Thierry PHULPIN, Didier RENAUT, Hervé ROQUET and Claude CAMY-PEYRET.

The vast majority of scientific data on the atmosphere today comes from observation satellites in these two types of orbits:

– a low orbit (from 700 km to 860 km for current satellites), giving very precise images of the entire Earth with two daily visits, one by day, one by night;

– a high orbit (36,000 km), called the geostationary Earth orbit, giving precise images of a large portion of the Earth, constantly (an image at least every quarter of an hour).

At first glance, the range of altitudes for the orbits of artificial Earth satellites is very wide: from 150 km for Soviet or American spy satellites during the Cold War era to 100,000 km for Vela satellites for the observation of nuclear tests (again, the Cold War!). Looking more closely, we can classify circular orbits into three groups according to their altitude and therefore their period (see Box 8.2). These three categories of satellites are typically denoted LEO*, MEO* and GEO* (respectively, Low, Medium, Geostationary Earth Orbit).

Elliptical orbits are generally classified in the HEO (Highly Eccentric Orbit) category, which is less extensive than LEO. For meteorological and Earth observation satellites, the main categories are LEO and GEO. We will briefly note their occurrence in the HEO and MEO categories.

8.2. Preliminaries

The motion of an artificial satellite in orbit around the Earth is governed initially by Kepler's three laws, which we briefly recall below.

1) The trajectory of the planets is represented by an ellipse with the Sun at one of the foci.

2) The areas covered by the Sun-planet radius vector for equal times are equal.

3) The square of the periods T is proportional to the cube of the semi-major axes a, where T^2/a^3 = constant.

This "ideal" movement, known as Keplerian, corresponds to the case where the Earth, the attracting body, has spherical symmetry and where the Earth and satellite are isolated in space, "alone in the world". In fact, the "real" movement is perturbed: the Earth is not strictly spherically symmetrical and other bodies (Sun, Moon, etc.) act on the satellite by virtue of gravity. In addition, other forces act on the satellite, such as the non-conservative forces like atmospheric drag and solar radiation pressure.

Box 8.1. *Laws of orbits: Keplerian orbit and perturbed orbit*

*See Glossary.

The laws established by Kepler in the 17th century, which govern orbital mechanics, are briefly recalled here (see Box 8.1). Two variables are essential to define the orbits that will be chosen for Earth observation: the periodicity and the crossing time, which requires clearly specifying how time is defined.

The *orbital period*, or in short the *period*, is the time taken by a body to complete one revolution in its orbit. We define several periods.

– Keplerian period T: the period calculated by Kepler's third law; it only depends on the value of the semi-major axis of the orbit (and the mass of the attracting body). For this period, we can take the origin of time at any point of the orbit.

In the real case, that is, perturbed, we define two other periods.

– Anomalistic period T_a: time elapsed between two consecutive passages at its perigee.

– Draconitic period T_d: the time that elapses between two passages of the object through its ascending node.

The *ascending node* is the point in the orbit where the satellite intersects the equatorial plane going from south to north; the *descending node* when it intersects the equatorial plane going from north to south.

These three periods, T, T_a and T_d, are slightly different from one another.

The *mean motion*, denoted n, is the angular speed required for a body to complete one orbit, $n = 2\pi/T$.

Box 8.2. *Orbital period*

Some useful definitions of "time" are given to define the orbits.

– TAI (International Atomic Time): obtained by a network of atomic clocks and decorrelated from the rotational movement of the Earth.

– UTC (Coordinated Universal Time): tied to TAI. Thanks to the appropriate skipping of whole seconds, it makes it possible to link the rotation of the Earth with the time of laboratory clocks.

– LMT (Local Mean Solar Time): obtained by the relation: LMT = UTC + $\lambda/15$, where the times are expressed in hours and where λ is the longitude expressed in degrees, negative in the west and positive in the east. LMT is local solar time. At the center of each time zone, LMT and LT (legal time) are combined, without taking into account the administrative shifts (+1 hour in winter or +2 hours in summer, for France, for example).

– LAT (Local Apparent Solar Time): related to the hour angle H, the angle that the direction of the Sun makes with the local meridian plane:

$$LAT = 12 + H/15$$

LAT in hours, true solar noon being 12:00. Angle H, noted in degrees, is negative in the morning and positive in the afternoon. This is the time given by sundials. The difference between LMT and LAT is called the equation of time, ET, a quantity whose average over the year is zero and whose maximum amplitude is 16 minutes:

$$LMT - LAT = ET$$

The ET equation of time is the sum of two astronomical effects: the heliocentric orbit of the Earth is not circular (but elliptical, $e = 0.0167$) and the equatorial plane is not to be confused with the ecliptic plane (but makes an angle of 23.5°, called obliquity).

– Mean day, D, corresponds to the time interval between two true solar noons, averaged over a year. It measures the diurnal rotational motion of the Earth by implicitly taking into account its annual revolution motion around the Sun.

D = 24 h 00 min 00 s

– Sidereal day, SD, corresponds to the period of rotation of the Earth in a fixed reference frame with respect to the stars (in fact, with respect to the vernal point, and thus includes the precession of the equinoxes).

SD = 23 h 56 min 04.1 s

Box 8.3. *Definitions of time*

8.3. Satellites in low Earth orbit

8.3.1. *Orbital characteristics*

The ideal, perturbation-free Keplerian orbit of the satellite is an ellipse contained in a fixed plane with respect to the stars. One of the foci of the ellipse is located at the center of mass of the Earth. The shape of the ellipse and its position in space (in a so-called Galilean reference frame, linked to the stars) are defined by five Keplerian parameters:

a, the semi-major axis of the ellipse;

e, its eccentricity;

i, its inclination on the terrestrial equatorial plane;

Ω, the right ascension of the ascending node;

ω, the argument of perigee.

The position of the satellite S is defined by the sixth Keplerian element:

v, the true anomaly, which locates the satellite on the ellipse and makes it possible to calculate the commonly used parameter, M, the average anomaly, which is actually a representation of time.

The relationships between these six parameters are classic and can be found in orbital mechanics books (Zarrouati 1987; Courtois 1998; Montenbruck and Gill 2000; Vallado 2007).

For conventional LEO satellites, the period is between $T = 93$ min for $h = 400$ km and $T = 103$ min for $h = 900$ km. By taking an average value of 100 minutes, we have a good approximation by saying that a LEO satellite makes approximately 14 revolutions in one day. In its fixed orbit with respect to the stars (Galilean frame of reference), the LEO satellite therefore circles the Earth approximately 14 times while the Earth circles around itself. This is how the satellite can "see" (as long as its sight is wide enough, say 60° on either side of the vertical, of the nadir) all the longitudes of the Earth. For the latitudes, it depends on the parameter i, the inclination, whose definition we specify here.

The inclination i is the dihedral angle made by the plane of the satellite's orbit with the Earth's equatorial plane xOy (Figure 8.1); i is between 0 and 180°. First, we define the ground track of the satellite S as the intersection of the segment OS with the surface of the Earth, O being the center of the Earth. The maximum latitude (north and south) reached by the ground track will be equal to the value of the inclination when i is acute (prograde orbit) and its supplement if i is obtuse (retrograde orbit).

Maximum latitude = i if $i < 90°$

Maximum latitude = $180° - i$ if $i > 90°$

An inclination close to 90° is necessary if we want to have a global view of the planet and Earth observation satellites will favor this so-called "polar" orbit inclination. But these missions will, for the most part, look for an additional property: to always observe the same areas at the same solar time. These are Sun-synchronous satellites ("synchronized" with the Sun).

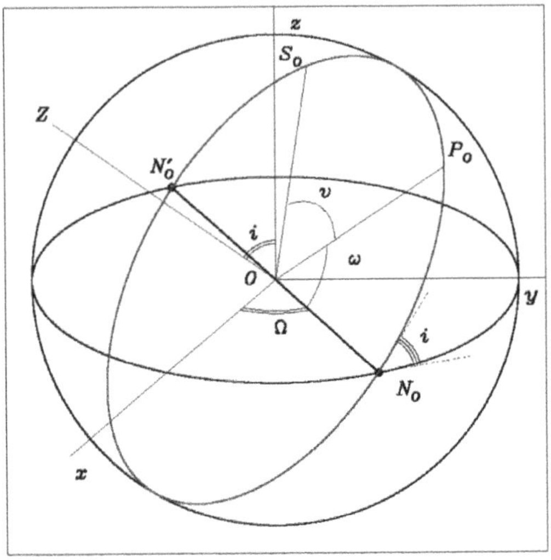

Figure 8.1. *Representation of the Earth and the ground track (in red) of the satellite. Plot of various points: S_O (for the satellite S), P_O (for the perigee), N_O and N'_O (for ascending and descending nodes). The angles i,* **Ω**, **ω** *and v are explained in the text. The terrestrial equatorial plane: xO_y; pole axis: O_z*

8.3.2. *Sun-synchronous satellites*

This property, Sun-synchronism (areas observed at the same time), is very important, and even decisive for remote sensing and Earth resource observation satellites. For a few consecutive days, for a given place, the solar incidence remains practically the same (important for the visible channels), and the thermal conditions are therefore more or less similar depending on the cloud cover (infrared). During the year, the incidence varies with the seasons (solar declination), but this variation is less than if the crossing times were at all times of the day.

8.3.2.1. *Calculating the perturbations in a real orbit*

For a Keplerian orbit, if the plane of the satellite's orbit remains fixed with respect to the stars, the crossing time to the ascending node will vary by 24 hours during the course of a year since the Earth takes a year to make a revolution around the Sun. But the orbit of a satellite is not "ideal" (it is not strictly Keplerian). Perturbation, in celestial mechanics, is an action that slightly alters the motion as it is classically calculated, but is small enough that it can be treated as a small adjustment to be made.

In the case of the satellite in orbit, we observe a main perturbation (whose action is, roughly, one thousandth of the central gravitational attraction) and secondary perturbations (which act one millionth at most). For the study of Sun-synchronism, we will only consider the main perturbation. This is due to the fact that the attracting body, the Earth, is not a sphere but can be considered, initially, as an ellipsoid of revolution. This deviation from sphericity is defined by a dimensionless parameter, conventionally denoted J_2, called *dynamic flattening*, which depends on the geometric flattening of the ellipsoid and the speed of rotation of the planet.

For Earth, $J_2 = 1.0826\ 10^{-3}$.

The application of the perturbation theory (Lagrange equations) separates the six Keplerian elements into two classes:

– metric elements: a, e, i;

– angular elements: Ω, ω, M.

The metric elements are not affected by the perturbation due to J_2. The shape of the ellipse remains the same, a and e are unchanged, and its inclination i does not vary. The angular elements (three angles) each undergo a variation. The derivative with respect to time of each of the angular elements is constant, so the variation is said to be *secular*.

– $(\mathrm{d}\Omega/\mathrm{dt})$ represents the motion of *nodal precession*. The plane of the orbit rotates around the axis of the poles and “glides” on the equatorial plane while maintaining a constant inclination.

– $(\mathrm{d}\omega/\mathrm{dt})$ represents the movement of *apsidal precession* (motion of perigee).

– $\Delta n = (\mathrm{d}M/\mathrm{dt}) - n$ gives the difference between the exact value of the period and that given by Kepler’s third law.

8.3.2.2. *Conditions of Sun-synchronism*

The calculation of the time the satellite intersects the equator, for an ideal orbit, is based on the fact that the orbit is fixed in a Galilean frame of reference. But, under the influence of the main perturbation, the orbital plane rotates around the axis of the poles, with a regular movement, at angular speed $(\mathrm{d}\Omega/\mathrm{dt})$, which is an expression of nodal precession. It is useful to take advantage of this inexorable variation to accompany the Earth in its annual revolution, whose angular velocity, noted $(\mathrm{d}\theta/\mathrm{dt})$, is equal to 360/365.242 degree/day. We understand that if we ensure that the precession velocity of the orbit $(\mathrm{d}\Omega/\mathrm{dt})$ remains equal to the angular velocity of the Earth $(\mathrm{d}\theta/\mathrm{dt})$, the angle between the plane of the orbit and the rays of the Sun will be constant. And so, the crossing time of the ascending node will always be the same. Satellites that use such an orbit are said to be Sun-synchronous.

The condition of Sun-synchronism (equality between these angular velocities) results in the relationship between the semi-major axis a and the inclination i, for quasi-circular orbits:

$$(a/R)^{7/2} = -\, k_h \cos i$$

with k_h = 10.10949 for Earth, a dimensionless constant function with three variables: J_2, (dθ/dt) and the density of the planet (Capderou 2012). In practice, we choose the altitude, therefore a, and we deduce the inclination i, with R denoting the equatorial radius of the Earth:

$$i = \text{arc cos}\,[\,-\,(a/R)^{7/2} \,/\, k_h\,]$$

We first note that the sign – implies a negative cosine, therefore an inclination greater than 90°. A Sun-synchronous satellite must be in a retrograde orbit.

The study of the possible values of (a, i) shows:

– the minimum value of i is i = 95.7° for h = 0 (and so a = R);

– the maximum value of a, for a circular orbit, is a = 12,331 km, that is, h = 5,964 km, for i = 180°. There cannot be a Sun-synchronous satellite above this altitude.

For operational Sun-synchronous satellites, therefore:

– i = 98° for h = 600 km;

– i = 99° for h = 900 km.

8.3.2.3. *Crossing time*

Transit times (local solar time) referred to here are LMT times. The difference between LAT and LMT (mean solar time) is given by the equation of time ET, which is a function of the day of the year (see Box 8.3).

For a Sun-synchronous satellite, the crossing time over a given location depends only on the latitude. When the ground track of such a satellite crosses the equator, in its ascending part (from south to north), at 21:30 for example for MetOp satellites, it will always cross it at the same time, regardless of the longitude of the point from the equator. And in its descending part, it will cross the equator with a 12-hour shift, at 09:30. But in this case, at what time will the satellite fly over the latitude of 60°, for example?

Latitude	Δτ (minutes)	Crossing time (ascending)	Crossing time (descending)
81.3° N	-360	15:30	15:30
75° N	-139	19:11	11:49
60° N	-61	20:29	10:31
45° N	-35	20:55	10:05
30° N	-20	21:10	09:50
15° N	-9	21:21	09:39
0° (equator)	0	21:30	09:30
15° S	+9	21:39	09:21
30° S	+20	21:50	09:10
45° S	+35	22:05	08:55
60° S	+61	22:31	08:29
75° S	+139	23:49	07:11
81.3° S	+360	03:30	03:30

Table 8.1. *Nominal crossing times (LMT) for the three Sun-synchronous MetOp satellites (-A, -B and -C) with regard to their latitude. The deviation $\Delta\tau$ is given for $\sigma = +1$*

It is possible to obtain (Capderou 2012), by spherical trigonometry, the relation between the latitude φ flown over and the deviation Δτ with crossing the equator:

$$\Delta\tau = \sigma/A \text{ arc sin } (\tan \varphi / \tan i_H)$$

where i_H is the Sun-synchronous inclination. A is a unit adjustment coefficient: if the times are in hours and the angles in degrees, A=15 (since 1 hour corresponds to 15°); σ characterizes the sign assigned to the part of the orbit: $\sigma = +1$ for the so-called ascending part (from south to north), $\sigma = -1$ for the so-called descending part (from north to south).

This deviation, noted in LMT, between the crossing times in a place of given latitude and the crossing of the equator (ascending node), is presented in Table 8.1, for the MetOp satellite (-A, -B or -C), h = 817 km, i_H = 98.7°. With the following relation, we obtain $\Delta\tau = -61$ minutes for $\varphi = 60°$ N in the ascending part of the orbit

($\sigma = +1$) which gives a crossing time of 20:29 LMT (since 21:30 – 01:01 = 20:29). On the descending pass, we get the time using the opposite value of $\Delta\tau$, therefore +61 minutes, which gives a crossing time of 10:31 LMT (since 09:30 + 01:01 = 10:31). The maximum latitude reached (in absolute value) is $180 - i_H = 81.3$ in the case considered here, that of the MetOp satellites. The values in Table 8.1 remain valid for any orbit close to that of the MetOp satellites.

8.3.2.4. *Choosing the crossing time*

The local crossing time at the ascending node is determined by the objectives of the mission. It is chosen as a compromise between various constraints, which we will number here from C1 to C5 (where C stands for constraint):

– C1: to obtain the best solar lighting conditions for the regions observed;

– C2: to take local meteorological factors into account (e.g. a certain region may be under cloud cover every morning);

– C3: to reduce the risks of Sun glint (reflection of sunlight from water surfaces can dazzle the satellite's instruments);

– C4: to limit periods of a solar eclipse (when the satellite, hidden by the Earth, is no longer illuminated by the Sun);

– C5: to take into account the crossing time of another Sun-synchronous satellite carrying out the same type of mission.

The optimal hours are chosen according to the type of mission.

8.3.2.4.1. Remote sensing and environmental research satellites

Constraints C1 and C3 have priority. The crossing time must be close to noon, for C1, but not too much, because of C3. Additional considerations, with C2, show that the best viewing conditions are around 11:00, as a first choice. The choice of 13:00 is also possible. For a flight over mid-latitudes (between 30° and 50° north latitude) around 11:00 LMT, it is preferable to pass the ascending node at 22:30 (see Table 8.1). The northern hemisphere is also preferred because it contains more land than the southern hemisphere. This is important because most of the Nations that finance the satellites are in the northern hemisphere.

This "22:30 ascending node" choice defines the most commonly used orbit for remote sensing satellites: we can cite the Landsat series, then Terra (United States), the SPOT and Pléiades series (France), Envisat and the Sentinel satellites of this domain (Europe), ADEOS and ALOS (Japan), Resurs-P (Russia), CBERS (China–Brazil), ResourceSat (India), etc. These satellites must be maintained in orbit, which means that the altitude and especially the inclination (the deviation of the crossing time depends mainly on the inclination) must be controlled. This is what the satellite's engines are for: every one or two months they release a little pulse of hydrazine.

8.3.2.4.2. Meteorological satellites

For these satellites, the crossing time is set by the coordination of the CGMS (Coordination Group for Meteorological Satellites) in order to have measurements compatible with the needs of meteorologists. It is the C5 constraint that comes into play: when we already have a "morning satellite" (i.e. one that flies over the average areas of the northern hemisphere around 10:00 LMT), for the next one an afternoon then evening schedule will be chosen, 13:30 then 19:00, for example.

NOAA alternates morning and early afternoon satellite launches. The first 19 (NOAA-1 to NOAA-19) are not maintained and the crossing times deviate after a few years, which is not a problem for meteorological satellites but becomes so for their use in climate monitoring. The last satellite in the series, NOAA-20 (originally named JPSS-1), is maintained at its nominal time, 13:25 at the ascending node.

To give an idea of this deviation, let us take the example of the NOAA-18 satellite. When it was launched in 2005, its ascending node crossing time was 14:00 LMT. It slowly drifted to 13:30 in 2007 and 2008. It then continued its "parabolic" drift (the representation of the hour at the ascending node as a function of time has the shape of a parabola – which is simply explained in space mechanics), which brought it to 13:50 in 2010, 14:30 in 2012, 15:50 in 2014, 17:40 in 2016, 19:30 in 2018, 21:00 in 2020, 22:00 in 2021 and so on.

The ESA MetOp and EUMETSAT satellites are strictly maintained. They are morning satellites (21:30 at the ascending node, therefore 09:30 at the descending node, for a crossing over Europe and North America around 10:00 LMT) (see Table 8.1). Figure 8.2 shows an example of the MetOp-B satellite, which is maintained in orbit. Each inclination rectification maneuver corresponds to an adjustment of the crossing time, which is thus extremely well controlled (between 21:28 and 21:32, a difference well below the ET).

For remote sensing or meteorological satellites whose main instrument is a radar, the C4 constraint (the solar panels must be illuminated for maximum time over a complete revolution to supply the radars) leads to a crossing at the equator at 06:00 or 18:00 LMT. These orbits are called dawn–dusk orbits.

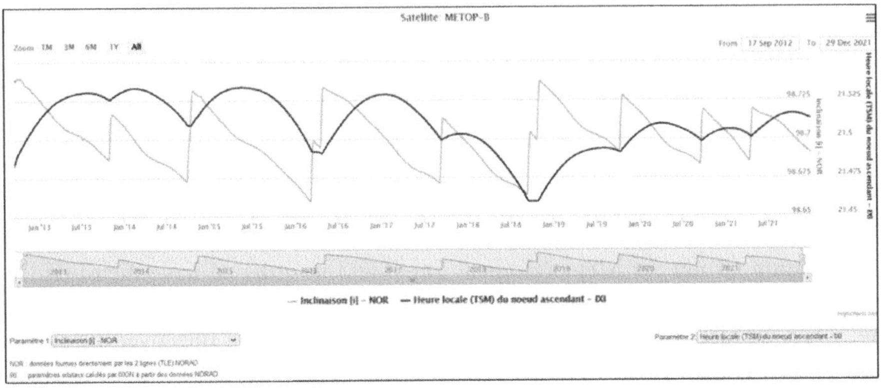

Figure 8.2. *Satellite MetOp-B. History over 9 years of two orbital parameters: inclination (in blue) and crossing time (LMT) at the ascending node (in black). A linear variation of the inclination leads to a parabolic variation in the crossing time*

8.3.3. *Non-Sun-synchronous satellites*

In the field of missions that study the atmosphere or the oceans, LEO satellites are Sun-synchronous. However, there are exceptions when the mission focuses on a given region of the globe or a latitude band. Thus, by favoring a fairly low inclination, these orbits necessarily trade Sun-synchronism for other advantages.

In the French–Indian mission (CNES-ISRO) Megha-Tropiques, the satellite instruments observe the areas at different times each day. The difference is about half an hour per day, the cycle with respect to the Sun being $C_s = 51$ days. The main advantage of this orbit is its observation frequency. A given place, near tropical latitudes, will be seen six times a day by Megha-Tropiques compared to one to two times by a Sun-synchronous satellite. We will return to this point in the section devoted to spatio-temporal sampling.

8.3.4. *Recurrent satellites*

8.3.4.1. *Definition of recurrence*

Recurrence is an important property for a satellite orbit: the ground track of the satellite exactly retraces itself after a certain number of days, noted C_T.

For a Sun-synchronous satellite, the recurrence cycle C_T with respect to the Earth is a whole number of days, since the satellite always intersects the equator at the same time. For a non-Sun-synchronous satellite, the recurrence cycle C_T is not a whole number of days. The recurrence sub-cycle is the number of days after which the satellite passes "not very far" from its initial ground track, this term is to be specified.

Establishing the recurrence for the orbit of a given satellite is not an immediate approach (Capderou 2014). It requires passing from the *anomalistic* period T_a in the *draconitic* period T_d (see Box 8.2), taking into account the terms (dω/dt) and (Δn) seen above. Depending on the semi-major axis a and inclination i, these quantities, a and I, can only be determined by an iterative method.

8.3.4.2. *Recurrent satellites*

8.3.4.2.1. Recurrence for Sun-synchronous satellites

N_T denotes the number of revolutions during the cycle of C_T days. The draconitic period, for Sun-synchronous satellites, is obtained from these values by:

$$T_d = N_T \,/\, C_T$$

All remote sensing satellites are recurrent and Sun-synchronous. Whether the satellite is observing below its ground track or at an angle, the areas are always observed from the same angle (since it is recurrent – the satellite goes back over its ground track – the satellite-pixel view geometry is identical from one cycle to another, for a given place) and always at the same solar time (since it is Sun-synchronous). Some meteorological satellites are recurrent and Sun-synchronous, such as the MetOp satellites (ESA-Eumetsat). After launching 19 non-recurrent Sun-synchronous satellites, NOAA decided that NOAA-20 would be recurrent.

In addition, recurrence allows for the establishment of a *grid*. This grid contains the ground tracks, over C_T days, of all N_T revolutions. The satellites of the same family can thus use the same grid. Let us give three examples as follows:

– A-Train and Terra satellites complete 233 revolutions in 16 days;

– the three MetOp satellites complete 412 revolutions in 29 days;

– NOAA-20, Suomi-NPP and Sentinel-5P complete 227 revolutions in 16 days.

From an operational point of view, the recurrence helps with the programming of missions.

8.3.4.2.2. Recurrence for non-Sun-synchronous satellites

For the Megha-Tropiques meteorological satellite, the recurrence is 97 revolutions in 7 days. But after 97 revolutions, the satellite moves to its ascending node after 6 d 20 h 47 m 23 s. The cycle $C_T = 7$ days is actually a cycle of $C'_T = 6.866$ days.

8.3.5. *Spatio-temporal sampling*

Apart from satellites for altimetry or those whose main instrument is a lidar, meteorological or Earth observation satellites do not just observe their ground track. Their instruments "see" either side of it, with a certain field of view*, depending on the type of scan*. Some examples of spatio-temporal sampling[1] are given for two meteorological satellites, one Sun-synchronous, MetOp-C and the other non-Sun-synchronous, Megha-Tropiques.

8.3.5.1. *Sampling for a Sun-synchronous satellite*

The European MetOp-C satellite is Sun-synchronous (crossing at the ascending node at 21:30) and recurrent (29-day cycle, five-day sub-cycle). The chosen field of view is 110.7° (half-aperture f = 55.37°), which corresponds to the orthogonal scanning of the AVHRR/3** instrument on board MetOp-C. The monthly table of overpasses is given for a point located at the equator (Figure 8.3), and another at latitude 60° N (Figure 8.4). The month chosen here is October 2021, but the general appearance of the table does not change if we consider another month (as long as the satellite is maintained in its nominal orbit). The overpasses take place around the same time, a result of Sun-synchronism.

At the equator (Figure 8.3), the overpasses are:

– around 09:30, in the descending part of the orbit (from north to south);

– around 21:30, in the ascending part of the orbit (from south to north).

With such a scan, it clearly appears that the crossing time varies by about 40 minutes around two central values 12 hours apart, 09:30 LMT and 21:30 LMT. These fixed central values are the hallmark of Sun-synchronism.

*See Glossary.

**See Appendix 3.

1. Here, the spatio-temporal sampling is established, over a defined time, for a given place P, for an instrument I characterized by its field of view, on board the satellite S. The monthly sampling table indicates, for each day of the month, at what instant the instrument I on board S saw the point P. We also note, for P, the direction PS, with the two classic angles of the spherical coordinates, the zenith angle of sight and the azimuth angle (relative to north).

As seen above (Table 8.1), these times vary with latitude. For latitude 60° N (Figure 8.4), the overpasses are:

– around 10:30, in the descending part of the orbit;

– around 20:30, in the ascending part of the orbit.

For the descending overpass, the central value is 10:30 LMT, the crossing time is able to vary between 08:40 and 12:20, that is, a variation of 110 minutes around the central value (it was only 40 minutes at the equator; this increase is explained by the narrowing of the meridians closer to the poles). Similarly for the ascending crossing, there is a variation of 110 minutes around the central value, 20:30. We can highlight the 29-day recurrence cycle for MetOp-C as well as the five-day sub-cycle.

The number of overpasses increases from the equator to higher latitudes. Equatorial and tropical regions are poorly sampled (64 overpasses in a month, i.e. two daily overpasses with the chosen field of view), the mid-latitudes not much more. From the latitude of 60°, the sampling becomes more abundant (139 per month, i.e. four or five daily overpasses). The polar regions are very frequently observed and with a slightly wider field of view they can be seen at every revolution (14 daily overpasses). These monthly tables (Figures 8.3 and 8.4) illustrate two notable points for Sun-synchronous satellites:

– a "weak point", with a restricted observing frequency for the most populated areas of the Earth;

– a "strong point" compared to geostationary satellites, with very good coverage of the polar regions thanks to a high observing frequency.

8.3.5.2. *Sampling for a non-Sun-synchronous satellite*

The Megha-Tropiques satellite is non-Sun-synchronous (crossing at the ascending node with precession) and recurrent (seven-day cycle). The chosen field of view is 85.9° (half aperture f = 42.95°), which corresponds to the orthogonal scanning of the Saphir** instrument on board. The monthly table of overpasses is given for a place located at latitude 12° N (see Figure 8.5). This table notes the times of crossing for each day of the chosen month, here December 2011.

With the inclination of the satellite being i = 20°, its ground track covers the latitudes between 20° N and 20° S and, thanks to their field of view, the instruments cover wider latitudes, between 28° N and 28° S for Saphir. At the equator, there is a

**See Appendix 3.

frequency of four to five overpasses per day; the maximum is reached around the latitude of 12° with about six overpasses per day (166 monthly overpasses in the example chosen here, Figure 8.5). The cycle with respect to the Sun is $C_S = 51.3$ days, which corresponds to a shift of 28 minutes per day, since 1,440/51.3 = 28. The recurrence cycle, $C_T = 7$ days, appears clearly (e.g. days 1 and 8 have exactly the same configuration, but are shifted by 3 h 16 min).

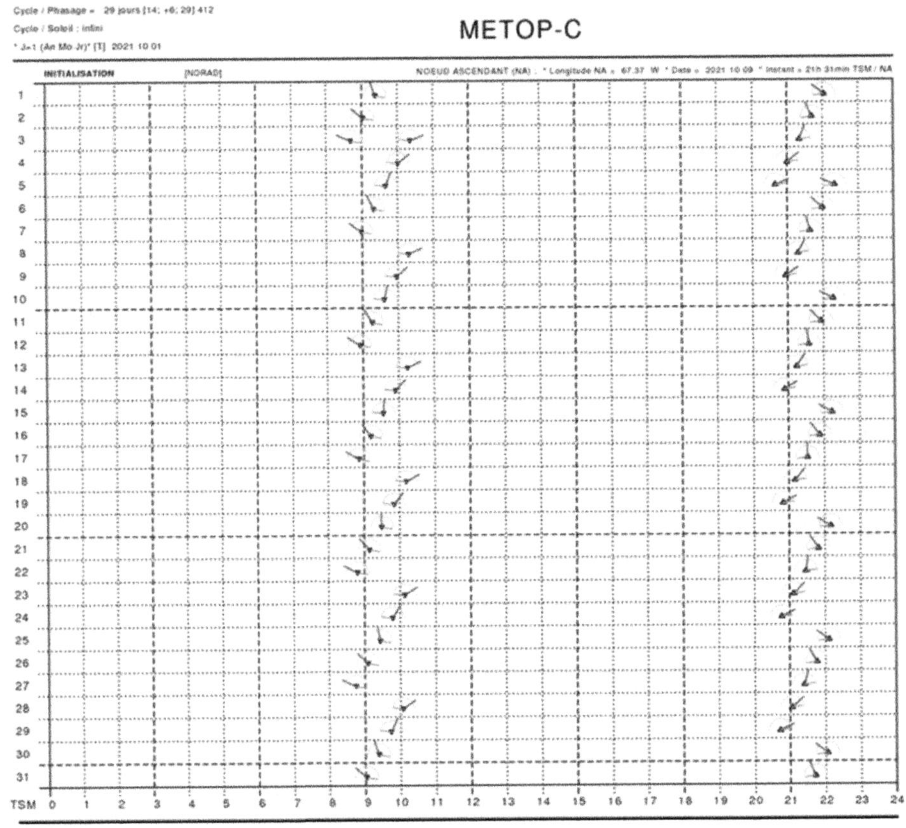

Figure 8.3. *Crossings for a location P located at the equator, on the prime meridian. The y-axis gives the days of the month, from top to bottom, from 1 to 31. Along the x-axis is the MST time, from 0 to 24 for the MetOp-C Sun-synchronous satellite. Field of view: 110.7° (AVHRR/3 instrument). Point P: latitude = 0°, longitude = 0°. Month of October 2021. 64 crossings per month for point P*

This type of orbit is of great interest as it provides a continuous sequence of six overpasses in the tropical zone for the monitoring of cyclones. An example is given

(Figures 8.6 and 8.7) with cyclone Thane, the evolution of which can be monitored over six consecutive overpasses on December 29, 2011. The cyclonic zone is located in southern India. By establishing the spatio-temporal sampling for this zone (centered at 12° N, 80° E) for the month of December 2011, the crossing times of the satellite in this zone can be obtained.

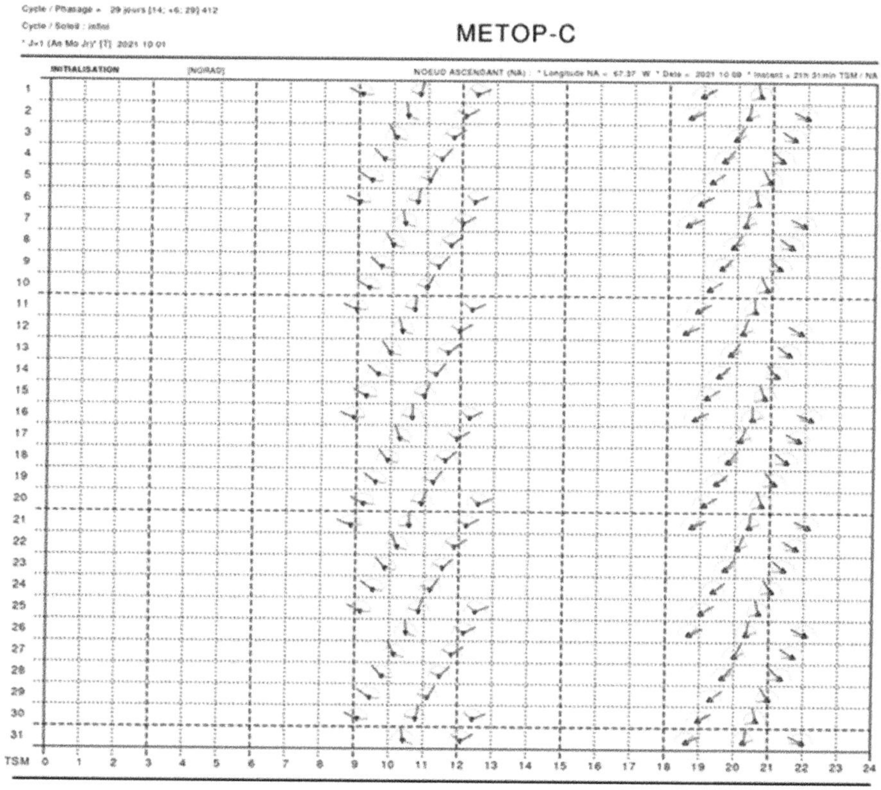

Figure 8.4. *Overpasses for a location at latitude 60° N. See legend to Figure 8.3. MetOp-C Sun-synchronous satellite. Field of view: 110.7° (AVHRR/3 instrument). Point P: latitude = 60° N, longitude = 0°. Month of October 2021. 139 overpasses per month for point P*

The monthly sampling table highlights the strong point and the weak point of this type of orbit:

– six consecutive overpasses at a given location, for crucial latitudes for the study of cyclones and tropical phenomena;

– an observation “lack” of 13 hours.

Two satellites, of the Megha-Tropiques type, on identical but shifted orbits would allow almost permanent coverage of the tropical zone.

Each small black triangle indicates a crossing (the orientation of the triangle makes it possible to distinguish the crossing during the ascending or descending orbit). The two lines, red and green, starting from each small triangle note the direction of the satellite seen from the place P considered: the red line for the zenith angle and the green line for the azimuth angle (direct trigonometric sense with respect to direction of movement)

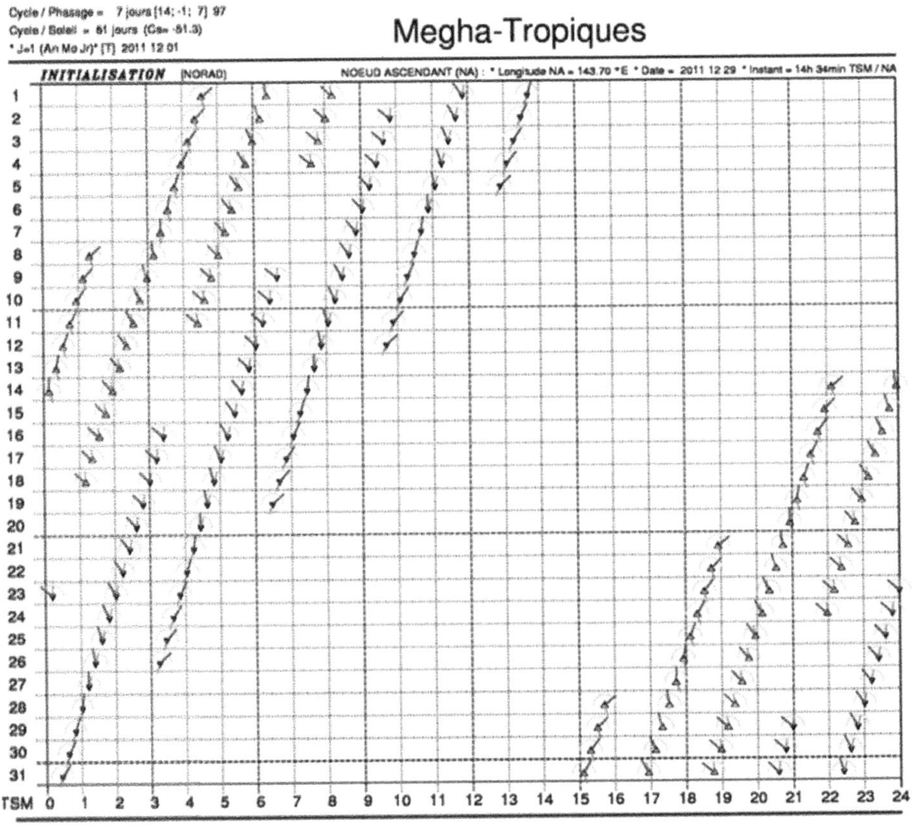

Figure 8.5. *Overpasses for a location at latitude 12° N. See legend to Figure 8.3. Megha-Tropiques non-Sun-synchronous satellite. Field of view: 85.9° (Saphir instrument). Point P: latitude = 12° N, longitude = 80° E (southern India, Chennai region). Month of December 2011. 166 overpasses per month for point P*

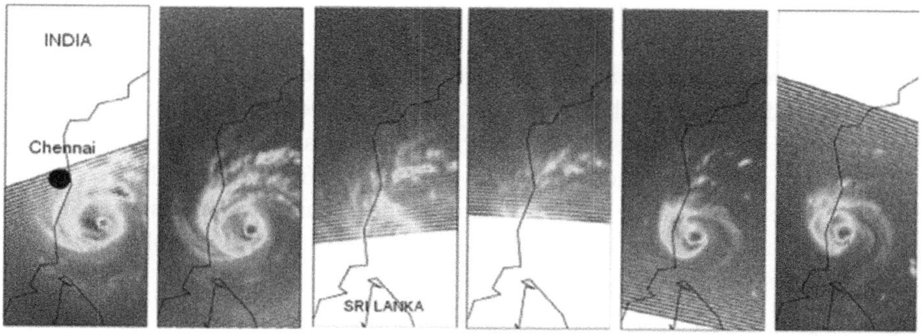

Figure 8.6. *Six consecutive views of cyclone Thane taken by the Saphir instrument aboard Megha-Tropiques, on December 29, 2011. The six overpasses, in LMT (as in Figure 8.5), correspond to 15:30, 17:20, 19:10, 21:00, 22:50 on December 29 and 00:35 on December 30 (© CNES-ISRO)*

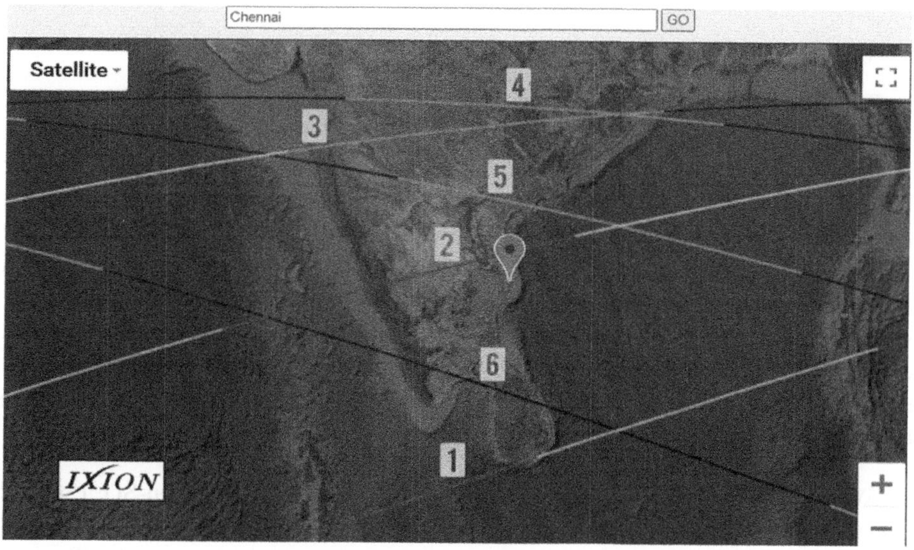

Figure 8.7. *The orbit of the Megha-Tropiques satellite during six consecutive revolutions above the Chennai region, on December 29, 2011, during cyclone Thane. The six consecutive overpasses are numbered from 1 to 6. At each of the overpasses, the city of Chennai (red marker) is seen by the Saphir instrument on board the satellite*

8.3.6. *Collaboration with LEO satellites*

NOAA sought collaboration for meteorological satellites between those "in the morning" and those "in the afternoon". An agreement made between the two organizations in 2015 reserves the morning satellites for EUMETSAT (MetOp, 09:30 LMT descending node) and those of the afternoon for NOAA (JPSS, 13:30 LMT ascending node). All of these satellites, Sun-synchronous and recurrent, must be in a maintained orbit. Another type of synergy is implemented with satellites that share the same LEO orbit.

8.3.6.1. *Co-orbital satellites*

When the MetOp-A, -B and -C satellites were operational at the same time, they were positioned equidistant in their common circular orbit, separated by 120° from each other (i.e. a difference in time of 34 minutes, one-third period). After MetOp-A stopped operating, the other two continued with a diametrically opposite 180° separation (and a gap of 51 minutes, half a period).

8.3.6.2. *Satellite train*

The first "train" of satellites comes from a very ambitious NASA project called the A-Train (for Afternoon Train), started in 2002 with the launch of the Aqua satellite. Several satellites then joined this orbit (including some from CNES or JAXA), separated from each other by a few minutes (see Chapter 2). The crossing time at the equator (ascending node) is around 13:30 LMT for all satellites. The A-Train has seen up to seven active satellites in orbit. Apart from a failed launch (Glory), this audacious experiment was successfully completed. This was a great scientific and operational success because it required close coordination between agencies (US, France, Japan) over 20 years. The synergy comes from the fact that the instruments on board the satellites are different and complementary and observe the same scenes almost simultaneously.

Without being able to speak of a "train", the future MetOp-SG (Second Generation) satellites will fly in close formation: a type A satellite (visible, infrared sounder and microwave imaging) and a type B satellite (microwave and millimeter imaging), thus analyzing the same scenes simultaneously. The MetOp-SG orbit will be strictly the same as that of MetOp, Sun-synchronous and recurrent.

8.4. Satellites in geostationary orbits

8.4.1. *Orbit characteristics*

8.4.1.1. *Orbit parameters*

Like any satellite, a geosynchronous satellite is characterized by its mean motion n, but in this case, it is such that n is equal to the Earth's rotation rate, that is, one turn (360°) in one sidereal day (23 h 56 m 04.0905 s).

$$n = 7.29211467\ 10^{-5}\ \text{rad.s}^{-1}.$$

If, however, its orbit is circular ($e = 0$) and in the equatorial plane of the Earth ($i = 0$), it is said to be *geostationary*. In this case, its trace is reduced to a point located on the equator. This point defines its *parking longitude* (or *longitude of the sub-satellite point*) and the satellite can be considered fixed with respect to the Earth (while, of course, it is in permanent motion in a Galilean frame of reference). We obviously see the benefit of such an orbit for Earth observation.

Application of Kepler's third law immediately gives the value of the semi-major axis a. This value must be slightly modified to take into account the gravity perturbations which, at this altitude, come mainly from three sources whose action is roughly equivalent, the Moon, the Sun and the dynamic flattening J_2 of the Earth. The following is obtained:

$$a = 42{,}165.7\ \text{km}$$

so, by noting R the equatorial radius of the Earth, the altitude h is:

$$h = a - R = 35{,}788\ \text{km}$$

Also note:

$$a/R = 6.611$$

8.4.1.2. *Stationkeeping*

When placed at the altitude defined above and the chosen longitude, the satellite remains at a fixed position relative to the Earth, for a certain period of time. It must be equipped with motors that allow its position to be frequently adjusted. The satellite is subject to two types of drift.

– A drift in latitude, because the Moon and the Sun tend to move the orbit from the equatorial plane to the ecliptic plane (these two planes make an angle of 23.5°, the terrestrial obliquity); the ground track is curved in a figure of 8 on either side of the equator, since the inclination of the satellite is no longer zero. For an operational satellite, an inclination of up to 2° is tolerated. Beyond that, the inclination is corrected.

– A drift in longitude, because the geoid differs slightly from the ellipsoid and does not have perfect rotational symmetry. These irregularities (well defined today by geopotential models) determine points on the equator that gently attract or repel geostationary satellites. The two points that "attract", known as stable points, have longitudes of 75.1° E and 104.9° W. The two points that "repel" are at 162.1° E and 11.4° W. Meteosat-11 (MSG-4), stationed at 0° longitude, tends to go east as it is attracted by the stable point at 75.1° E. Another example is the two operational American geostationary satellites. GOES-East (currently GOES-16) is stationed at longitude 75° W and GOES-West (currently GOES-17) at 135° W. Both are drawn to longitude point 104.9° W, located in between them.

Stationkeeping involves returning the satellite to its nominal position. These maintenance operations require maneuvers involving fuel and limit the life of the satellite. North–south control results in much more fuel consumption than east–west control.

8.4.2. *Observation conditions*

Located at an altitude of more than five times the radius of the Earth, the satellite delivers an image of the planet that covers almost half of its surface. This image, this disk (*slot*), does not cover the poles of the Earth. The cone of observation of the Earth from the geostationary satellite has a half-angle at the top equal to:

$$\arcsin (R/a) = 8.7°.$$

With the complement of this arc being 81.3°, the area limits are therefore:

– latitude: between 81.3° N and 81.3° S;

– longitude: between $\lambda_S - 81.3°$ and $\lambda_S + 81.3°$, with λ_S being the parking longitude.

In practice, when the viewing angles deviate too far from the central view, the resulting transformation of the pixel, accentuated by the roundness of the Earth, gives images that are too distorted. The images are not considered usable beyond 55° or 60° (in latitude or longitude) on either side of the central point.

With three or four satellites well distributed around the equator, we can have a good permanent view of the Earth, excluding the poles.

The United States, European countries, India, China, Korea or Japan quickly saw the importance of these satellites, while Russia (and previously, the USSR) did not. The coordination of these meteorological satellites is globally managed by the CGMS (see Chapter 7).

For example, Metéo-France uses, for its missions in metropolitan France and overseas, five main geostationary satellites distributed along the equator:

– two European satellites: Meteosat (0°) and Meteosat Indian Ocean (57° E);

– a Japanese satellite: MTSAT (145° E);

– two American satellites: GOES-W (135° W) and GOES-E (75° W).

8.5. Other types of orbits used

Satellites for meteorology or Earth observation mainly use LEO or GEO orbits. However, below the cases of satellites using HEO orbits are discussed as well as the contribution that satellites in MEO orbits can provide.

8.5.1. *Satellites in HEO orbits*

As explained above, Russia is not hugely motivated by GEO orbits. Despite this, the Russians have used them for a long time, since the beginning of the space age, which started so brilliantly for the USSR. To overcome these intangible geographical position problems, Soviet engineers invented the *Molniya* ("lightning" in Russian) orbit.

This orbit is a very eccentric ellipse (HEO orbit), $a = 26{,}553$ km, $e = 0.75$. The period is 12 hours, exactly half a sidereal day, or 717.72 minutes. Another property of the Molniya orbit is that it is recurrent with a 1-day cycle. The application of Kepler's second law shows that the satellite moves very slowly near the apogee and very quickly near the perigee. The Molniya orbit was calculated so that when the satellite is near apogee, the speed of its ground track will be close to that of the Earth at those latitudes. Thus, for several hours, the satellite can be considered geostationary.

But an important point comes into play here. The calculation of the perturbation shows that, under the effect of the term J_2, the perigee, identified by the angle ω, is subject to a secular variation, the apsidal precession, ($d\omega/dt$) seen above. The apsidal precession causes the perigee to "circulate". With an eccentric orbit, the "interesting" part, which is the apogee in the case here, is at a given time over Russia and can be found later over Australia! The equation giving ($d\omega/dt$) involves the factor ($5 \cos^2 i - 1$). By canceling out this term, we get ($d\omega/dt$) = 0 and the perigee, like the apogee, is fixed. This condition is obtained with $i = 63°$ (or its supplement $i = 117°$). This inclination, $i = 63°$, is called the *critical inclination*.

As much as Russia could not take advantage of geostationary orbits, it is favored by the gods of celestial mechanics. The critical inclination (resulting from a simple arc cosine) is of great interest to Russia, whose territory extends over ten time zones in longitude, on either side of latitude 63° N. Since 1964, Molniya satellites have been launched regularly, especially during the Cold War. Including the Molniya-1, -2 and -3 series, a total of 200 satellites can be counted. They were officially satellites for internal communication in this vast country. Given the sheer number of them, they must have had other uses.

Russia has long used LEO orbits for meteorological satellites. With the launch of Arktika-M1, it inaugurated its first Molniya orbit in 2021 for such a mission. It is calculated that, over a period of 12 hours, the Arktika-M1 satellite (altitude at apogee: 40,000 km; altitude at perigee: 300 km) spends 11 hours in the Northern Hemisphere (including 10 hours above latitudes of interest to Russia).

8.5.2. *Uses of satellites in MEO orbit*

The contribution of satellites to meteorology does not stop at observation satellites, whether in LEO, GEO or even HEO orbits. For about 20 years, with the advent of GNSS (Global Navigation Satellite System) navigation satellites, meteorology has been using these GPS-type satellites for its benefit.

8.5.2.1. *Ground measurement of GPS signals*

The method is based on the reception of GPS signals by ground stations. These fixed reference stations, whose position is well defined, simultaneously receive the signal from several GNSS satellites. The continuous comparison between the real position and the position obtained by GPS (very close) makes it possible to quantify this difference (due to the delay taken by the signal while crossing the atmosphere), which depends mainly on the temperature and humidity in the troposphere. With a fairly dense network of stations, the evolution of the amount of water vapor in the troposphere can be mapped in 3D over time. This data is assimilated in weather forecasting models (see Chapter 1 of Volume 2).

8.5.2.2. *Radio occultation: LEO–MEO combination*

With a LEO satellite equipped with a GPS receiver, the following physical phenomenon is observed: when it goes from the MEO-GPS satellite to the LEO satellite, the electromagnetic signal crosses layers of the Earth's atmosphere and is slightly deflected. This deviation, around one tenth of a degree, is accurately measured (since the position of the two satellites is very well known). Its value is a function, as seen previously, of the temperature, the pressure and the integrated water vapor content of the atmosphere crossed. These physical parameters, obtained by these radio occultations, are assimilated by the forecasting models (see Chapters 1 and 4 of Volume 2).

The advantage of this method is that it provides a lot of data. Let us quickly evaluate this. A LEO satellite like Megha-Tropiques, equipped with a GPS receiver called ROSA**, makes about 14 revolutions per day and a GPS satellite makes two. Both orbit in the same direction, the signal between the two satellites is hidden 12 times (12 = 14 – 2) per day by the Earth. The signal therefore crosses the atmosphere 24 times. With the 24 satellites in the GPS constellation (counting up to 30 with the reserve ones), this therefore gives more than 500 observations per day for the LEO satellite, and the possibility of 15,000 observations per month. This number is further increased if the MEO satellites of the Galileo, Glonass or Beidou constellation are included.

A GPS receiver is added to many LEO satellites now. The Taiwanese FormoSat-3/Cosmic mission (in collaboration with the United States), launched in 2006, is a constellation of six small LEO satellites whose sole purpose is to receive and analyze GPS signals. During the mission, 1,800 usable data were collected per day. The experiment was so conclusive that it was repeated in 2019 with the FormoSat-7/Cosmic-2 constellation.

The table and all the figures (except Figure 8.6) were obtained using the IXION software (Capderou 2012 and 2014).

8.6. References

Capderou, M. (2012). *Satellites – De Kepler au GPS*. Springer, Paris and Berlin.

Capderou, M. (2014). *Handbook of Satellite Orbits*. Springer, New York.

Courtois, M. (ed.) (1998). *Technique et technologie des véhicules spatiaux*. CNES/Cepadues, Toulouse.

**See Appendix 3.

Montenbruck, O. and Gill, E. (2000). *Satellite Orbits: Models, Methods and Applications*. Springer, Berlin.

Vallado, D.A. (2007). *Fundamentals of Astrodynamics and Applications*. Springer, New York.

Zarrouati, O. (1987). *Trajectoires spatiales*. CNES/Cepadues, Toulouse.

9

Measurement Physics

Clémence PIERANGELO[1], Fatima KARBOU[2] and Claude CAMY-PEYRET[3]

[1] *CNES, Toulouse, France*

[2] *Météo-France, CNRM, Centre d' Etudes de la Neige, Saint-Martin-d'Hères, France*

[3] *IPSL, Paris, France*

This chapter presents the main notions of the physics of measurement that describe the physical mechanisms on which the observation of the atmosphere by satellite is based. After a general description of the basics of atmospheric remote sensing (absorption and scattering, gas spectroscopy, etc.), the radiative transfer equation is introduced, analyzing its practical consequences for space observation. Finally, the principles and advantages of the various sensors on board satellites, whether passive or active, in the optical or microwave domain are described.

9.1. Physical principles of observation of the atmosphere by satellite

9.1.1. *Basic principles of remote sensing*

Space observation of the atmosphere uses remote sensing, i.e. remote measurement, in this case from a satellite; it is based on the detection of electromagnetic radiation by a sensor, after this radiation has traveled through the atmosphere and has therefore interacted with the various atmospheric components.

For a color version of all the figures in this chapter, see www.iste.co.uk/phulpin/satellites1.zip.

Satellites for Atmospheric Sciences 1,
coordinated by Thierry PHULPIN, Didier RENAUT, Hervé ROQUET and Claude CAMY-PEYRET.

The source of this radiation can be natural (the sun, the Earth, a star, etc.) or artificial (a laser or an antenna). In the first case, we speak of passive observations, in the second case, of active observations.

The electromagnetic wave emitted and/or backscattered by the Earth's surface can be characterized by its amplitude, its phase, its direction of propagation, its polarization and its wavelength. Here, the radiation has a spectral range extending from ultraviolet (UV) to microwaves. The UV, visible and infrared (IR) domains correspond to the optical domain, while microwaves correspond to the radiofrequency domain.

From a geometric point of view, by introducing the line of sight of the sensors, we distinguish two cases depending on whether the lines of sight intercept the Earth's surface or not. The geodetic nadir is the direction of sight that intercepts the Earth's surface perpendicular to the reference ellipsoid. By extension, we can speak of nadir sounding for the first case, and of limb sounding for the second case, in which the lines of sight cross the atmosphere with no intersection with the surface (Figure 9.1).

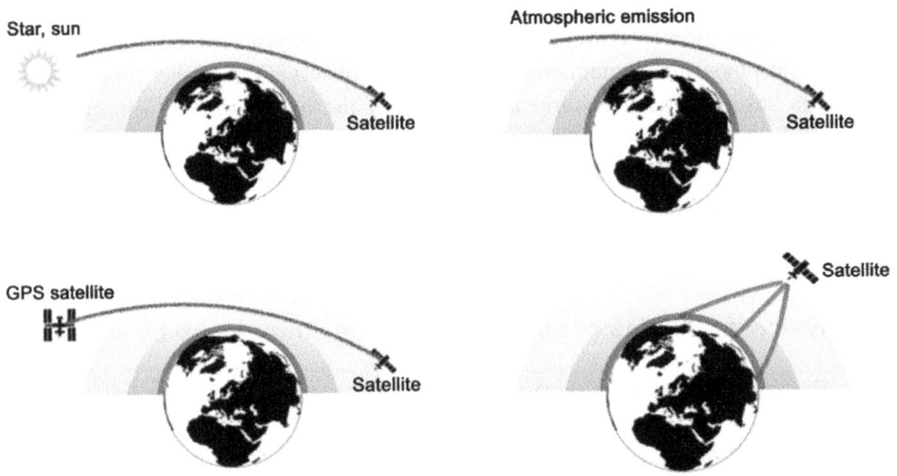

Figure 9.1. *Observation geometry for a sensor in limb sounding (top left: radio occultation, top right: transmitting, bottom left: GPS radio occultation) or nadir sounding (bottom right)*

Figure 9.2 schematically presents the interactions between radiation, surface and atmosphere. A distinction is made between passive methods, for which the source of radiation is natural (the sun, the Earth, or, at the limb, a star), from active methods, for which the source of radiation is on board the satellite and that measures the radiation, backscattered by the surface and the atmosphere (lidar or radar). Diffusion, absorption and emission will be detailed in section 9.1.2.

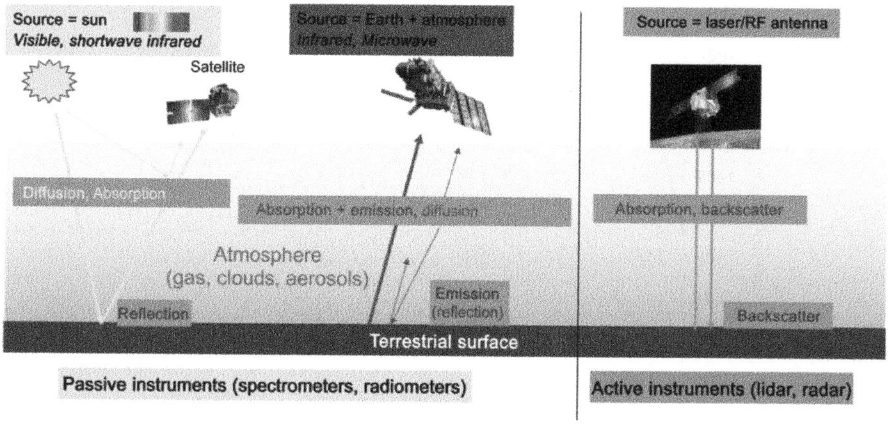

Figure 9.2. *Principle of implementation of atmosphere remote sensing from space*

Before going into detail, it is necessary to specify the spectral variable used to describe the spectral distribution of this radiation. For both historical and physical reasons, it may be necessary to use either the frequency ν (in MHz or GHz in the microwave domain), the wavenumber $\tilde{\nu}$ (in cm^{-1} in the infrared domain) or the wavelength λ (in μm or in nm) in the visible or ultraviolet range. By approximating the emission of natural environments as that of a black body at the local temperature[1] T, the emitted radiation has a spectrally variable intensity (or energy spectral radiance), which essentially depends on its temperature and is given by the Planck function (Figure 9.3) defined by the following formulas:

$$B(\nu,T) = \frac{2h\nu^3}{c^2[e^{\frac{h\nu}{kT}}-1]} \text{ or } B(\tilde{\nu},T) = \frac{2hc^2\tilde{\nu}^3}{e^{\frac{hc\tilde{\nu}}{kT}}-1} \text{ or } B(\lambda,T) = \frac{2hc^2\lambda^{-5}}{e^{\frac{hc}{\lambda kT}}-1} \quad [9.1]$$

1. We will see later that it is necessary for material surfaces to introduce the notion of emissivity, which quantifies the deviation from the black body. Similarly for dilute media (here the atmosphere), we will assume the local thermodynamic equilibrium between the molecules and the radiation and what is called the source function will be identical to the Planck function.

The quantities $B(\text{v},T)\,d\text{v}$, $B(\tilde{\text{v}},T)d\tilde{\text{v}}$ and $B(\lambda,T)d\lambda$ are equal and represent the power radiated per unit area and per unit solid angle according to the spectral variable used. The quantities B are expressed, respectively, in W m^{-2} sr^{-1} MHz^{-1}, W m^{-2} sr^{-1}/cm^{-1} and W m^{-2} sr^{-1} µm^{-1}.

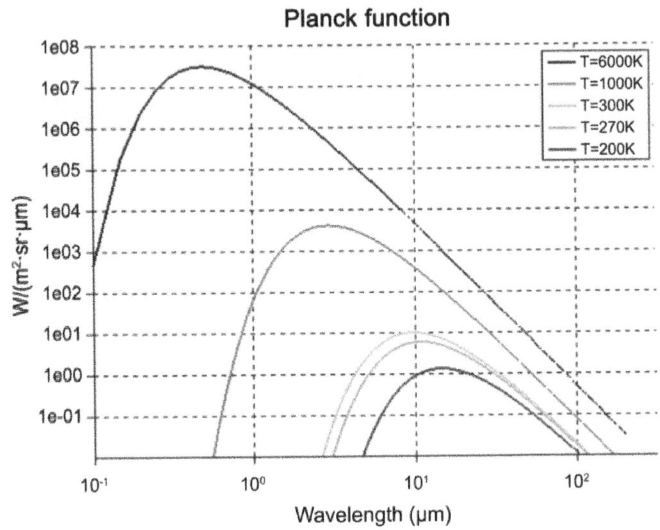

Figure 9.3. *Spectral luminosity emitted by a blackbody as a function of wavelength for a temperature of 200 to 6,000 K*

9.1.2. *Absorption, scattering, emission*

We define the extinction coefficient α_{ext} as the proportion of energy lost by the wave by absorption or by scattering, when passing through a medium, per unit length. In other words:

$$\frac{dL}{dl} = -\alpha_{ext}L \qquad [9.2]$$

where *dl* is the infinitesimal thickness of the medium crossed by the wave.

Similarly, the absorption coefficient α_{abs} is defined as the probability, per unit length, that a photon is absorbed by the medium and the scattering coefficient α_{diff} as the probability, per unit length, that the photon is scattered in any direction.

The formulation of equation [9.2] shows that the attenuation of the radiation depends on the product between the extinction coefficient and the length of the medium. This product, called "optical thickness" and denoted τ, is unitless. The optical thickness in the interval of physical length [*z,z'*] can be calculated for an inhomogeneous medium by the relation:

$$\tau = \int_{z}^{z'} \alpha_{ext}(l)dl \qquad [9.3]$$

Finally, the transmission, denoted *t*, is between 0 and 1 and quantifies the ratio between the amount of energy propagating in the incident direction at the output of the medium and at the input of the medium. By combining the two previous equations, the transmission is written:

$$t = exp\ (-\tau) \qquad [9.4]$$

9.1.3. *Spectroscopy of gaseous species*

Gaseous absorption lines are visible on the top of the blackbody radiation as shown in Figure 9.4. They are linked to the energy transitions caused by the movements of atoms in polyatomic molecules (vibration-rotation) (ranging from the visible to microwaves) or to electronic transitions (in the UV and visible domains). A line is characterized by its position, intensity and shape. Its frequency is related to the gap between the energy levels. The intensity of the line is related to the probability of transition between two energy levels.

The shape of the line depends on the properties specific to each molecular species which themselves depend on the temperature, the pressure of the medium and the concentration of the absorbing molecule. Classical modeling involves a Voigt profile, convolution of a Doppler broadening linked to the movement of the molecules, depending on the temperature, and a Lorentz profile linked to the collisions between the molecules, depending on the pressure and the temperature (see Appendix 2).

When modeling in detail the absorption or emission of molecules in the atmosphere, spectroscopic parameters are needed to calculate the corresponding spectrum of the molecule in the gaseous phase. The spectroscopic parameters and the equations giving the cross-sections (i.e. the absorption coefficient related to a molecule) and line profiles are detailed in Appendix 2.

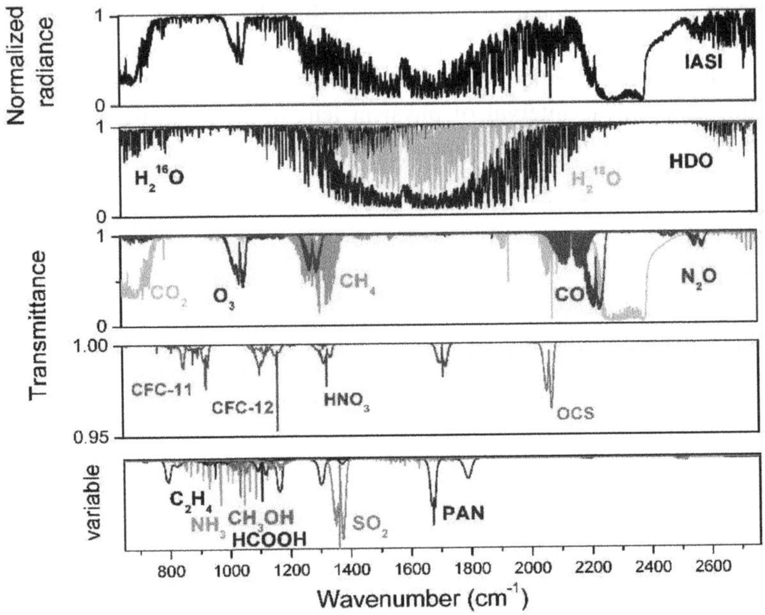

Figure 9.4. *Transmission between the surface and the top of the atmosphere in the thermal infrared spectral range and identification of the main absorbers (source: Clerbaux et al. (2009))*

9.1.4. *Optical properties of particles*

In addition to gases, the atmosphere also contains solid or liquid particles in suspension (residing there from a few hours to a few days), which are called atmospheric aerosols (Pierangelo 2005). Clouds, which are generally dealt with separately, are a special kind of aerosol, made up of water, either in the form of liquid droplets or in the form of ice crystals. Some examples of atmospheric aerosols, which can be of natural or anthropogenic origin, are desert dust, volcanic smoke, sulfate aerosols resulting from the conversion of gases emitted by human activities, etc.

From a microphysical point of view, aerosols are characterized by their mineralogical/chemical composition, their size distribution (of the order of one μm) and their shape distribution. The composition of the aerosol is reflected in its refractive index, denoted m, real part and imaginary part, which varies with wavelength.

From an optical point of view, aerosols have the effect of absorbing part of the incident radiation (soot, for example, strongly absorbs in the visible range) and also of scattering part of the incident radiation. Aerosols are thus characterized by the following optical properties:

– extinction coefficient, α_{ext}, related to the extinction cross-section σ_{ext}, by the number of particles per unit volume N: $\alpha_{ext} = N\ \sigma_{ext}$;

– single-scattering albedo, $\varpi_0 = \frac{\sigma_{diff}}{\sigma_{ext}}$, which is 0 for a solely absorbing aerosol and 1 for a solely scattering aerosol;

– phase function, which describes the angular distribution of scattered radiation as a function of the scattering cross-section σ_{diff}:

$$p(\theta, \phi, \theta', \phi') = \frac{4\pi}{\sigma_{diff}} \frac{d\sigma_{diff}}{d\Omega}.$$

In the case of spherical particles (which is a good approximation for liquid aerosols, much less so for solid aerosols), the Mie theory makes it possible to calculate these optical properties as a function of the refractive index and the ratio $2\pi R/\lambda$, with R being the particle radius. For spheroidal particles, so-called "T-Matrix" calculations also allow an analytical calculation. For more complex shapes, there are numerical alternatives, such as discrete dipole approximation (DDA) techniques, which are very expensive, however, in terms of computation time (Draine and Flatau 1994).

A few remarks about the phase function:

– in the case of a homogeneous medium for randomly oriented particles, the phase function depends only on the relative angle between the direction of the incident radiation and that of the emerging radiation;

– the phase function of aerosols generally has one or two preferred directions: a very strong forward peak and sometimes a backward peak (Figure 9.5, left);

– a more rigorous description requires introducing the polarization of the radiation; the phase function is then replaced by a phase matrix (see section 9.2.3);

– in the case where the particles are very small compared to the wavelength of the radiation, the Mie theory tends to the Rayleigh theory; this is the case for scattering by molecules (negligible at long wavelengths), whose phase function is relatively simple (Figure 9.5, right).

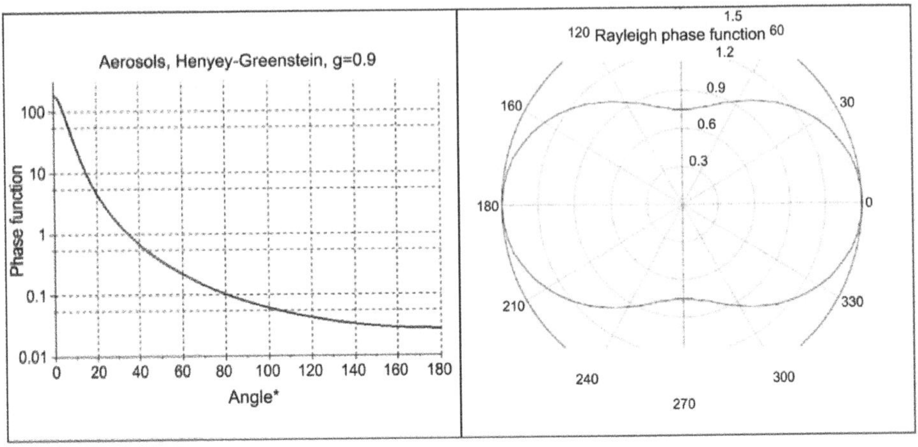

Figure 9.5. *Typical phase function of aerosols, modeled by a Henyey–Greenstein function (left) and phase function of Rayleigh scattering by molecules (right)*

Readers wishing to go further on the question of remote sensing of aerosols can refer to Lenoble et al. (2013).

In the microwave range, and outside of the water vapor and oxygen absorption windows, the clear-sky atmosphere is almost transparent to electromagnetic radiation. In the presence of clouds, the interaction of radiation with hydrometeors (droplets, water drops and ice crystals) results in the mechanisms of absorption and scattering. By considering hydrometeors as spherical particles, the Mie theory can be used to define the scattering and absorption coefficients. These coefficients vary according to the radius of the particle, its complex refractive index and the frequency or wavelength. These are complex relationships that make the interpretation and use of microwave measurements in rainy, cloudy skies very difficult. To simplify and in terms of information content, taking into account the frequencies used for the observation of the atmosphere, below 50 GHz, the main phenomenon is absorption by water drops; indeed, it is often imagers at 19 and 37 GHz that are typically used. The phenomenon of scattering by ice mainly occurs at high frequencies (beyond 85 GHz).

9.1.5. *At the surface: reflection and emission*

The Earth's surface absorbs part of the radiation coming from the sky, reflects part of it and emits thermal radiation.

In the domain where the emission from the surface is negligible (UV–visible and near-infrared domain), for a given wavelength, the surface is characterized by its albedo, that is, the ratio between the integral of the reflected radiation and the hemispherical integral of the received radiation. The albedo, denoted a_ν, is close to 1 for very shiny surfaces (e.g. snow in the visible range), and 0 for very dark surfaces. However, it is sometimes necessary to go into more detail by not only considering the hemispherical radiation but also by looking at the angular aspects of the reflectance via the BRDF (see Glossary). A surface for which the BRDF is uniform is said to be Lambertian.

In the thermal infrared (TIR) and microwave domains, emissivity, denoted ε_ν, is the ratio between the energy emitted at a given frequency and the energy that a blackbody would emit at the same temperature. It is a dimensionless number between 0 and 1. In the thermal infrared, the emissivity is close to 1 for terrestrial surfaces (typically 0.98), but can take lower values, for example, over deserts (Péquignot et al. 2008). The emissivity of surfaces for microwaves is highly variable depending on soil conditions, surface roughness and type of vegetation (Karbou et al. 2005). An example of a monthly emissivity map at 89 GHz is shown in Figure 9.6.

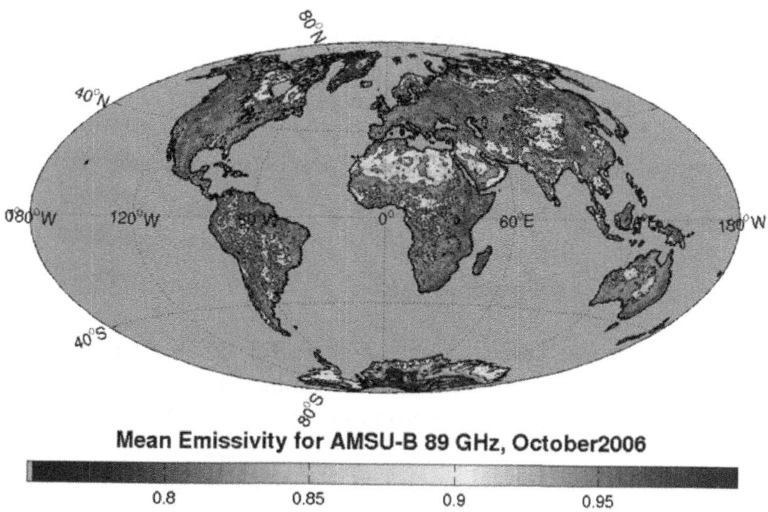

Figure 9.6. *Map of surface emissivity in the microwave domain, at 89 GHz, for the month of October 2006*

Albedo and emissivity are linked: with the Earth's surface being an opaque body, we have for each frequency $a_\nu = 1 - \varepsilon_\nu$.

All of the parameters described above are inputs necessary to calculate the incident radiation at the level of the satellite. Before describing the radiative transfer equations, some useful databases are introduced below.

9.1.6. *Spectroscopic parameter database*

See Appendix 2.

9.1.7. *Aerosol and cloud databases*

With regard to aerosols and clouds, there are several databases whose purpose is to allow for upstream calculation of optical properties from their microphysical properties (refractive indices, size distribution, shape, etc.). For example, the OPAC database (Hess et al. 1998) considers nine species of aerosols (desert dust, sulfate aerosols, soot, etc.) for which Mie calculations have been carried out for given size distributions. This database is supplied with software to make external mixtures of aerosols. There are also climatologies giving a typical aerosol profile depending on the place and the season, for example, the GADS database associated with the OPAC models (Global Aerosol Data Set) (Koepke et al. 1997).

Data provided by CAMS (Copernicus Atmosphere Monitoring Service) is also widely used[2].

9.1.8. *Atmospheric profile databases*

In order to carry out simulations to prepare for future space missions, it is necessary to calculate pseudo-observations (simulated atmospheric spectra) for various atmospheric situations. These profiles of pressure, temperature and humidity (and possibly concentration of other gases such as ozone, etc.) can be either standard profiles, for example, the five so-called "U.S. standard" profiles, or taken from wider databases (see Appendix 1).

For calculations requiring many more cases or spatially and temporally distributed data, the use of data from atmospheric transport models is quite

2. https://atmosphere.copernicus.eu/.

widespread, in particular CAMS data, for example, to have profiles of trace gases that are more representative of a given area or period.

9.1.9. *Surface databases*

The surface must be described in the emissive domain (thermal infrared and microwaves) by the surface temperature and the surface emissivity. The surface temperature is often taken to be equal to the temperature of the air near the surface, and is obtained in this case from the supporting databases described above. The spectral values of the surface emissivity can be taken from libraries, such as the ECOSTRESS database[3], or modeled by empirical formulas, for example the formula of Cox and Munk over the oceans. We also use the combination of surface emissivity maps and integrated empirical models, for example, in the distributions of the radiative transfer model RTTOV[4]. It should be noted that the surface emissivity spectra were measured on restricted areas allowing for the evaluation of surface emissivity models. For example, estimates of soil moisture and its roughness have been used to derive soil emissivity, as well as for the canopy by exploiting a simplified radiative transfer model. Physical approaches (based on the radiative transfer equation) have also made it possible to estimate large-scale surface emissivity from space remote sensing observations (Prigent et al. 1997; Karbou et al. 2005).

9.2. Radiative transfer equation

Although the radiative transfer equation (RTE) can be linked to Maxwell's equations (Mishchenko et al. 2002), we limit ourselves here to an empirical description of the radiative transfer.

9.2.1. *Differential RTE*

The simplest case is that of a medium without scattering.

In the thermal infrared and beyond, the self-emission of the medium is represented by a source function J_{emit}. Equation [9.2] becomes:

$$\frac{dL}{dl} = -\alpha_{abs} L + \alpha_{emit} J_{emit} \quad [9.5]$$

3. https://speclib.jpl.nasa.gov/.
4. https://nwpsaf.eumetsat.int/site/software/rttov/download/emissivity-data/.

By making the assumption of local thermodynamic equilibrium, the source function is given by the Planck function B, at the local temperature T, while Kirchhoff's law[5] implies $\alpha_{abs} = \alpha_{emit}$ and we therefore obtain:

$$\frac{dL}{dl} = \alpha_{abs}(B(T)) - L) \qquad [9.6]$$

In this equation, σ_{abs} and L depend on the frequency of the radiation; λ_{abs}, T and L depend on l, the position along the beam path. Depending on the needs, the dependence on l can sometimes be replaced by a simple change of variable by a dependence on transmission, optical thickness or pressure for a nadir atmospheric sight.

In the case where scattering must be taken into account, an additional source term must be introduced at each level of depth τ (variable designating the optical thickness). It takes into account all the scattering occurring in the direction (μ, φ) and is written (if $= cos\theta$):

$$J_{scat}(\tau, \mu, \varphi) = \frac{\varpi(\tau)}{4\pi} \int_0^{2\pi} \int_{-1}^{1} p(\tau, \mu, \varphi, \mu', \varphi') L(\tau, \mu', \varphi') d\mu' d\varphi' \qquad [9.7]$$

Finally, in the case where the irradiance of the sun must be taken into account, an additional source term is introduced which corresponds to the solar radiation, reflected once by the scatterers located at the depth τ, in the satellite direction:

$$J_{sun}(\tau, \mu, \varphi) = \frac{\varpi(\tau)}{4\pi} p(\tau, \mu, \varphi, \mu_0, \varphi_0) E_0 exp(-\tau/\mu_0) \qquad [9.8]$$

In the visible or near-infrared range, where the emission can be neglected, we therefore obtain the following form for the differential RTE:

$$\mu \frac{dL}{d\tau} = L - J_{scat}(\tau, \mu, \varphi) - J_{sun}(\tau, \mu, \varphi) \qquad [9.9]$$

9.2.2. *Integration of the RTE*

The vertical integration of the radiative transfer equation provides the radiation reaching the sensor, which is therefore the quantity measured.

5. This relation makes it possible to discriminate the different atmospheric layers by infrared or microwave sounders. We will come back to this in section 9.5.

The "plane-parallel" atmosphere hypothesis is commonly used: the thickness of the atmosphere is considered to be very small compared to the radius of curvature of the Earth. The atmosphere is therefore divided into horizontal virtual "layers".

9.2.2.1. *Beer–Lambert equation*

From equation [9.2], called the Beer–Lambert equation, it is easy to establish a radiative transfer equation in the integral form, useful in obtaining the luminance at the output of the medium:

$$L = L_0\, exp\left(- \int_0^l N(l')\sigma_{abs}(l')dl'\right) \quad [9.10]$$

However, this formulation is of little practical use: in the visible or near-infrared range, scattering by molecules or aerosols is generally not negligible; while in the longer wavelengths, the thermal emission of the medium must be taken into account.

9.2.2.2. *In the emissive range, without scattering*

It is possible to carry out an integration of the RTE without scattering in the thermal infrared domain: by taking into account the boundary conditions at the surface and at the top of the atmosphere, and by carrying out a change of variable of transmission to atmospheric pressure[6], obtaining the following:

$$I(sat) = t_{surf}\epsilon_{surf}B(T_{surf}) + \int_{-\infty}^{ln\,(P_{surf})} B(T(P))\frac{dt}{dlnP}dln(P)$$

$$+t'_{surf}(1 - \epsilon_{surf}) \int_{ln(P_{surf})}^{-\infty} B(T(P))\frac{dt\prime}{dlnP}dln(P) \quad [9.11]$$

When we consider the RTE after vertical integration from the surface to the sensor, we see that the quantity measured is the sum of three terms:

– a term emanating from the surface attenuated by the total transmittance along the path;

– a term corresponding to the direct contribution of the atmosphere. It may result from the addition of emissions specific to the atmospheric layers. This term is weighted by a function called "weighting function" ($\frac{dt}{dlnP}$);

6. The radiative transfer equation can also be expressed as a function of transmission or optical thickness. Here, we choose to express it in terms of a purely thermodynamic quantity, the pressure. This corresponds to a more intuitive and universal vertical scale. This also allows for easy calculation of gaseous absorption cross-sections.

– finally, a term due to the surface reflection of the radiation emitted by the atmosphere downwards. Since the emissivity of the surface is close to 1, this term is often disregarded. In the microwave domain, this term also takes into account the contribution of the cosmic background.

9.2.2.3. *In the solar domain without scattering*

The equation in the solar domain can be solved analytically only in the case without scattering and is then simply given by:

$$I = a_{surf} E_{sun} \frac{cos(SZA)}{\pi} \tau_{atm}(sun \rightarrow surf)\, \tau_{atm}(surf \rightarrow sat) \qquad [9.12]$$

where *SZA* represents the solar zenith angle, a_{surf} is the surface albedo and E_{sun} is the solar irradiance at the top of the atmosphere.

9.2.2.4. *In the presence of scattering*

In the presence of scattering, with the equation being an integro-differential equation, it is necessary to use numerical techniques to solve it.

Among the widely used algorithms, the discrete ordinate technique (Chandrasekhar 1960), DISORT code written in FORTRAN (Stamnes et al. 1988), involves splitting the phase function into Legendre polynomials, and then the total intensity and the phase function into Fourier series according to the azimuths. The equation obtained for each azimuth is rewritten by approximating the integral over μ’ by a Gaussian quadrature. The discretization of the calculation for two hemispheres (ascending/descending) is realized for a given number of polar angles *N* called “streams”. Finally, the system is solved through the calculation of the eigenvalues of a matrix of dimension (2N·L)·(2N·L), *L* denoting the number of vertical layers in the atmosphere.

Another algorithm, that of successive orders of scattering (Lenoble et al. 2007), is based on a first Fourier decomposition then a decomposition on the orders of scattering (simple, order 2, 3, etc.) and is solved by iteration. The advantage of this approach is that it can be quite fast; however, it only works for low scattering optical thicknesses.

9.2.3. *Polarized RTE*

The RTE was presented above in its scalar form. In reality, electromagnetic radiation should be described in vector form, that is, in polarized mode. The impact of scatterers on the polarization of radiation explains the benefit of polarized measurements in characterizing aerosols from space.

In the polarized mode, the description of the radiation in the atmosphere is done by the Stokes vector [I,Q,U,V]: I is the total intensity of the radiation (the radiance in scalar mode), Q and U are the linearly polarized components of the radiation and V is the circularly polarized component. In the atmosphere, component V is negligible, it will be considered zero. Coming back to a plane called the polarization plane, the U component is zero. In usual geometric views from space, this plane is generally equal to or very close to the simple diffusion plane.

Mie codes are used to calculate aerosol phase matrices for polarized components of radiance (I, Q, U, V) as a function of the angle between the incident ray and the scattered ray. For further information on radiative transfer, and in particular polarization, readers can refer to the work of Lenoble (1993).

9.2.4. *Recent advances for radiative transfer*

Recent work in the field of radiative transfer relates, on the one hand, to techniques allowing for faster calculations: citing, for example, calculations on graphics cards (GPU) or the low stream interpolation (LSI) method (O'Dell 2010), which is based on the fact that the calculations are generally carried out for many frequencies and present redundancies. This technique involves first making a rough calculation of the scattering for all the spectral samples, then a very accurate calculation for a few well-chosen samples and then deducing a corrective term for all the samples. Many research studies also focus on so-called "3D" calculations in which the atmosphere and/or the surface are described by three-dimensional fields. This is particularly relevant if we are interested in small-scale cloud fields (Cornet et al. 2010). An overview of the latest scientific work on the subject can be found in the conclusions of the Trattoria 2020 workshop[7].

9.2.5. *RTE analysis and implications for space-based remote sensing of the atmosphere*

The integral equation of the radiative transfer in the emissive domain shows that the radiation reaching the satellite carries information about the temperature of each atmospheric layer, via the term $B(T(P))$, as well as the amount of absorbing gas, via the transmittance t, which is a function of the optical thickness, itself a function of the gas concentration. If the gas concentration is known, then it is possible to retrieve the temperature. In practice, the absorption lines of CO_2, a relatively

7. https://www.meteo.fr/cic/meetings/2020/trattoria.

uniform and constant gas at short time scales, allows temperature sounding in the thermal infrared, and oxygen lines in the microwave range.

An important concept to understand the ability of sounders to differentiate vertical levels in the atmosphere is that of the weighting function (the factor $\frac{dt}{dlnP}$), which we will first explain physically. The weighting functions are used to characterize the transmission properties changing with altitude, thus making it possible to better identify the atmospheric layers having contributed the most to the signal measured by the satellite. Due to the absorption caused by the various atmospheric gases, the transmission of rising radiation between a given level of the atmosphere and the satellite decreases when the altitude of this level decreases. At the top of the atmosphere, the transmission reaches its maximum, with a value of 1; indeed, the radiation emitted by the top of the atmosphere no longer encounters absorbing gas.

A physical phenomenon complementary to absorption is the thermal emission of each layer, which increases as the absorption increases (Kirchhoff's law). For a given temperature, a strongly absorbing layer emits more than a weakly absorbing layer. On the other hand, with the same absorption power, a colder layer emits less radiation than a warmer layer.

By analyzing the radiation reaching the satellite at several spectral frequencies, corresponding to varying levels of transmittance, it is possible to distinguish different vertical levels. Indeed, at frequencies where the absorption is very low, the radiation interacts little with the atmosphere, and we essentially observe the surface. At frequencies with high absorption, the transmission between the surface and the satellite can be almost zero: the surface is therefore not seen. However, higher up, the transmission takes intermediate values between 0 and 1. There is therefore a certain level in the atmosphere (not too high for there to remain a sufficient molecular density for emission, but not too low either for the emitted radiation to be transmitted) for which the emitted radiation will be collected by the satellite. By scanning the different absorption levels in the spectrum, the different vertical levels in the atmosphere are scanned (Figure 9.7).

The weighting function gives the contributions of each level of the atmosphere to the observed radiance (Figure 9.8). The pressure level at which the weighting function reaches a maximum is the one whose emitted radiation will contribute the most to the radiation measured at the satellite. Pressure levels where the weighting function is zero will not contribute to the observed radiance; either because they are levels above the maximum of the weighting function and for which no gas absorbs or emits the radiation at the wavelength considered, or because these are levels located below the maximum of the weighting function and for which all the radiation emitted upwards will be absorbed before reaching the satellite.

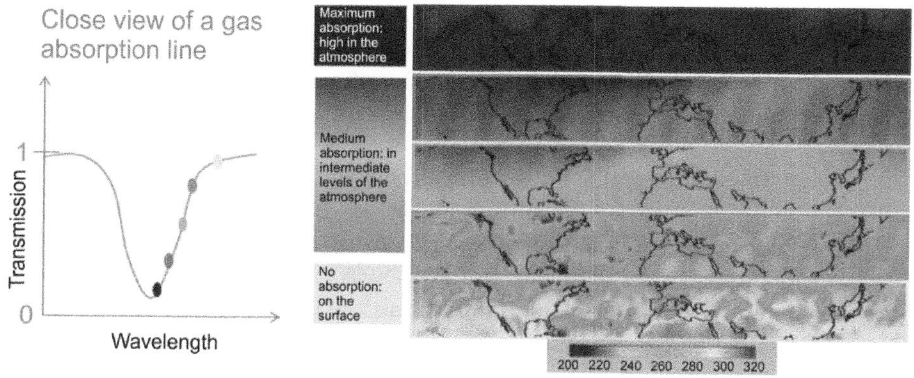

Figure 9.7. *Ability of a sounder to differentiate between several layers of the atmosphere: theoretical example of five spectral channels in the vicinity of a line. On the right, the high layers have lower brightness temperatures (in K, IASI measurements) because the temperature decreases with altitude in the troposphere*

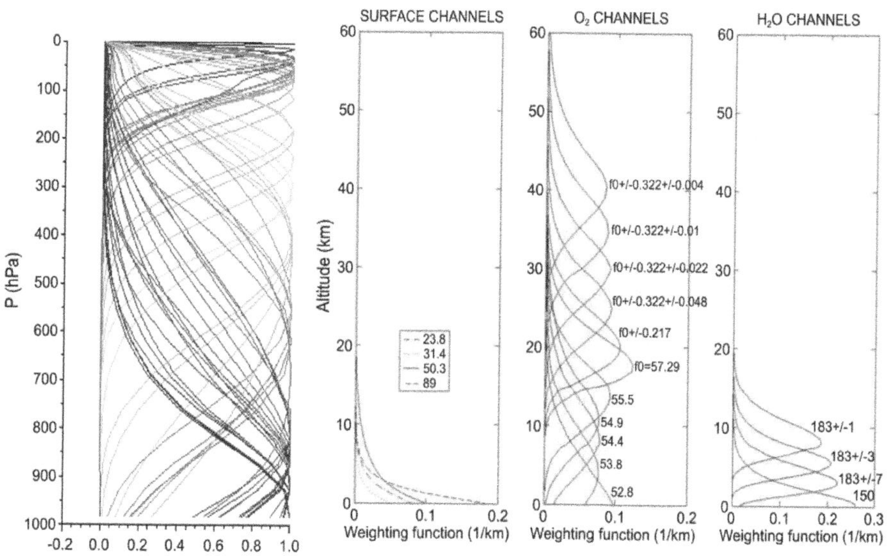

Figure 9.8. *Weighting functions of some channels of the IASI infrared instrument (left) and the AMSU-A microwave sounding instrument (right)*

It is this notion of atmospheric "sounding" that makes it possible, by inversion, to obtain vertical profiles, on several levels, of temperature or water vapor (see Chapter 10). For example, the AMSU-A instrument includes molecular resonance frequencies (O_2 and H_2O) and window channels for which the atmosphere is almost transparent to radiation. The combination of channels at frequencies more or less close to the absorption lines allows the sounding of water vapor and temperature at different altitudes from the surface.

In the microwave range, the impact of hydrometeors is characterized by emission and scattering phenomena. A simplified model has been proposed (Wilheit et al. 1977) to simulate an empirical relationship between brightness temperatures and rain intensity. More advanced models (RTTOV-SCAT, CRTM) are used to simulate brightness temperatures in a cloudy atmosphere, taking into account the vertical distribution of hydrometeors.

The forms of the RTE without proper emission (e.g. in the visible range) do not allow such vertical sensitivity. On the other hand, the spectral signatures of the absorption make it possible to retrieve a vertically integrated column for different gases. The impact of scattering by aerosols is stronger in the shorter wavelengths (due to the small size of the aerosols) and the analysis of the RTE with scattering also makes it possible to retrieve the optical thickness of the aerosols.

9.2.6. *Example: the 4A/OP source code*

Many computer tools, called radiative transfer codes, help solve the RTE. These tools vary according to the application; they can work line by line or with broader spectral band models, take into account solar and/or terrestrial radiation, consider scattering or not, gaseous absorption or not, or be scalar or polarized. For example, the 4A/OP code is a line-by-line code developed several decades ago (Scott and Chédin 1981) and whose specificity is that the line shape calculations are made for different temperature and pressure conditions, with the gaseous optical thicknesses being stored in tables. The code entries are shown in Figure 9.9. This code, working from the visible to the far-infrared domain, includes several scattering modules (DISORT, LIDORT, VLIDORT and SOS) and allows the calculation of Jacobians (useful for retrievals, see Chapter 10)[8].

8. The software is free and available at https://4aop.aeris-data.fr/.

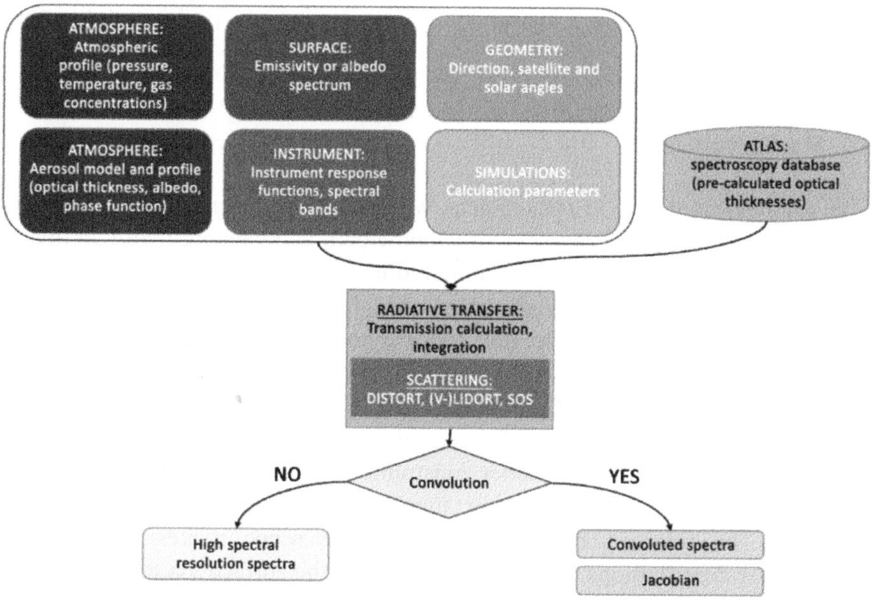

Figure 9.9. *Block diagram of radiative transfer calculation with 4A/OP software*

9.3. Passive optical sensors: radiometers and spectrometers

Satellite-based passive optical sensors have different names depending on their characteristics. Without this being really codified, we can refer to the glossary for the definitions of radiometers, spectrometers and multi-spectral or hyper-spectral imagers.

These categories are not exclusive: radiometers are very often imagers, there are spectro-imagers, and sounders are often spectrometers. The name chosen reflects more the functionality put forward in the instrumental concept and the performance of the system in terms of radiometry (signal-to-noise ratio), spectral resolution and spatial resolution. For the observation of the atmosphere from space, it is the radiometric and spectral aspects that take precedence (high-spatial-resolution imaging from space instead meets the needs of surface mapping).

Some satellites take continuous measurements and thus make it possible to obtain global data for the atmosphere, while others will observe according to a programming plan and focus on certain areas. Upstream of the instrument, a sighting mirror can be used to target an area that is not located under the satellite track, or to keep the on-ground pointing. This function can sometimes be performed directly by controlling the platform.

9.3.1. *Radiometers*

We are interested here in radiometers, which are very often multi-spectral (with "wide" spectral bands, of a few tens of nanometers), and generally with an imaging function (typically pixels from a few hundred meters to a few kilometers for a total swath of a few hundred kilometers). Although generally designed for several applications, some of which do not concern the atmosphere, these instruments provide information on clouds (cloud cover, optical thickness of the cloud) and aerosols, with the possibility of measuring their optical thickness and characterizing their size distribution. The choice of several well-positioned spectral bands for these instruments makes it possible to better constrain, for example, the spectral variations of extinction by aerosols (reflecting their size distribution via an Angström parameter). The frequent revisiting of the same scene makes it possible to distinguish the contributions of the atmosphere and the surface via temporal aspects (the reflectance of the aerosols varies from one day to another while that of the surface changes slowly (Hagolle et al. 2010)).

Illustrations of cloud observations are given in Chapter 14 of Volume 2 and those of particles in Chapters 8–11 of Volume 2.

If they include one or more bands absorbed by water vapor, these radiometers also provide the integrated content of water vapor (Vesperini et al. 1999).

A radiometer generally includes the following elements: collecting optics (telescope), filters to select only the useful spectral band(s), optics (lenses, mirrors, diaphragm) to focus the flux up to the detector(s), which are generally cooled passively (by coupling them with the cold space) or actively by a cryogenic machine, to limit thermal noise. The imaging function is obtained by the use of 1D detectors orthogonal to the satellite track and a temporal acquisition along the track (the so-called "push-broom" mode) or by the use of 2D detector matrices. Onboard calibration devices (lamps, blackbodies, etc.) can complete the payload. Electronics complete the detection chain and allow control of the instrument and the acquisition, or even onboard processing of data.

Some examples of radiometers providing atmospheric products are SEVIRI aboard the geostationary satellite MTG, MODIS aboard the A-Train, MERIS, etc.

Among the radiometers are the polarimeter radiometers that make it possible to measure the polarization of the radiation. The POLDER instrument was a precursor for this type of observation. Thanks to a spectral filter wheel also including polarizers, it measures radiation in nine spectral bands, three of which are polarized. This is of great interest for the detection of atmospheric aerosols over land, and also for the speciation of aerosols. This type of instrument will soon be part of the family of operational instruments for meteorology, thanks to the carriage of 3MI on Post-EPS (Fougnie et al. 2018).

First-generation infrared sounders, such as HIRS, are also radiometers whose specificity is to use relatively narrow filters to position spectral bands in the absorption lines of different gases (CO_2 for the temperature sounding, H_2O and O_3).

9.3.2. *Spectrometers*

A spectrometer has several hundred or even thousands of spectral channels, that is, the radiation received is collected in many, very fine "bands" of wavelengths. An important quantity is the spectral resolution, that is, the narrowness of the spectral channels, generally measured in cm^{-1}. The instrumental spectral response function (ISRF) gives the spectral response of the instrument to a monochromatic wave, and its width at half-maximum is the spectral resolution. In addition to the signal-to-noise ratio of the measurement, the higher the spectral resolution and the wider its spectral coverage, the more efficient the spectrometer.

Spectrometers allow fine observation of gaseous absorption lines. Their main use is therefore to measure gas concentrations in the atmosphere over the entire spectrum, from UV to infrared, as well as temperature and humidity sounding in the thermal infrared, because the higher the spectral resolution, the sharper the weighting functions (Figure 9.8) (Huang et al. 1992). Figure 9.10 shows this effect related to spectral resolution.

The two main families of spectrometers are dispersive element spectrometers and Fourier transform (FT) spectrometers.

Dispersive element spectrometers are based on the refraction or diffraction of light by an optical element, usually a prism, or, for better spectral resolving power and a more compact instrument, a grating. An entrance slit oriented perpendicular to the satellite track collects the light flux, which then passes through a collimator optic, then into the dispersive element and is finally imaged on a detection matrix. Figure 9.11 shows that an image is obtained, one dimension being spectral and the other spatial (according to the swath in the direction perpendicular to the trace). The along-track dimension is given by the acquisition time and the measurement integration step. The spectrum is finally obtained by a binning of the signal in the spatial dimension over all or part of the height of the matrix, according to the number of fields of view desired along the swath.

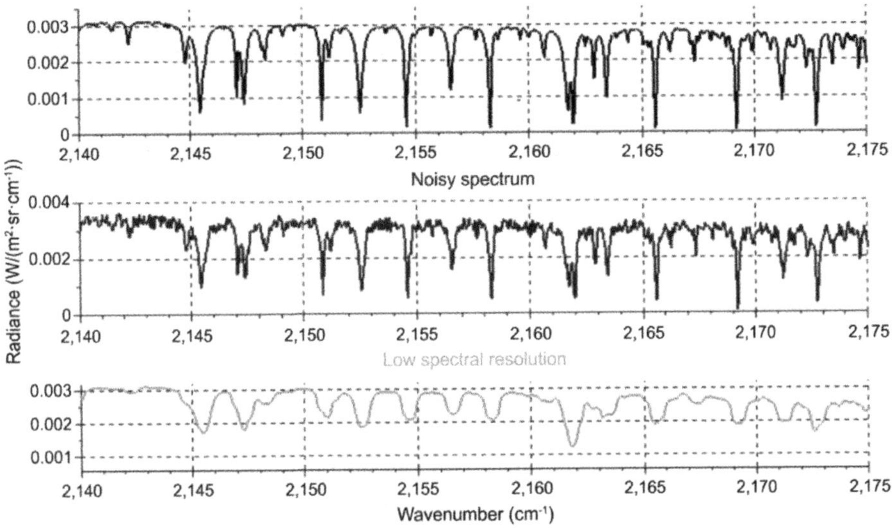

Figure 9.10. *Simulations of the same atmospheric spectrum with high noise (in blue) or degraded spectral resolution (in green)*

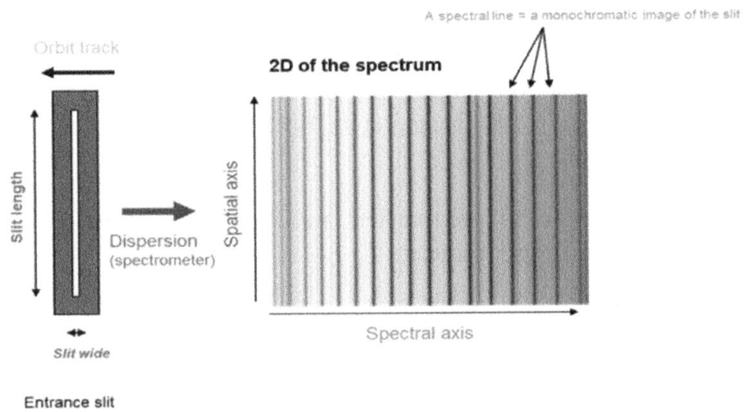

Figure 9.11. *Measurement of a grating spectrometer on a satellite; the entrance slit is placed perpendicular to the track of the satellite (© CNES)*

An example of a grating spectrometer is given by the MicroCarb mission (Pascal et al. 2017) intended for the measurement of carbon dioxide by spectrometry in the near infrared range (see Chapter 6).

Fourier transform spectrometers are based on two-wave interference. The Michelson interferometer (Figure 9.12) is the simplest implementation. The flow entering the instrument is separated into two by a beamsplitter and sent to two mirrors, one of which is fixed and the other moving. The waves reflected by the mirrors are recombined by the beamsplitter and the total flux is recorded by a detector. Depending on the position of the moving mirror, the optical path difference (OPD) between the two waves varies and their recombination is successively constructive (waves in phase) or destructive (waves in phase opposition). The instrument thus optically realizes the FT of the spectrum (the interferogram), as indicated in the following equation as a function of the OPD (in cm) between the two optical paths passing through the fixed and mobile mirrors:

$$I(OPD) = \int_{-\infty}^{+\infty} L(\sigma)d\sigma + \int_{-\infty}^{+\infty} L(\sigma)cos(2\pi\sigma.OPD)d\sigma \qquad [9.13]$$

The spectrum can then be obtained by the inverse FT of the interferogram, with a spectral resolution linked to the maximum OPD.

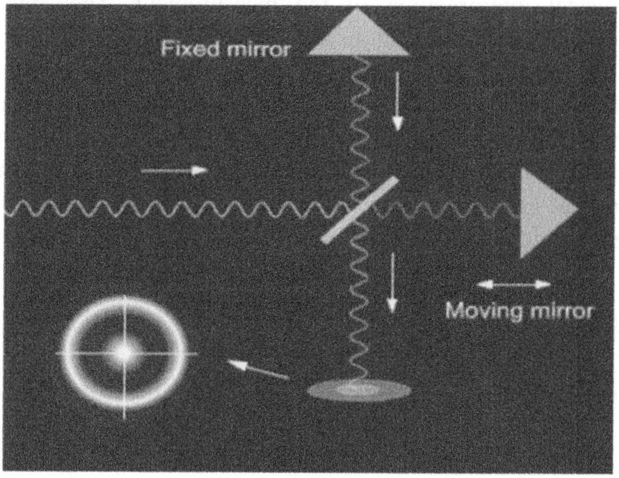

Figure 9.12. *Measurement of a Michelson interferometer (© CNES)*

Table 9.1 presents the advantages and disadvantages of the two types of instruments.

	Dispersive spectrometer	FT spectrometer
Advantages	No moving part Simultaneous acquisition of the entire spectrum Compact design (immersed grating, echelette, etc.)	Very high-spectral-resolution possible (only depends on the mirror path) Multiplex gain: the detector sees the integrated intensity Possibility of single-pixel detector, easier to regulate thermally Easy spectral calibration
Disadvantages	Detector size (as long as the spectrum) Distortions Variability of ILS (Instrument Line Shape) High aperture f/1 to f/3	Temporal acquisition of samples: stabilization of the line of sight (7 km/s) Moving parts: critical qualification and lifetime, sensitivity to microvibrations Reduced field of view to avoid self-apodization Large calculations (FFT)

Table 9.1. *Comparison of the two families of spectrometers*

9.3.3. *Level 1 processing*

Level 1 processing is computer processing applied to the raw measurement in order to remove any instrumental errors. These processing operations are mandatory for the geophysical interpretation of the measurement. They include at least radiometric calibrations (associating an absolute physical unit of luminosity with a signal level), spectral calibrations (associating a central wavelength and an ISRF with a spectral sample), geometric calibrations (associating a latitude and longitude with a measurement point, possibly resampling the image in a terrestrial reference frame). Depending on the type of instrument, other processing is necessary, for example the inverse Fourier transformation for a Michelson interferometer, or the correction of geometric distortions for a grating spectrometer.

9.3.4. *The sensors of the future*

Beyond the classic sensors briefly described here, developments of innovative sensors are in progress. These are already developments allowing for greater radiometric or spectral performance (e.g. the Mertz interferometer for IASI-NG[9]. It is also the addition of an imaging function to spectrometers or high spectral resolution for imagers. For the observation of the atmosphere using spectrometers, the possibility of descending to spatial scales of a few kilometers or less offers new possibilities, as already shown by the TROPOMI instrument (Veefkind et al. 2012) for monitoring polluting gas emissions.

9. https://iasi-ng.cnes.fr/fr/.

Let us also mention developments that allow for a reduction in the size of the instrument and therefore affordable constellations (e.g. static FT interferometers with a long integration time), or new measurement principles, for example, the tomography of the atmosphere by constellations.

9.4. Active optical sensors: lidars

9.4.1. *Lidar principle*

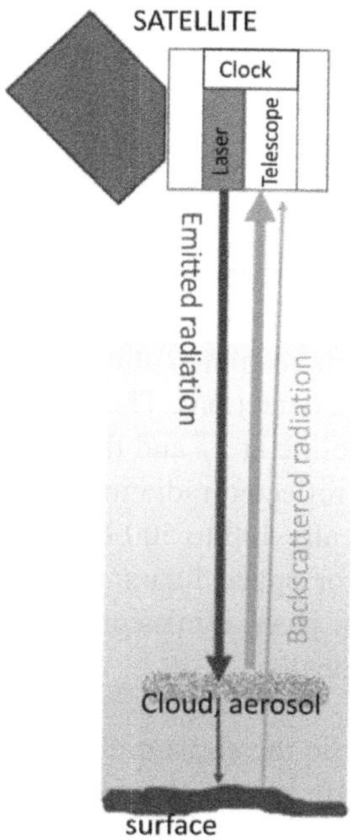

Figure 9.13. *Measurement using a spatial Lidar. The onboard clock separates the signal backscattered by the aerosols (thick arrows) from that reflected by the surface (thin arrows)*

A lidar is an active optical instrument; the source of the radiation is a laser on board the platform, which illuminates the scene observed by a telescope located on the same platform (Figure 9.13). The main advantage of lidar comes from the fact that the emission of radiation by the laser is in the form of short pulses emitted at a known time thanks to an onboard clock. The signal backscattered by the atmosphere or the surface and collected by the telescope is recorded by a detector temporally using the same clock. Since the speed of light is known, measuring the distance is possible by measuring time of flight. This allows either a fine vertical analysis of the signal and therefore highly resolved vertical profiles (a few hundred or tens of meters) (section 9.4.3.1) or a filtering of the radiation on the ground-echo allowing for knowledge of the optical path followed, with no impact of scattering (section 9.4.3.3).

9.4.2. *Lidar equation*

The intensity of the radiation collected by the lidar corresponding to the backscatter by an altitude z is given by the following lidar equation:

$$P(z) = KF_O A \frac{\beta(z)}{(z_T - z)^2} exp\left(-2\int_{z_T}^{z} \alpha(z)dz\right)$$

In this equation, the first factors depend on the instrument; F_0 designates the flux emitted by the laser (limited by the electrical power available on board), A is the area of the telescope and K is an absolute calibration parameter (which depends for example on the efficiency of the detection). The denominator $(z_T - z)^2$ is the square of the distance between the satellite at z_T and the reflective altitude z and represents the spherical dispersion of the reflected radiation. It is therefore advantageous to fly a lidar as low as possible (typically 400 to 500 km). Nevertheless, the distance to the target is a strong constraint for space lidars, whose measurements have a much lower signal-to-noise ratio than passive missions. This often requires averaging the signals horizontally and/or vertically.

Finally, the last terms contain the signature of the atmospheric medium crossed: $\beta(z)$ is the backscattering coefficient of the particles or molecules at altitude z (in km^{-1}), and we recognize the transmission of the atmosphere in the exponential term. The factor 2 in front of the integral indicates a round trip. $\alpha(z)$ is the extinction coefficient, and its integral gives the optical thickness of the medium crossed.

9.4.3. *Different types of spatial Lidar*

9.4.3.1. *Backscatter lidar*

The objective of backscatter lidars is to characterize the scatterers (aerosols and clouds) in the atmosphere, and in particular their vertical distribution. They acquire successive vertical profiles along the trace, as shown in Figure 9.14.

Some lidars operate at multiple wavelengths. Generally based on a source that is an Nd:YAG laser, they emit and collect radiation at 1,064 nm and, using an optical device called a doubler, at 532 nm (formerly CALIPSO) or, with a tripler, at 355 nm (EarthCare).

On the other hand, since the radiation emitted by the laser is naturally in a state of linear polarization, it is possible to collect the return signal in two ways, one corresponding to the parallel polarization (no depolarization with respect to the radiation emitted), the other to the perpendicular polarization. The depolarization thus measured provides information on the microphysics of scatterers, it allows us, for example, to distinguish ice clouds from clouds of liquid water.

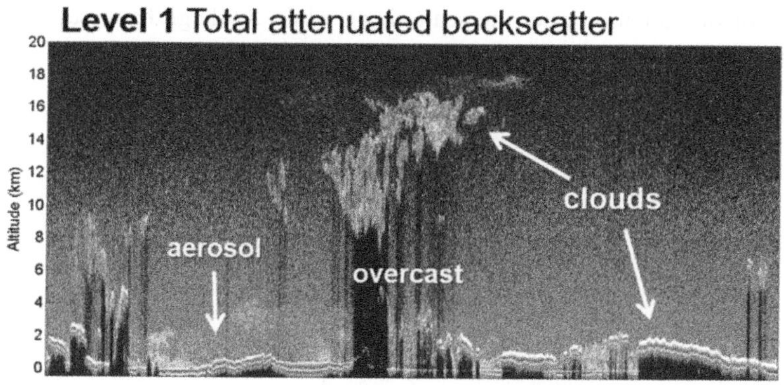

Figure 9.14. *Example of a signal obtained by a space lidar; the horizontal axis represents the track of the satellite, the vertical axis represents the altitude above the Earth's surface and the color indicates the intensity of the backscattered signal (source: NASA)*

The extraction of geophysical information from a backscatter lidar is based on the lidar equation given above. The two elements to be retrieved to characterize aerosols or clouds, in addition to the altitude of the layers, are the coefficients $\alpha(z)$ and $\beta(z)$. However, the challenge comes from the fact that we have only one equation and two unknowns. A strong hypothesis, not always valid, is then made: the ratio $\beta(z)/\alpha(z)$, called the aerosol lidar ratio, is assumed to be known and is assigned a fixed value. The inversion then takes place, for example, iteratively layer by layer starting from the top of the atmosphere (Klett 1981). However, this technique propagates and amplifies errors and appears to be unstable.

Another use of lidar measurements is to compare scattering ratio (SR) histograms acquired over certain areas with synthetic histograms obtained from lidar simulations carried out for scenes from a global climate model (Chepfer et al. 2008).

$$SR(z, \lambda) = \frac{\left(\beta_{part}(z,\lambda)+\beta_{mol}(z,\lambda)\right)exp\left(-2\int_z^{TOA} \eta\alpha_{part}(l,\lambda)+\alpha_{mol}(l,\lambda)dl\right)}{\left(\beta_{mol}(z,\lambda)\right)exp\left(-2\int_z^{TOA} \alpha_{mol}(l,\lambda)dl\right)} \quad [9.14]$$

To solve the difficulty described above with respect to the inversion of the coefficients $\alpha(z)$ and $\beta(z)$, the high-spectral-resolution lidar (HSRL) technique makes it possible to obtain two independent equations. It is based on the fact that Mie backscattering from aerosols is spectrally very sharp (no change in the frequency of the radiation after the scattering), while Rayleigh backscattering from molecules produces spectral broadening of the radiation (Figure 9.15).

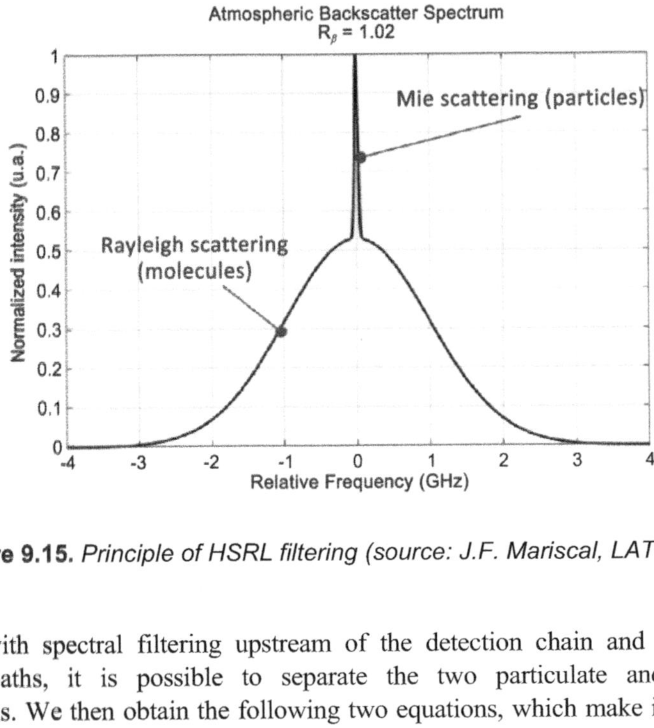

Figure 9.15. *Principle of HSRL filtering (source: J.F. Mariscal, LATMOS)*

Thus, with spectral filtering upstream of the detection chain and two distinct detection paths, it is possible to separate the two particulate and molecular contributions. We then obtain the following two equations, which make it possible to go back to the quantities $\alpha(z)$ **and** $\beta(z)$ without assumption on the aerosol lidar ratio. This therefore allows for a more accurate characterization of aerosols (type of aerosol).

$$P_{mol}(z) = K_{mol}\frac{1}{R^2}\beta_{mol}(z)exp\left(-2\int_z^{TOA} \eta\alpha_{part}(l) + \alpha_{mol}(l)dl\right) \quad [9.15]$$

$$P_{part}(z) = K_{mol}\frac{1}{R^2}\beta_{part}(z)exp\left(-2\int_z^{TOA} \eta\alpha_{part}(l) + \alpha_{mol}(l)dl\right) \quad [9.16]$$

This HSRL function is implemented in the UV on the EarthCare lidar, which should be launched into orbit in 2023 (Illingworth et al. 2015).

9.4.3.2. *Doppler lidar*

In order to measure the wind speed in the atmosphere, the Doppler lidar offers unique possibilities. This technique is based on the frequency modification of the radiation during its backscattering when the target (the molecule or the backscattering particle) is moving along the axis of radiation propagation. The frequency variation $\Delta\nu$ associated with a radial velocity V is written:

$$\frac{\Delta\nu}{\nu} = -2\,\frac{V}{c} \quad [9.17]$$

Thus, a very precise measurement of the emitted frequency and the backscattered frequency makes it possible to retrieve the radial velocity of atmospheric compounds. It should be noted that the speed of the satellite in its orbit and the speed of the Earth's rotation must be taken into account, but these elements are deterministic and can be calculated easily. For terrestrial winds, depending on the wavelength considered, the frequency shift is typically 0.1 to 10 MHz. A very inclined sight is preferred in order to get a wind component close to the horizontal.

We saw above the difference in spectral terms between backscatter on particles and on molecules (Figure 9.15). If we wish to exploit molecular backscatter (short wavelengths are favored, ideally UV) to get atmospheric winds, only direct detection is possible. An interference filter or an interferometer placed upstream of the detection makes it possible to measure the spectral shift (Garnier and Chanin 1992). For particle backscatter, the detection can be direct or heterodyne. Heterodyne detection consists of beating the backscattered wave with an onboard continuous laser, called a local oscillator. The electric current produced by the interference between the two signals is written in the form of a sinusoidal function of frequency equal to the difference between the backscattered frequency and that of the local oscillator.

Rayleigh scattering depends on pressure and temperature conditions; the processing algorithms use meteorological data to obtain a more accurate measurement of winds.

The measurements of the ADM-Aeolus instrument, launched in 2018, exploit both Mie backscatter and Rayleigh backscatter (Tan et al. 2008) by implementing a Fizeau interferometer for the HSRL function. This mission is described in more detail in Chapter 4 of Volume 2.

9.4.3.3. *DIAL lidar*

A DIAL (Differential Absorption Lidar) is a lidar that emits at least two frequencies, at least one of which corresponds to an absorption line of the gas whose concentration is to be measured. For satellite use, due to signal-to-noise constraints, only the IPDA (Integrated Path Differential Absorption) category is relevant. No IPDA lidar has yet been launched in space, but the MERLIN project (Ehret et al. 2017), an IPDA lidar for methane, is in the development phase and should be in orbit from 2026. The principle of such a lidar is shown in Figure 9.16.

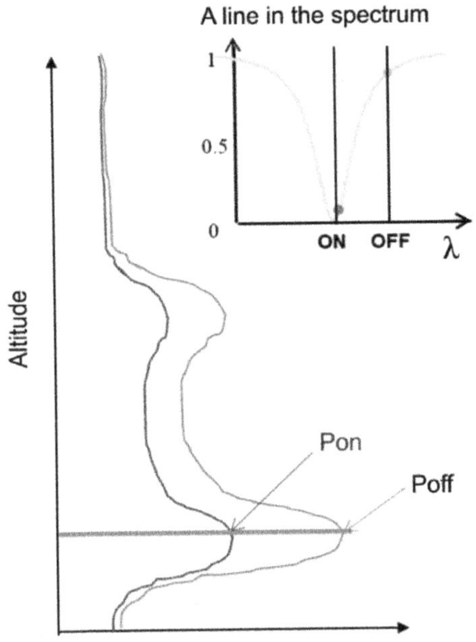

Figure 9.16. *Principle of a spatial IPDA lidar; the orange curve corresponds to the unabsorbed backscattered signal, and the blue to the signal absorbed by the target molecule and backscattered*

The processing of the measurements of an IPDA space lidar requires the calculation of the differential absorption optical depth (DAOD) by combining the signals corresponding to the ground echo P_{on} and P_{off}:

$$DAOD = \frac{1}{2} ln \frac{P_{off}}{P_{on}} \quad [9.18]$$

It is possible to calculate a weighting function (WF) from exogenous data (temperature and humidity profiles provided by meteorological analyses, absorption cross-sections calculated from spectroscopic databases):

$$WF(P) = \frac{\sigma_G(\lambda_{on},P,T)-\sigma_G(\lambda_{off},P,T)}{g(m_{dry}+m_{H2O}x_{H2O}(P))} \quad [9.19]$$

where σ_G denotes the cross-section of the target gas *G*, *P* denotes the pressure, *T* denotes the temperature in the pressure layer *P*, m_{dry} and m_{H2O} are the molecular masses of dry air and water, respectively, and x_{H2O} denotes the vertical profile of mixing ratio of water in dry air.

The variable ultimately retrieved for an IPDA lidar is the dry-air mixing ratio of the gas *G*, vertically integrated with the weighting function, and denoted XG. Therefore, it can be shown that we get the following equation:

$$XG = \frac{DAOD}{\int_0^{P_{surf}} WF(P)dP} \quad [9.20]$$

We therefore obtain the mixing ratio XG and its associated weighting function by combining the measurements of the ground echoes P_{on} and P_{off} with exogenous data, and also using the distance measurement between the satellite and the ground echo provided by the time of flight of the signal. Indeed, this distance measurement makes it possible, combined with accurate orbitographic data, to know the altitude of the scattering surface. This then allows an accurate estimate of P_{surf} by taking the surface pressure data (provided by meteorological analyses) and correcting it from the discrepancy between the topography of the meteorological model and the altitude of the scattering surface. A more precise description of data processing is described in Kiemle et al. (2011).

9.4.3.4. *Other lidars*

On the ground, the Raman-type lidar uses Raman scattering to measure, for example, the water vapor profile. However, due to the low flux of Raman scattering, a long integration time (several minutes) is required. It is therefore not suitable for a space application due to the speed of the satellite (typically 7 km/sec).

More recent developments include frequency comb lidar (Rieker et al. 2014). The frequency combs make it possible to simultaneously emit thousands of regularly spaced wavelengths. Nevertheless, given the optical flow constraints for satellite lidars, it is more appropriate to limit the number of frequencies emitted in order to obtain a sufficient signal-to-noise ratio in each tooth of the comb. Mini frequency combs can be generated by radio frequency techniques based on acousto-optic modulators. For greenhouse gas measurement missions, this concept under study at CNES would make it possible to combine the advantages of active techniques (no bias linked to scatterers) with the advantages of passive techniques (fine description of the line, and the possibility of spectral calibration to relax the stability constraints of the laser emission) (Hébert and Lemaître 2018).

Finally, for more details on the various lidars for the atmosphere, the reader can refer to the book by Weitkamp (2005).

9.4.4. *Comparison of optical sensors*

	Optical imagers	Passive TIR spectrometers	Passive UV/VIS/NIR/SWIR spectrometers	DIAL (Differential Absorption Lidar)	Backscatter lidar
Main target species	Aerosols, clouds, humidity	Temperature Gaseous species (humidity, ozone, greenhouse gases, etc.) Clouds/aerosols	O_3, NO_2, CO_2, CH_4, CO	CH_4 CO_2, H_2O, O_3	Particles (clouds, aerosols)
Strengths	High-spatial-resolution imaging, sometimes polarization	T, H_2O, O_3: true vertical profile with high DOFS Number of species Day and night	Weighting function down to the ground, well suited for determining surface emission fluxes or monitoring pollution in the boundary layer	If DIAL resolved (not IPDA): vertical resolution Weighting function down to the ground Day and night Few regional biases	Very good vertical resolution Day and night
Weaknesses	Little information on the vertical	Poorly resolved spatially Upper troposphere only for CO_2, CH_4	Aerosol impact: regional biases Daytime measurements only	Low SNR requires averaging. Measurement along-track only	Low SNR requires averaging. Measurement along-track only
Examples	MODIS, Polder, S3, 3MI	AIRS, IASI, CrIS, IASI-NG, MTG/IRS	GOME, TROPOMI, GOSAT, OCO-2, MicroCarb, S4, S5	MERLIN, ASCENDS, WALES, SCALE	CALIPSO, EarthCare, MESCAL, etc.

Table 9.2. *Comparison of different atmospheric sounding techniques*

9.5. Passive and active microwave sensors

9.5.1. *Specificities of microwave sensors*

Microwave remote sensing complements visible and infrared remote sensing capabilities, which have shorter wavelengths. Microwave radiation has the ability to penetrate clouds and to some extent rain, and is independent of sunlight, allowing the Earth to be observed in any weather, day or night. Another particularity of microwaves is their ability to penetrate vegetation and different types of surface, such as bare ground and snow (at varying depths depending on the observation frequency and the surface type).

9.5.2. *Passive microwave sensors*

Passive microwave radiometers are generally of two types: imagers, which have so-called "window" channels to observe the surface and precipitation, and sounders, which have channels in the more "opaque" spectral regions to probe temperature, humidity and hydrometeors at several altitudes. All current and planned radiometers are onboard polar low-orbit satellites, although geostationary orbit mission projects are gradually emerging.

A microwave radiometer is a very sensitive receiver for measuring low levels of electromagnetic radiation. All matter emits electromagnetic radiation. A classic example of a total power radiometer is that of a heterodyne receiver, for which the received signal, called a radio frequency (RF) signal, is converted into a different signal and is generally of a lower frequency (the intermediate frequency (IF)); it is easier to use than the initial frequency, before being detected. A detailed description of microwave systems and components is available (Doghri 2014).

By selecting observation frequencies, the radiometer can measure atmospheric or surface-integrated physical quantities or surface properties. As illustrated above, the interaction between the different components of the atmosphere and electromagnetic radiation forms the physical basis for the exploitation of satellite measurements. In the microwave domain, the molecules that have a significant impact in terms of absorption of electromagnetic radiation are oxygen (50–70 GHz, 118 GHz) and water vapor (22 GHz, 183.31 GHz). Apart from the pure rotation resonance lines of H_2O and O_2, another type of signal attenuation exists and is commonly called the "continuum". It is an excess of absorption not predicted by the theoretical models, which has been observed in situ and in the laboratory. The contribution of the continuum increases slightly with frequency and becomes greater in the presence of water vapor. Thus, the microwave transmission spectrum exhibits a wide range of conditions ranging from almost total transparency to near total opacity, as shown in Figure 9.17.

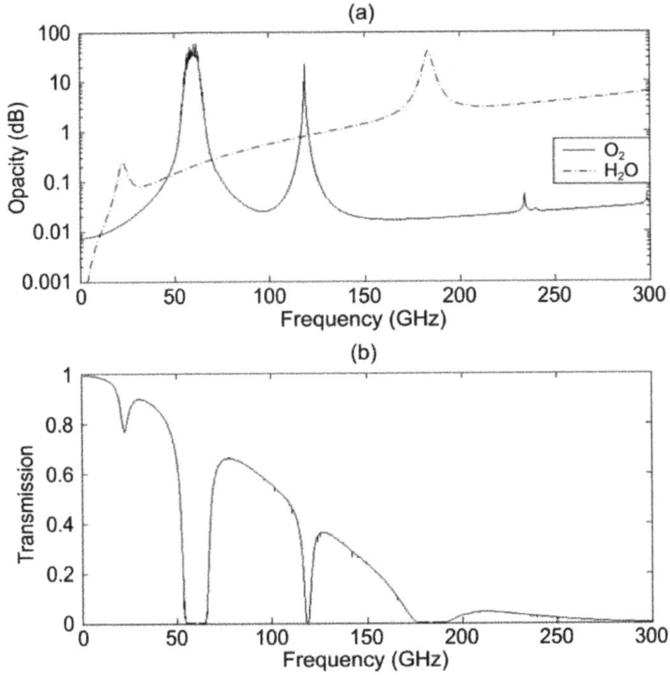

Figure 9.17. *Opacity and transmission of the atmosphere in microwaves*

Frequencies below 15 GHz, for which the atmosphere is almost transparent to radiation, are often used for the observation of land and ocean surfaces (by active remote sensing). Measurements close to the absorption lines of water vapor or oxygen, by passive sounders, make it possible to measure atmospheric humidity and temperature at different altitudes.

The first operational attempts to assimilate the brightness temperatures of passive microwave sensors in rainy cloudy skies date back to 2010 (Bauer et al. 2010) with AMSR-E and SSMI observations. Subsequently, increasingly sophisticated modeling was implemented to assimilate more observations from imagers and microwave sounders (Geer et al. 2017, 2018).

9.5.3. *Active microwave sensors*

Active microwave sensors include two families of instruments: non-imaging radars such as scatterometers, altimeters and weather radars, and imaging radars such as SAR (Synthetic Aperture Radar) and InSAR (Interferometric Synthetic Aperture Radar) systems. An active sensor of this type generates its own energy to illuminate the target. The electromagnetic wave is emitted by an antenna towards the target, the reflected radiation is measured by the sensor. All of these remote sensing instruments are used particularly for observing precipitation, clouds and ocean surface winds. The wide range of microwave frequencies exploited makes it possible to obtain three-dimensional (3D) information on the microphysical properties of clouds and rain. For example, the precipitation radar (in Ku-band, 13.8 GHz) of the TRMM mission was the first weather radar to be launched into space to map rainfall distributions in 3D.

9.5.4. *List of microwave instruments*

Appendix 3 summarizes the main use of microwave measurements according to the frequency range, as well as a non-exhaustive list of current or future instruments[10]. Depending on the observation frequencies, the exploitation of these data makes it possible to obtain useful information on the atmospheric column (vertical distribution of temperature, humidity, hydrometeors) as well as on the physical quantities integrated into the surface (surface temperature, soil humidity, sea surface wind, integrated water vapor content, precipitation rate).

9.6. References

Bauer, P., Geer, A.J., Lopez, P., Salmond, D. (2010). Direct 4Dvar assimilation of all-sky radiances. Part I: Implementation. *Quart. J. Roy. Meteor. Soc.*, 136, 1868–1885.

Chandrasekhar, S. (1960). *Radiative Transfer*. Dover, Mineola, New York.

Chepfer, H., Bony, S., Winker, D., Chiriaco, M., Dufresne, J.-L., Sèze, G. (2008). Use of CALIPSO lidar observations to evaluate the cloudiness simulated by a climate model. *Geophys. Res. Lett.*, 35, L15704.

Clerbaux, C., Boynard, A., Clarisse, L., George, M., Hadji-Lazaro, J., Herbin, H., Hurtmans, D., Pommier, M., Razavi, A., Turquety, S. et al. (2009). Monitoring of atmospheric composition using the thermal infrared IASI/MetOp sounder. *Atmos. Chem. Phys.*, 9, 6041–6054.

10. An overview of satellites and instruments is available at https://space.oscar.wmo.int/.

Cornet, C., Labonnote, L.C., Szczap, F. (2010). Three-dimensional polarized Monte Carlo atmospheric radiative transfer model (3DMCPOL): 3D effects on polarized visible reflectances of a cirrus cloud. *J. Quant. Spectrosc. Radiat. Transfer*, 111(1), 174–186.

Doghri, A., Djerafi, T., Ghiotto, A., Wu, K. (2011). Broadband substrate-integrated-waveguide six-port applied to the development of polarimetric imaging radiometer. *Proceedings of the 41st European Microwave Conference (EuMA)*, Manchester.

Draine, B.T, and Flatau, P.J. (1994). Discrete-dipole approximation for scattering calculations. *J. Opt. Soc. Am. A*, 11, 1491–1499.

Ehret, G., Bousquet, P., Pierangelo, C., Alpers, M., Millet, B., Abshire, J.B., Wirth, M. (2017). MERLIN: A French-German space lidar mission dedicated to atmospheric methane. *Remote Sensing*, 9(10), 1052.

Fougnie, B., Marbach, T., Lacan, A., Lang, R., Schlüssel, P., Poli, G., Couto, A.B. (2018). The multi-viewing multi-channel multi-polarisation imager – Overview of the 3MI polarimetric mission for aerosol and cloud characterization. *J. Quant. Spectrosc. Radiat. Transfer*, 219, 23–32.

Garnier, A. and Chanin, M.L. (1992). Description of a doppler rayleigh lidar for measuring winds in the middle atmosphere. *Appl. Phy.*, B55, 35–40.

Geer, A., Baordo, J.F., Bormann, N., Chambon, P., English, S.J., Kazumori, M., Lawrence, H., Lean, P., Lonitz, K., Lupu, C. (2017). The growing impact of satellite observations sensitive to humidity, cloud and precipitation. *Quart. J. Roy. Meteor. Soc.*, 143, 3189–3206.

Geer, A.J., Lonitz, K., Weston, P., Kazumori, M., Okamoto, K., Zhu, Y., Schraff, C. (2018). All-sky satellite data assimilation at operational weather forecasting centres. *Quart. J. Royal Meteorol. Soc.*, 144(713), 1191–1217.

Hagolle, O., Huc, M., Pascual, D.V., Dedieu, G. (2010). A multi-temporal method for cloud detection, applied to FORMOSAT-2, VENμS, LANDSAT and SENTINEL-2 images. *Remote Sens. Environ.*, 114(8), 1747–1755.

Hébert, P.-J. and Lemaitre, F. (2018). SCALE: Validations and prospects for a novel type of sounding lidar using short frequency combs. *ICSO 2018*, Chania.

Hess, M., Koepke, P., Schult, I. (1998). Optical properties of aerosols and clouds: The software package OPAC. *Bull. Am. Meteorol. Soc.*, 79(5), 831–844.

Huang, H.L., Smith, W.L., Woolf, H.M. (1992). Vertical resolution and accuracy of atmospheric infrared sounding spectrometers. *J. Appl. Meteorol. Climatol.*, 31(3), 265–274.

Illingworth, A.J., Barker, H.W., Beljaars, A., Ceccaldi, M., Chepfer, H., Clerbaux, N., Van Zadelhoff, G.J. (2015). The EarthCARE satellite: The next step forward in global measurements of clouds, aerosols, precipitation, and radiation. *Bull. the Am. Meteorol. Soc.*, 96(8), 1311–1332.

Karbou, F., Prigent, C., Eymard, L., Pardo, J.R. (2005). Microwave land emissivity calculations using AMSU measurements. *IEEE Trans. Geosci. Remote Sens.*, 43(5), 948–959.

Kiemle, C., Quatrevalet, M., Ehret, G., Amediek, A., Fix, A., Wirth, M. (2011). Sensitivity studies for a space-based methane lidar mission. *Atmos. Meas. Tech.*, 4(10), 2195–2211.

Klett, J.D. (1981). Stable analytical inversion solution for processing lidar returns. *Appl. Opt.*, 20(2), 211–220.

Koepke P., Hess M., Schult I., Shettle E.P. (1997). Global aerosol dataset. Report no. 243, Max-Plank-Institut für Meteorologie, Hamburg, September.

Lenoble, J. (1993). *Atmospheric Radiative Transfer*. A. Deepak Publishing, Hampton, VA.

Lenoble, J., Herman, M., Deuzé, J.L., Lafrance, B., Santer, R., Tanré, D. (2007). A successive order of scattering code for solving the vector equation of transfer in the earth's atmosphere with aerosols. *J. Quant. Spectrosc. Radiat. Transfer*, 107(3), 479–507.

Lenoble, J., Remer, L., Tanré, D. (2013). *Aerosol Remote Sensing*. Springer, Berlin and Heidelberg.

Mishchenko, M.I., Travis, L.D., Lacis, A.A. (2002). *Scattering, Absorption, and Emission of Light by Small Particles*. Cambridge University Press, Cambridge.

O'Dell, C.W. (2010). Acceleration of multiple-scattering, hyperspectral radiative transfer calculations via low-streams interpolation. *Journal of Geophysical Research: Atmospheres*, 115(D10).

Pascal, V., Buil, C., Loesel, J., Tauziede, L., Jouglet, D., Buisson, F. (2017). An improved MicroCarb dispersive instrumental concept for the measurement of greenhouse gases concentration in the atmosphere. In *International Conference on Space Optics – ICSO 2014*, Vol. 10563, p. 105633K. International Society for Optics and Photonics.

Péquignot, E., Chédin, A., Scott, N.A. (2008). Infrared continental surface emissivity spectra retrieved from AIRS hyperspectral sensor. *Journal of Applied Meteorology and Climatology*, 47(6), 1619–1633.

Pierangelo, C. (2005). Apports du sondage infrarouge à l'étude des aérosols atmosphériques. Thesis, Université Pierre et Marie Curie (Paris VI), 22 September.

Prigent, C., Rossow, W.B., Matthews, E. (1997). Microwave land surface emissivities estimated from SSM/I observations. *J. Geophys. Res. Atmos.*, 102(D18), 21867–21890.

Rieker, G.B., Giorgetta, F.R., Swann, W.C., Kofler, J., Zolot, A.M., Sinclair, L.C., Newbury, N.R. (2014). Frequency-comb-based remote sensing of greenhouse gases over kilometer air paths. *Optica*, 1(5), 290–298.

Scott, N.A. and Chedin, A. (1981) A fast line-by-line method for atmospheric absorption computations: The Automatized Atmospheric Absorption Atlas. *J. Appl. Meteorol.*, 20, 802–812.

Stamnes, K., Tsay, S.C., Wiscombe, W., Jayaweera, K. (1988). Numerically stable algorithm for discrete-ordinate-method radiative transfer in multiple scattering and emitting layered media. *Appl. Op.*, 27(12), 2502–2509.

Tan, D.G., Andersson, E., Kloe, J.D., Marseille, G.J., Stoffelen, A., Poli, P., Nett, H. (2008). The ADM-Aeolus wind retrieval algorithms. *Tellus A: Dyn. Meteorol. Oceanogr.*, 60(2), 191–205.

Veefkind, J.P., Aben, I., McMullan, K., Förster, H., de Vries, J., Otter, G., Claas, J., Eskes, H.J., de Haan, J.F., Kleipool, Q. et al. (2012). TROPOMI on the ESA Sentinel-5 Precursor: A GMES mission for global observations of the atmospheric composition for climate, air quality and ozone layer applications. *Remote Sens. Environ.*, 120, 70–83.

Vesperini, M., Breon, F.M., Tanré, D. (1999). Atmospheric water vapor content from spaceborne POLDER measurements. *IEEE Trans. Geosci. Remote Sens.*, 37(3), 1613–1619.

Weitkamp, C. (2005). *Lidar: Range-resolved Optical Remote Sensing of the Atmosphere.* Springer, Berlin and Heidelberg.

Wilheit, T.T., Chang, A., Rao, M.V., Rodgers, E., Theon, J.S. (1977). A satellite technique for quantitatively mapping rainfall rates over the oceans. *J. Appl. Meteorol.*, 16(5), 551–560.

10

The Inverse Problem and Techniques for Atmospheric Variable Retrieval

Clémence PIERANGELO
CNES, Toulouse, France

While the previous chapter presented the direct problem, that is, the calculation of the atmospheric spectrum depending on the state of the atmosphere and the surface via the radiative transfer equation, the inverse problem involves estimating atmospheric variables (temperature and humidity profile, concentration of absorbing gases, properties of clouds or aerosols, etc.) from a measured spectrum. The purpose of this chapter is to briefly present the main techniques used.

10.1. General remarks on the inversion of atmospheric parameters

While the direct model is physical and allows for an analytical calculation, the inverse model does not allow for the analytical formulation of a solution. The approach is therefore purely mathematical. The only exceptions are the inversion of lidar data (e.g. the inversion of a gas concentration by an IPDA lidar, Chapter 9), or the inversion of the Beer–Lambert equation for a homogeneous medium, described below, not adapted to the atmosphere, heterogeneous in pressure, temperature and concentration.

For a color version of all the figures in this chapter, see www.iste.co.uk/phulpin/satellites1.zip.

Satellites for Atmospheric Sciences 1,
coordinated by Thierry PHULPIN, Didier RENAUT, Hervé ROQUET and Claude CAMY-PEYRET.

In a homogeneous medium (e.g. a gas cell of length l), with a single spectral channel and no emission, the radiative transfer equation is given by the Beer–Lambert law: $I = I_0 exp\left(-\int_0^l N(l')C_{ext}(l')dl'\right)$.

And the number of absorbent particles or molecules per unit volume can be calculated very simply: $N = \frac{1}{l\, C_{ext}} ln\left(\frac{I_0}{I}\right)$.

But for a typical spatial observation of the atmosphere, we have:

– several spectral channels: it is then necessary to combine the measurements taking into account their respective precision (probabilistic aspect);

– a medium heterogeneous in pressure, temperature and concentration, hence integrals and nonlinearity: the practical resolution requires a linearization of the problem.

The inversion then involves adjusting a simulated measurement to an observed measurement. This therefore implies the ability to simulate a measurement and therefore the use of a radiative transfer code.

10.2. Matrix expression of the direct problem

10.2.1. *Matrix expression*

In a very schematic way, the measurement can be written via the following equation:

$$\boldsymbol{y} = \boldsymbol{f}(\boldsymbol{x}, \boldsymbol{b}) + \boldsymbol{\varepsilon} \qquad [10.1]$$

where $\boldsymbol{y}$ is the observation or measurement vector containing the measured elements (typically radiances), $\boldsymbol{f}$ is the function (or observation operator) of the direct model (here the radiative transfer equation, Chapter 9), x is the state vector containing the elements to be retrieved (geophysical quantities), $\boldsymbol{b}$ is the vector of the "parameters" of the model (the quantities that we are not trying to retrieve but that we know from elsewhere) and ε is the error vector on the measurement. In these equations and thereafter, vectors or matrices have bold symbols.

This error vector contains the noise linked to the instrument (radiometric noise linked to the flux in a so-called "photon-limited" case), noise linked to the detection chain (such as the dark current or background noise), which is usually well characterized, as well as the noise linked to errors on the radiative transfer model (e.g. the imperfectly known spectroscopic parameters, or the failure to take into account fine effects such as polarization or refraction in the atmosphere). These are difficult to assess so sometimes an arbitrary value is applied for the model error.

In order for this formalism to apply, it is crucial that the measurement be previously well calibrated radiometrically and spectrally, which is the subject of so-called "level 1" processing. The retrieval is the calculation of the best estimate of x, the so-called "level 2" product.

10.2.2. *Linearization of the problem*

To estimate the state vector, some methods are based on the first-order linearization of equation [10.1], which involves introducing the Jacobian matrix **K**, defined by $K_{i,j} = \frac{dy_i}{dx_j}$:

$$\boldsymbol{y} = \boldsymbol{Kx} + \boldsymbol{\varepsilon} \quad [10.2]$$

to simplify the equations, the parameter vector **b** is not included anymore in the following.

This assumes that the problem is linear or weakly nonlinear, which is not always valid (e.g. in a cloudy situation). There are techniques to examine the degree of nonlinearity of the problem, described in Rodgers (2000).

Linearization is sometimes done around a reference point $(\boldsymbol{x_0}, \boldsymbol{y_0})$ (for which $\boldsymbol{y_0}$ is the value calculated by applying the function **f** on vector $\boldsymbol{x_0}$) and gives:

$$\boldsymbol{y} - \boldsymbol{y_0} = \boldsymbol{K}(\boldsymbol{x} - \boldsymbol{x_0}) + \boldsymbol{\varepsilon} \quad [10.3]$$

10.2.3. *Typical dimensions of the problem*

In order to visualize the dimensions of the problem, let us take the example of the inversion of a temperature T and humidity H_2O profile from the radiance measurements of a hyperspectral infrared sounder. Let m be the number of spectral channels selected and n be the number of vertical levels that describe the atmosphere.

Equation [10.1] is then written in more detail:

$$\begin{bmatrix} I(v_1) \\ \vdots \\ I(v_m) \end{bmatrix} = f\left[\begin{bmatrix} T(P_1) \\ \vdots \\ T(P_n) \end{bmatrix}, \begin{bmatrix} H_2O(P_1) \\ \vdots \\ H_2O(P_n) \end{bmatrix}, b\right] + \begin{bmatrix} \varepsilon(v_1) \\ \vdots \\ \varepsilon(v_m) \end{bmatrix}$$

and in a linearized form:

$$\begin{bmatrix} I(v_1) \\ \vdots \\ I(v_m) \end{bmatrix} = \begin{bmatrix} \frac{dI(v_1)}{dT(P_1)} & \cdots & \frac{dI(v_1)}{dT(P_n)} & \frac{dI(v_1)}{dH_2O(P_1)} & \cdots & \frac{dI(v_1)}{dH_2O(P_n)} \\ \vdots & \ddots & \vdots & \vdots & \ddots & \vdots \\ \frac{dI(v_m)}{dT(P_1)} & \cdots & \frac{dI(v_m)}{dT(P_n)} & \frac{dI(v_m)}{dH_2O(P_1)} & \cdots & \frac{dI(v_m)}{dH_2O(P_n)} \end{bmatrix} \begin{bmatrix} T(P_1) \\ \vdots \\ T(P_n) \\ H_2O(P_1) \\ \vdots \\ H_2O(P_n) \end{bmatrix} + \begin{bmatrix} \varepsilon(v_1) \\ \vdots \\ \varepsilon(v_m) \end{bmatrix}$$

The measurement vector, of dimension m, typically covers a few hundred to a few thousand spectral channels. The state vector, of dimension $N = 2n$, typically contains a few tens to a few hundred elements. The Jacobian matrix is therefore of dimension $m \times N$, which can be complicated to handle especially in an operational context with a lot of data to process. Hence, meticulous channel selection is usually implemented both for assimilation and retrieval of IASI spectra (Collard 2007).

10.3. Solutions to the inverse problem

10.3.1. *Least squares*

Note first that the matrix **K** is generally not square, so we cannot calculate its mathematical inverse. In general, $m > N$, the problem is therefore overdetermined. However, we can introduce its pseudo-inverse, which is written $(\boldsymbol{K}^T\boldsymbol{K})^{-1}\boldsymbol{K}^T$, where the exponent ^{T} stands for the transpose operator.

One solution of the inverse problem, denoted $\hat{\boldsymbol{x}}$, is the least squares solution, given by the following equation:

$$\hat{\boldsymbol{x}} - \boldsymbol{x}_0 = (\boldsymbol{K}^T\boldsymbol{K})^{-1}\boldsymbol{K}^T(\boldsymbol{y} - \boldsymbol{y}_0) \qquad [10.4]$$

An improved version consists of a version weighted by the inverses of the uncertainties on the measurement.

Nevertheless, this solution is often of little use in practice due to the poor conditioning of matrix $\boldsymbol{K}$, which makes $\boldsymbol{K}^T\boldsymbol{K}$ non-invertible or very badly conditioned, leading to correct solutions mathematically but not physically (see Figure 10.1). This is related to the fact that the problem is also underdetermined, in the sense that the number of unknowns is greater than the number of independent pieces of information (the information contained in the different spectral channels is highly redundant).

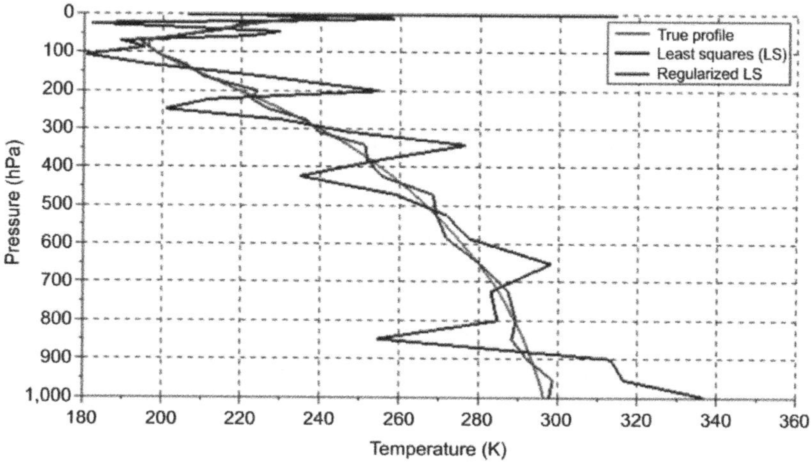

Figure 10.1. *Retrieval of a profile by least squares (in black) and with regularization (in blue). Oscillations in the least-squares solution are typical of poor matrix conditioning* $\boldsymbol{K}^T\boldsymbol{K}$*. For a color version of this figure, see www.iste.co.uk/phulpin/satellites1.zip*

It is therefore necessary to introduce information a priori. This can be done explicitly, to reduce the number of unknowns, by using exogenous data (e.g. a temperature profile provided by a meteorological analysis) or by taking into account the correlations between the unknowns (regularization techniques, see next section). This can also be done implicitly, by limiting the solution domains to be investigated, which is the case when building a basis for an inversion from regressions, Look-Up-Tables, or neural networks.

The prior decorrelation of the measurements, whether by channel selection or by principal component analysis, also makes it possible to improve the conditioning of the problem. Finally, another possibility is the introduction of other independent measurements, such as adding microwave sensor channel radiance to a retrieval of infrared radiance to constrain the temperature profile (Chédin et al. 2003).

10.3.2. *Probabilistic methods*

The least squares solution fits the measurements to the data. In the following methods, we always minimize the attachment to the data, but we also introduce an additional constraint. The simplest regularization is a norm 2 constraint of the finite vector $\boldsymbol{x}$, which gives a smoother vertical profile (see Figure 10.1). A solution can then be written, by introducing a scalar variable λ and I the identity matrix:

$$\hat{\boldsymbol{x}} - \boldsymbol{x}_0 = (\boldsymbol{K}^T\boldsymbol{K} + \lambda \boldsymbol{I})^{-1}\boldsymbol{K}^T(\boldsymbol{y} - \boldsymbol{y}_0) \tag{10.5}$$

It is observed that the poor conditioning of $\boldsymbol{K}^T\boldsymbol{K}$ can be bypassed as long as λ is high enough.

The most classic technique for spectra inversion relies on a Bayesian approach and is called optimal estimation. The reference work is the book by Rodgers (2000). In this context, the vectors $\boldsymbol{x}$ and $\boldsymbol{y}$ are considered to be random variables. Optimal estimation, sometimes also known as "1D-Var", involves maximizing the probability of the state vector given the measurement, that is, according to Bayes equality $p(\boldsymbol{x}|\boldsymbol{y}) = p(\boldsymbol{y}|\boldsymbol{x})\, p(\boldsymbol{x}) \frac{1}{p(\boldsymbol{y})}$.

The following equations are the result of this equality in the case where all the distributions are Gaussian and are obtained by identifications of the linear or quadratic terms in $\boldsymbol{x}$. This assumption of Gaussianity is sometimes questionable, especially for the distributions of species such as aerosols, for example.

10.3.2.1. *Linear or iterative inversion*

If the problem is linear, then for a priori Gaussian distributions, we obtain an a posteriori Gaussian distribution, with an average $\hat{\boldsymbol{x}}$:

$$\hat{\boldsymbol{x}} - \boldsymbol{x}_0 = (\boldsymbol{K}^T\boldsymbol{S}_\varepsilon^{-1}\boldsymbol{K} + \boldsymbol{S}_a^{-1})^{-1}\left(\boldsymbol{K}^T\boldsymbol{S}_\varepsilon^{-1}(\boldsymbol{y} - \boldsymbol{y}_0) + \boldsymbol{S}_a^{-1}(\boldsymbol{x}_a - \boldsymbol{x}_0)\right) \tag{10.6}$$

In this equation, we have introduced the a priori state vector x_a as well as variance–covariance matrices of measurement uncertainty $\boldsymbol{S}_\varepsilon$ and of the state vector a priori $\boldsymbol{S}_a$. The diagonal of this matrix represents the variance (square of standard deviation), while the off-diagonal elements are linked to the correlations between spectral channels for $\boldsymbol{S}_\varepsilon$ or between vertical levels, or other elements of the state vector for $\boldsymbol{S}_a$. An example of a priori covariance matrices on the state vector is given in Figure 10.2.

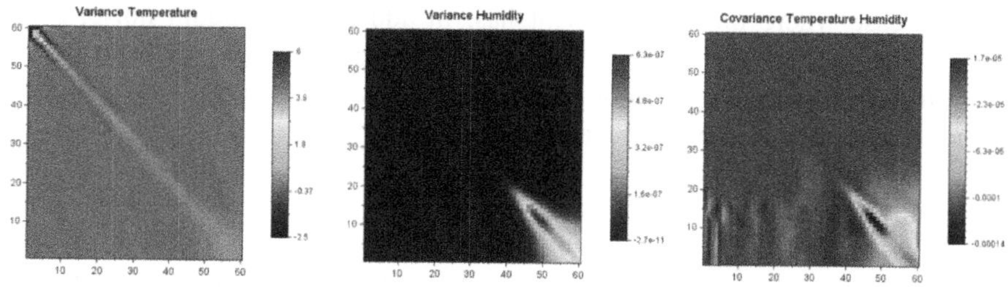

Figure 10.2. *Examples of variance–covariance matrix for temperature (left), humidity (center) and cross-covariance (right): the atmosphere is discretized here on 60 vertical levels; the bottom of the atmosphere corresponds to the first rows and columns of matrices. For a color version of this figure, see www.iste.co.uk/phulpin/satellites1.zip*

If the problem is weakly nonlinear, the inversion is then carried out by carrying out iterations (index i) on equation [10.6]: from an initial state x_0, the Jacobians are updated and the spectrum is calculated on each iteration:

$$\hat{x}_{i+1} - \hat{x}_i = (\boldsymbol{K}_i^T \boldsymbol{S}_\varepsilon^{-1} \boldsymbol{K}_i + \boldsymbol{S}_a^{-1})^{-1} \left(\boldsymbol{K}_i^T \boldsymbol{S}_\varepsilon^{-1}(\boldsymbol{y} - \boldsymbol{f}(\hat{x}_i)) + \boldsymbol{S}_a^{-1}(\boldsymbol{x}_a - \hat{x}_i)\right) \qquad [10.7]$$

We observe that radiative transfer code with the calculation of the Jacobians is necessary at each iteration, which has a strong impact on the calculation time.

10.3.2.2. *Transporting uncertainty*

Bayesian estimation also makes it possible to obtain the variance–covariance matrix of the a posteriori distribution of the state vector $\hat{\boldsymbol{S}}$, that is, via its diagonal, the best estimate of the a posteriori random uncertainty.

$$\hat{\boldsymbol{S}} = (\boldsymbol{K}^T \boldsymbol{S}_\varepsilon^{-1} \boldsymbol{K} + \boldsymbol{S}_a^{-1})^{-1} \qquad [10.8]$$

This uncertainty depends only on the Jacobians. It is therefore a very practical tool for estimating the uncertainty of the level 2 product according to the level 1 characteristics (choice and position of the spectral bands, resolution and spectral sampling, noise in the measurement) and is widely used in the dimensioning phase of a space mission to compare different instrumental concepts, or to make compromises between requirements.

The uncertainty on the parameters (vector $\boldsymbol{b}$) can be taken into account in two ways, either by replacing $\boldsymbol{S}_\varepsilon$ in the uncertainty transport equation with $\boldsymbol{S}_\varepsilon + \boldsymbol{K}_b\, \boldsymbol{S}_b\, \boldsymbol{K}_b^T$ or including $\boldsymbol{b}$ in the state vector. If $\boldsymbol{b}$ and $\boldsymbol{x}$ are independent, it can be shown that the two approaches are equivalent.

In the case where the measurement has, in addition to random noise, a bias (constant difference over time with respect to what is thought to be true), the formalism of the optimal estimation also makes it possible to carry out a transfer of bias measurement $\boldsymbol{\delta}_m$ towards the state vector.

$$\boldsymbol{\delta}_x = \left[\left(\boldsymbol{K}^T\boldsymbol{S}_\epsilon^{-1}\boldsymbol{K} + \boldsymbol{S}_a^{-1}\right)^{-1}\boldsymbol{K}^T\boldsymbol{S}_\epsilon^{-1}\right]\boldsymbol{\delta}_m \qquad [10.9]$$

This equation is used in particular to determine the calibration requirement specifications, which are generally more constrained than random errors.

10.3.2.3. *Averaging kernels*

The characterization of the retrieved quantity is essential in order to then be able to compare measurements from different sensors (Rodgers and Connors 2003) or to get level 4 (assimilation in models).

The averaging kernels[1] function gives the contribution of the different layers of the atmosphere to the product obtained. This is the matrix A defined by

$$\hat{\boldsymbol{x}} = \boldsymbol{A}\boldsymbol{x} \qquad [10.10]$$

A perfect averaging kernel would be the identity matrix but in practice the vertical resolution is limited, that is, the value of the estimated state vector at a given vertical level depends on the true values at that level as well as at neighboring levels. The matrix $\boldsymbol{A}$ is then rather band-diagonal at best. An illustration is given in Figure 10.3. Note that, if we estimate an integrated column, then A is a vector.

The optimal estimation formalism provides a simple way to compute averaging kernels:

$$\boldsymbol{A} = (\boldsymbol{K}^T\boldsymbol{S}_\varepsilon^{-1}\boldsymbol{K} + \boldsymbol{S}_a^{-1})^{-1}\boldsymbol{K}^T\boldsymbol{S}_\varepsilon^{-1}\boldsymbol{K} \qquad [10.11]$$

This characterization also makes it possible to know the number of degrees of freedom of the system, or DOFS (Degrees Of Freedom for Signal), which is equal to the trace of matrix $\boldsymbol{A}$.

In the SWIR (see Glossary), we generally obtain only one to two elements of information on the vertical but with a significant contribution from the lower layers.

1. Sometimes also called the "weighting function" when the retrieved quantity is an integrated column, but this can lead to confusion with the radiative weighting function defined in Chapter 9.

It is therefore a domain of choice for the estimation of gaseous concentrations close to the surface (e.g. to trace it back to surface fluxes).

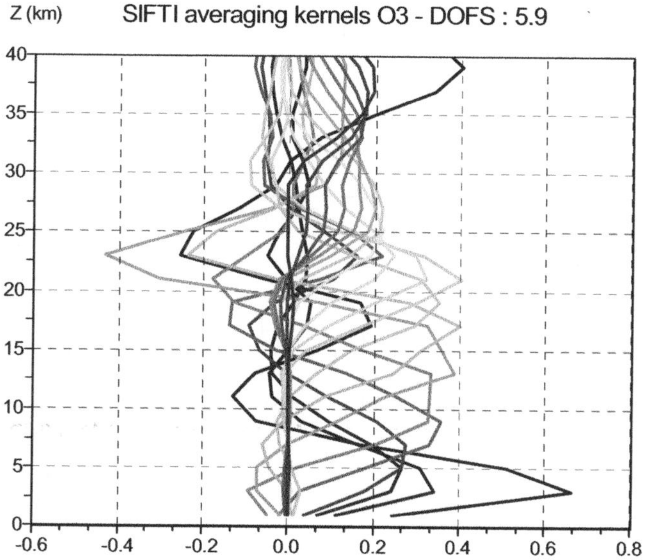

Figure 10.3. *Averaging kernel matrix examples for an inversion of the vertical profile of ozone. The blue functions correspond to the rows of the matrix for the lower levels of the atmosphere, the red functions for the upper atmosphere*

In the TIR (see Glossary), we obtain much more vertical information for temperature and humidity (typically between 5 and 10 DOFS for each of the two variables), as well as for ozone, or even carbon monoxide.

10.3.2.4. *Example: 4ARTIC software*

The 4ARTIC software is an example of an optimal estimation implementation. It is based on the 4A/OP code as a direct model (see Chapter 9) and was developed at CNES, both for design studies of new missions in the thermal infrared or near-infrared domains and to be implemented in the ground segment of the future MicroCarb space mission. Figure 10.4 shows that it can be used in "perfo" mode (uncertainty transport via equations [10.8] to [10.11]) or in "retrieval" mode (inversion of a spectrum by iterations, equation [10.7]).

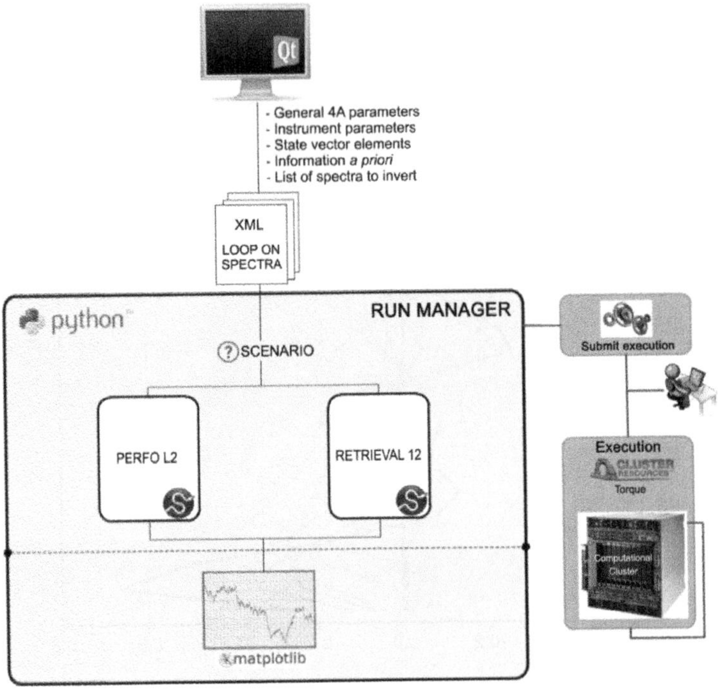

Figure 10.4. *Diagram of the 4ARTIC software. For a color version of this figure, see www.iste.co.uk/phulpin/satellites1.zip*

10.3.3. *Methods with pre-calculated bases*

Optimal estimation solves the under-constraint problem by introducing an explicit prior vector and covariance matrix, but there is another family of inversion techniques that implicitly introduce a constraint a priori via the use of a pre-computed basis, used either empirically (methods with Look-Up-Tables, LUT), or as a learning basis for a neural network retrieval.

Building a pre-calculated base first requires determining all the parameters to which the measurement is sensitive. This can be done, for example, by examining the Jacobians of the different gases or by analyzing the differences between simulated spectra with/without a given variable (Figure 10.5).

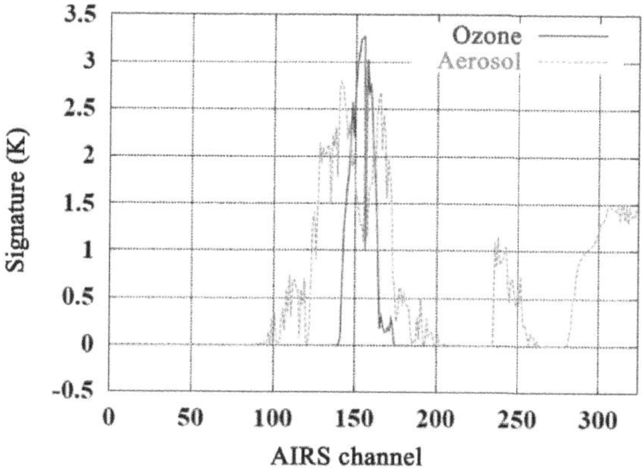

Figure 10.5. *Impact of ozone (in red) and desert aerosols (in green) on a spectrum from the NASA Aqua/AIRS sounder. For a color version of this figure, see www.iste.co.uk/phulpin/satellites1.zip*

Once the variable parameters have been set (atmospheric as well as surface parameters or related to the geometry of the sensor), the associated measurements (radiance, spectra, etc.) should be simulated using a radiative transfer code. The radiative transfer calculations, which take a long time, are thus done "once and for all" upstream of the retrievals. The base construction step is quite tricky as it is crucial to have an input base that is representative of the variety of atmospheric situations that the sensor may encounter.

10.3.3.1. *Empirical methods*

The retrieval is then based on the comparison between the observation and the simulated radiances. A distance measurement is therefore defined, which generally takes into account the expected noise for each wavelength. The result of the inversion is the atmospheric situation with the smallest distance between the simulated measurements in the table and the observation.

This approach is used, among others, for the remote sensing of aerosols (Tanré et al. 1997), for observations of the MODIS space radiometer (Pierangelo et al. 2004) or for AIRS spectrometer observations.

Beyond LUT methods, many other empirical approaches exploiting pre-computed tables exist, such as the use of regressions established in tables (difference or ratios between spectral channels, linear combinations of channels, polynomial adjustment, etc.).

10.3.3.2. *Neural network methods*

Methods based on artificial neural networks make it possible to perform nonlinear regressions and have been applied to the inversion of atmospheric parameters for several decades (Butler et al. 1996; Hadji-Lazaro et al. 1999). A network, represented in Figure 10.6, contains input neurons (in our case, radiance or brightness temperatures measured by the spatial sensor), one or more hidden layers of neurons, and an output layer containing the variables to be retrieved (e.g. the concentration of an atmospheric gas). A classic type of network is the multi-layer perceptron. Each neuron of the hidden layers performs an operation, which is a weighted sum of the outputs of the neurons of the previous layer followed by the application of an activation function. The activation function is generally a threshold function and is used to introduce nonlinearity (sigmoid, hyperbolic tangent, etc.).

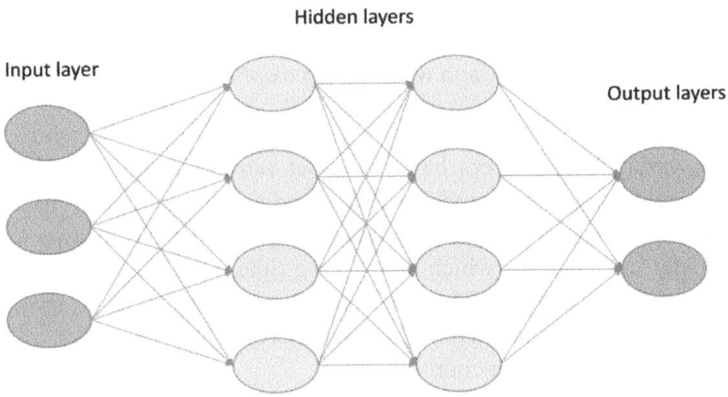

Figure 10.6. *Schematic representation of a neural network. For a color version of this figure, see www.iste.co.uk/phulpin/satellites1.zip*

The development of a neural network retrieval requires determining the architecture of the network (number of hidden layers, number of neurons per layer, choice of inputs and outputs, etc.), which is often done empirically, and the weights of the synapses, in the network learning step.

Usually supervised learning is done by different algorithms (e.g. gradient backpropagation). This learning requires a representative database of inputs and outputs. The construction of this database can be done:

– starting from an atmospheric database and calculating by radiative transfer the radiances of the sensor (Crevoisier et al. 2004);

– starting from a set of pre-existing real measurements co-located with ground truths or models of the atmospheric variables to be retrieved (David et al. 2021). However, this can only be done once the mission has been in flight for some time.

The precise management of learning is necessary to avoid over-learning. To do this, once the learning has been carried out, using a second database, called the "test database", we verify that the statistics obtained are consistent.

In addition to their ability to process nonlinear problems, the advantages of neural methods are their execution speed once the learning has been carried out, since it is then sufficient to present the measurement as input to the network and to perform the neural operations, which are simple operations, to get the outputs.

One of the drawbacks of neural methods lies in their difficulty to provide the error bars (confidence interval) of the retrieved products. However, possibilities exist by grouping the data into "bins", for example, whose uncertainty is characterized (Aires and Pellet, 2020) or via QRNNs (Quantile Regression Neural Networks) whose output is a probability distribution of the variables to be retrieved (Pfreundschuh et al. 2018).

Finally, we should mention deep learning, which is instead applied to image data (and not for an individual pixel) for classification or segmentation but whose applicability to atmospheric retrievals is the subject of current research (Aires et al. 2021).

10.4. References

Aires, F. and Pellet, V. (2020). Estimating retrieval errors from neural network inversion schemes – Application to the retrieval of temperature profiles from IASI. *IEEE Transactions on Geoscience and Remote Sensing*.

Aires, F., Boucher, É., Pellet, V. (2021). Convolutional neural networks for satellite remote sensing at coarse resolution. Application for the SST retrieval using IASI. *Remote Sensing of Environment*, 263, 112553.

Butler, C.T., Meredith, R.V.Z., Stogryn, A.P. (1996). Retrieving atmospheric temperature parameters from DMSP SSM/T-1 data with a neural network. *Journal of Geophysical Research: Atmospheres*, 101(D3), 7075–7083.

Chédin, A., Saunders, R., Hollingsworth, A., Scott, N., Matricardi, M., Etcheto, J., Crevoisier, C. (2003). The feasibility of monitoring CO_2 from high-resolution infrared sounders. *Journal of Geophysical Research: Atmospheres*, 108(D2).

Collard, A.D. (2007). Selection of IASI channels for use in numerical weather prediction. *Quarterly Journal of the Royal Meteorological Society*, 133(629), 1977–1991.

Crevoisier, C., Heilliette, S., Chédin, A., Serrar, S., Armante, R., Scott, N.A. (2004). Midtropospheric CO_2 concentration retrieval from AIRS observations in the tropics. *Geophysical Research Letters*, 31(17), L17106.

David, L., Bréon, F.M., Chevallier, F. (2021). XCO_2 estimates from the OCO-2 measurements using a neural network approach. *Atmospheric Measurement Techniques*, 14(1), 117–132.

Hadji-Lazaro, J., Clerbaux, C., Thiria, S. (1999). An inversion algorithm using neural networks to retrieve atmospheric CO total columns from high-resolution nadir radiances. *Journal of Geophysical Research: Atmospheres*, 104(D19), 23841–23854.

Pfreundschuh, S., Eriksson, P., Duncan, D., Rydberg, B., Håkansson, N., Thoss, A. (2018). A neural network approach to estimating a posteriori distributions of Bayesian retrieval problems. *Atmospheric Measurement Techniques*, 11(8), 4627–4643.

Pierangelo, C., Chédin, A., Heilliette, S., Jacquinet-Husson, N., Armante, R. (2004). Dust altitude and infrared optical depth from AIRS. *Atmospheric Chemistry and Physics*, 4(7), 1813–1822.

Rodgers, C.D. (2000). *Inverse Methods for Atmospheric Sounding: Theory and Practice*, Volume 2. World Scientific, Singapore.

Rodgers, C.D. and Connors, B.J. (2003). Intercomparison of remote sounding instruments. *Journal of Geophysical Research: Atmospheres*, 108(D3).

Tanré, D., Kaufman, Y.J., Herman, M., Mattoo, S. (1997). Remote sensing of aerosol properties over oceans using the MODIS/EOS spectral radiances. *Journal of Geophysical Research: Atmospheres*, 102(D14), 16971–16988.

Appendices

Appendix 1

Atmospheric Profiles

Claude CAMY-PEYRET
IPSL, Paris, France

This appendix on atmospheric profiles is intended to supplement information that could not be fully developed in several chapters of the book.

A1.1. AFGL profiles

A useful source of profiles that has been widely used for modeling radiative transfer in the Earth's atmosphere was developed by Anderson et al. (1986) to serve as a basis for code comparisons or simulations across the spectrum from microwave to UV through the infrared and visible. Simplified "typical" atmospheres were selected based on the format of the reference atmosphere known as the U.S. Standard Atmosphere, 1976. This reference covered only "average" thermodynamic profiles, focusing instead on the structure and composition of the upper atmosphere. For simulations in the so-called middle atmosphere, information on minor constituent profiles and seasonal and latitude variations was missing.

It is this type of data that has been published under the title "AFGL atmospheric constituent profiles (0–120 km)" by Anderson et al. (1986) where AFGL stands for Air Force Geophysical Laboratory. There are six AFGL profiles with an index i that identifies them with i = [1;6]. They correspond to the profiles in Table A1.1. New profile databases have been proposed as described below.

Satellites for Atmospheric Sciences 1,
coordinated by Thierry PHULPIN, Didier RENAUT, Hervé ROQUET and Claude CAMY-PEYRET.

AFGL profile index	AFGL profile name	Information on conditions
1	Tropical	15N, annual average, day
2	Mid-latitude summer	45N, July, day
3	Mid-latitude winter	45N, January, day
4	Sub-arctic summer	60N, July, day
5	Sub-arctic winter	60N, January, day
6	U.S. Standard Atmosphere, 1976	45° 32' 33", average, day

Table A1.1. *AFGL profiles*

The AFGL profiles are given on a grid of 50 levels in geometric altitude from 0 to 120 km with a discretization of 1 km between 0 and 25 km, 2.5 km between 25 and 50 km and 5 km above 50 km up to 120 km. The classification of the first five profiles is associated with different seasonal and latitudinal conditions. The last profile ($i = 6$) is an annual zonal average applicable for mid-latitudes. The temperature profiles (kinetic temperature) as a function of the geometric altitude between 0 and 120 km are shown in Figure A1.1.

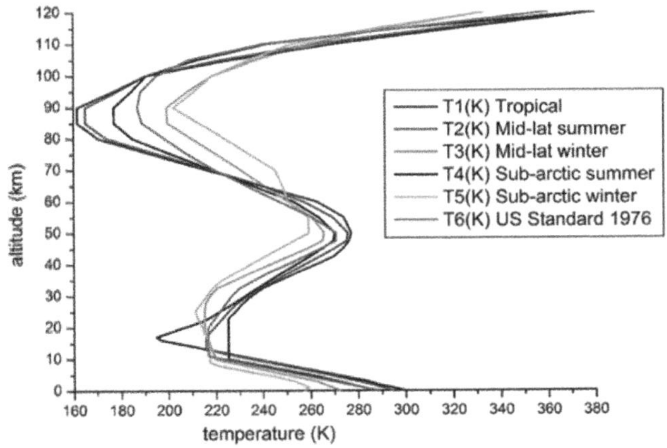

Figure A1.1. *Temperature variation as a function of altitude for the six AFGL profiles. For a color version of this figure, see www.iste.co.uk/phulpin/satellites1.zip*

We identify the different characteristic levels that correspond to the significant changes (value and/or sign) of the vertical thermal gradient from bottom to top:

– the tropopause with a marked minimum temperature located at 17 km for tropical atmosphere, a tropopause that varies with the season for other atmospheres with an altitude of 9 km for the atmosphere in the vicinity of the Arctic Circle in winter of the northern hemisphere;

– the stratopause with a maximum of approximately 50 km corresponding to maximum heating by photodissociation of ozone;

– the minimum temperature at the mesopause approximately 80 km;

– the increase of the temperature above this last altitude due to the photodissociation of molecular oxygen and the gradual transition to the thermosphere whose kinetic temperature is greatly modulated during the solar activity cycle.

Pressure variations as a function of altitude are exponential: the pressure decreases rapidly over eight decades between 0 and 120 km, with small modulations related to variations in the vertical temperature gradient as seen in Figure A1.2.

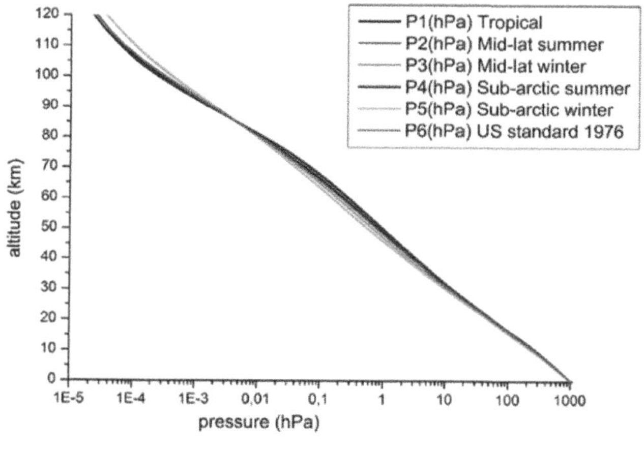

Figure A1.2. *Pressure variation as a function of altitude for the six AFGL profiles. For a color version of this figure, see www.iste.co.uk/phulpin/satellites1.zip*

In the following, the AFGL concentration profiles of H_2O, CH_4 and NO_2 are discussed (with associated comments) as examples of the variation with an altitude of three minor constituents typical of the Earth's atmospheric composition. Concentrations of the chemical species are expressed in volume mixing ratio (ppmv or parts per million by volume) which are noted here xX(z) with X = [H_2O, CH_4, NO_2]. The difference between the mixing ratio (without further specification) and mixing ratio with respect to dry air is important for precision measurements (in situ in general) when taking into account (or not) the presence of water vapor.

As seen in Figure A1.3 for H_2O in the vicinity of the surface, increasing values of humidity are observed when going from the sub-arctic winter profile (the driest) to the sub-arctic summer, mid-lat winter, mid-lat summer and finally to the tropical (the wettest). The cold trap (minimum of the H_2O) in the vicinity of the tropopause is characteristic and varies according to the altitude of this thermodynamic "barrier". The increase of the H_2O profile up to the stratopause comes from the oxidation and photodissociation of CH_4 which produces H_2O. At higher altitudes, the photodissociation of H_2O by UV radiation (less and less filtered by ozone) is explaining in turn the decrease of water vapor (the simplified AFGL model no longer differentiates between the six profiles above 80 km).

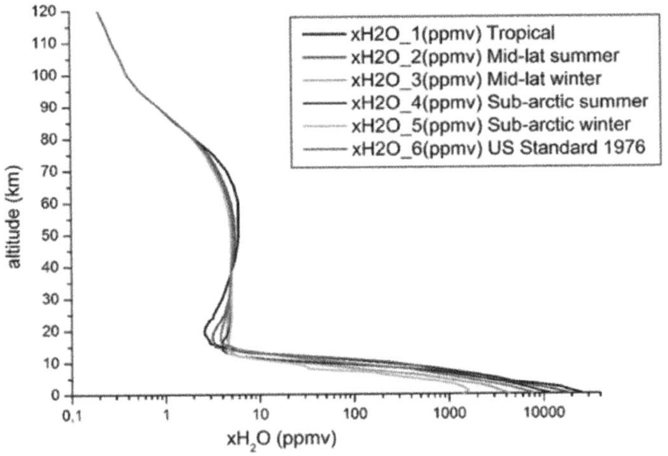

Figure A1.3. *Variation of the H_2O mixing ratio as a function of altitude for the six AFGL profiles. For a color version of this figure, see www.iste.co.uk/phulpin/satellites1.zip*

Compared to the original AFGL profiles (which date back to the 1980s), the CH_4 mixing ratio profiles of Figure A1.4 have been multiplied by a factor to reproduce the increasing trend of the average concentration of methane (a greenhouse gas) at the surface with a value close to 1.86 ppmv in 2016. Again, the AFGL model for CH_4 is a simplified version of reality: the six profiles are identical above 60 km, and the U.S. Standard profile is quite different from the other five. But the vertical distribution of methane is realistic: a constant vertical profile of CH_4, a species quite well mixed in the troposphere (the vertical resolution of the AFGL 1D model is

1 km in this altitude domain); above the tropopause, the CH_4 destruction processes by radical OH, by $O(^1D)$ and by direct photodissociation rapidly decrease the mixing ratio of CH_4.

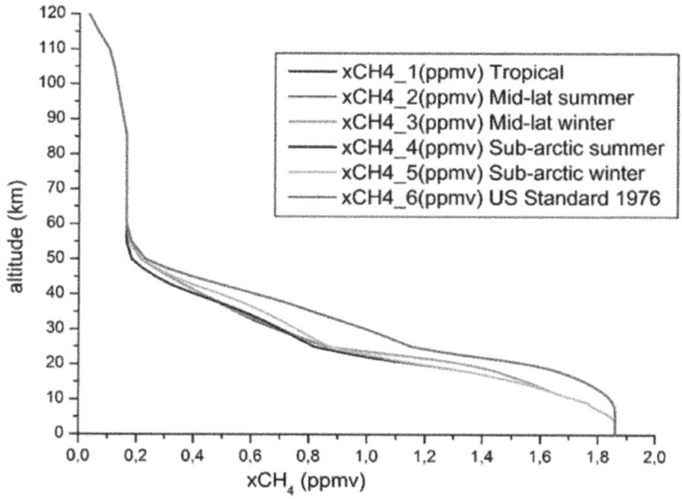

Figure A1.4. *Variation of the CH_4 mixing ratio as a function of altitude for the six AFGL profiles. For a color version of this figure, see www.iste.co.uk/phulpin/satellites1.zip*

For nitrogen dioxide NO_2, a reactive species, the AFGL profile is not provided for each of the $i = [1;6]$ profiles. Implicitly, it corresponds to a standard atmosphere with a very simplified stationary 1D model (for noon). The vertical profile nevertheless shows a very low NO_2 concentration in the unpolluted troposphere (which corresponds here to a background level of 2.3×10^{-5} ppmv or 0.023 ppbv), then a rise above the tropopause is associated with the conversion of the source of nitrogen oxides in the stratosphere, that is, nitrous oxide (N_2O). As a result of its photodissociation or reaction with $O(^1D)$, this source species is transformed into NO_2 and NO (its partner in photochemical equilibrium in the presence of ozone), both species being gathered as $NO_x = NO + NO_2$.

The photodissociation of NO_2 above the mesopause leads to its gradual transformation into NO which becomes the most abundant species during the day in the NO_x partition. Time-dependent 3D photochemical and dynamic models (via the solar zenith angle) are needed to go beyond the very "simplified" AFGL profile for NO_2 shown in Figure A1.5.

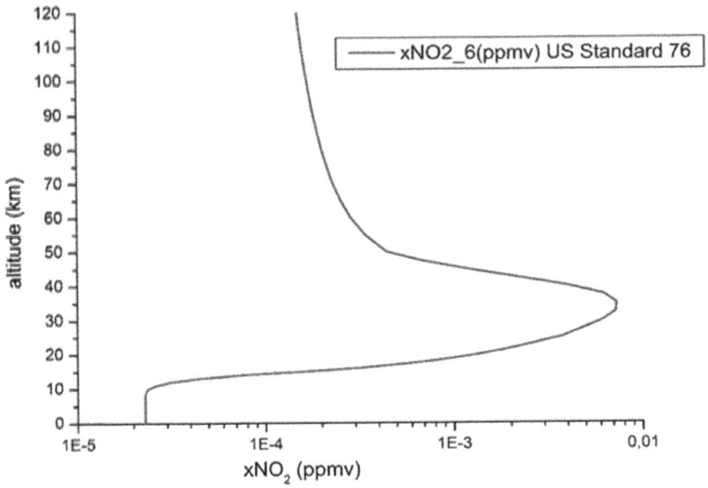

Figure A1.5. *Variation of the NO_2 mixing ratio as a function of altitude for the AFGL profile. For a color version of this figure, see www.iste.co.uk/phulpin/satellites1.zip*

A1.2. Other profile databases

To enlarge the profile base and better adhere to geophysical variability, we can use the TIGR database[1] (Thermodynamic Initial Guess Retrieval), Chevallier et al. (1998), which includes 2,311 profiles and where the most extreme situations are represented.

Other profile databases have been introduced for more specific objectives such as the SAFREE database for simulations of sea surface temperature restitution (François et al. 2002) or the GAPRI database for simulations of surface temperature restitution on land surfaces ((Mattar et al. 2015).

A1.3. References

Anderson, G.P., Clough, S.A., Kneizys, F.X., Chetwynd, J.H., Shettle, E.P. (1986). AFGL atmospheric constituent profiles (0-120 km). AFGL-TR-86-0110. Environmental Research Papers.

1. https://ara.lmd. polytechnique.fr/index.php?page=tigr.

Chevallier, F., Chéruy, F., Scott, N.A., Chédin. A. (1998). A neural network approach for a fast and accurate computation of a longwave radiative budget. *J. Appl. Meteorol.*, 37, 1385–1397.

François, C., Brisson, A., Le Borgne, L., Marsouin, A. (2002). Definition of a radiosounding database for sea surface brightness temperature simulations. Application to sea surface temperature retrieval algorithm determination. *Remote Sens. Environ.*, 81, 309–326. doi:10.1016/S0034-4257(02)00008-1.

Mattar, C., Durán-Alarcón, C., Jiménez-Muñoz, J.C., Santamaría-Artigas, A., Olivera-Guerra, L., Sobrino, J.A. (2015). Global atmospheric profiles from reanalysis information (GAPRI): A new database for earth surface temperature retrieval. *Int. J. Remote Sens.*, doi:10.1080/01431161.2015.1054965.

NOAA/NASA/USAF (1976). U.S. Standard Atmosphere. Technical memorandum, U.S. Government Publishing Office, Washington, DC.

Appendix 2

Spectroscopic Parameters for Radiative Transfer

Claude CAMY-PEYRET
IPSL, Paris, France

This appendix on spectroscopic parameters is intended to supplement information that could not be fully developed in several chapters of the book.

A2.1. Spectroscopic parameters and line profiles

When trying to model in detail the absorption or emission of molecules present in the atmosphere[1], spectroscopic parameters must be available to compute the corresponding spectrum of the molecule in the gas phase. For this reason, these parameters are based on the long-standing work in the field of experimental and theoretical molecular spectroscopy, which has made it possible to produce very useful spectroscopic databases as input to direct radiative transfer calculation models. We will first determine these properties specific to each molecular species which depend on the temperature, the pressure of the medium and the abundance of the absorbing molecule, and depending on the type of molecule concerned, we will use either a line-by-line description of the radiative properties of the molecule, or a description via absorption cross section for molecules whose spectrum is too dense

1. Except in the upper atmosphere where photodissociation phenomena break molecules into their atomic constituents. Here, we will not cover, for example, atomic oxygen lines in the submillimetric domain.

Satellites for Atmospheric Sciences 1,
coordinated by Thierry PHULPIN, Didier RENAUT, Hervé ROQUET and Claude CAMY-PEYRET.

or whose line-by-line modeling proves impracticable (and empirical cross sections, i.e. measured experimentally, will be used in this situation).

A2.1.1. *Line-by-line parameters*

In the terrestrial atmosphere that concerns us here (air for which N_2 and O_2 are the main constituents), in order to describe a molecular transition of species A in absorption " ← ' between a lower level identified by the index ' and an upper level identified by the index ", spectroscopic databases for the atmosphere use the following parameters (the spectral variable used in atmospheric spectroscopy is usually the wavenumber $\tilde{\nu}$ expressed in cm^{-1}):

– the position of the line $\tilde{\nu}_0$ (at zero pressure) in cm^{-1};

– the intensity of the line $S(T_{ref})$ at a reference temperature $T_{ref} = 296$ K that is expressed in $cm^{-1}/(molecule\ cm^{-2})$;

– energy E" of the lower level of transition in cm^{-1};

– the broadening coefficient $\gamma_{A\text{-}air}(T_{ref})$ in $cm^{-1}\ atm^{-1}$;

– the exponent n_{air} of temperature dependence $\gamma_{A\text{-}air}(T) = \gamma_{A\text{-}air}(T_{ref})\ (T_{ref}/T)^{nair}$;

– the self-broadening coefficient of the line $\gamma_{A\text{-}A}(T) = \gamma_{A\text{-}A}(T_{ref})\ (T_{ref}/T)^{nA}$. This latter coefficient is important to consider for A = H_2O since water vapor is the most abundant minor constituent, at least in the lower atmosphere where collisions H_2O–H_2O play a significant role;

– the shifting coefficient of the line resulting from collisions between the radiatively active molecule and air molecules noted $\delta_{A\text{-}air}(T_{ref})$ in $cm^{-1}\ atm^{-1}$. In general, this parameter is not always very well documented, as it has only been measured in the laboratory for a limited number of transitions. Otherwise, we take either a value by default, or a value resulting from theoretical calculations (which have their limitations);

– the rotational and vibrational (and electronic in some cases, such as O_2) quantum numbers of the upper and lower levels of the transition;

– nuclear statistical weights (related to the spin of equivalent nuclei where available) needed to calculate partition functions (see below) or to calculate emission spectra under conditions of non-local thermodynamic equilibrium (non-LTE);

– quality indicators for these different parameters;

– information about the origin of these parameters.

In general, the parameters $\tilde{\nu}_0$ and E” result from theoretical models (vibration–rotation Hamiltonian) and are the most precise considering the numerous works on the analysis of laboratory spectra recorded with high-performance instruments (Fourier transform, laser spectrometry, frequency comb technique).

The line intensity parameter $S(T_{ref})$ is more difficult to determine experimentally with precision (we reach 0.5% in the best cases), and therefore theoretical models (vibration–rotation transition moment operator) are used for this purpose. This is all the more necessary as the lines to be measured are numerous and often superimposed on each other, which makes it quite difficult to determine intensities. For example, the Ozone Atlas of Flaud et al. (1990) (calculated using the theoretical tools mentioned above but based on the best experimental data of the time) contains about 93,000 lines of O_3 from the microwave (0.3 cm^{-1}) to the mid-infrared (3,200 cm^{-1}) region.

A2.1.2. *Absorption cross sections*

When it is not possible to generate them reliably, the parameters of individual lines (in the case of a very dense spectrum) or when models are not available, laboratory spectra are recorded under very precise conditions of purity, temperature and pressure of the mixture (pure gas plus collision partner, in general, N_2 or synthetic dry air is used as a buffer gas in order to approach atmospheric conditions). The intensity transmitted through a cell of length L (in cm) containing the absorbing molecule A diluted in the buffer gas in homogeneous conditions of pressure P (in hPa) and temperature T (in K) is written as (Beer–Lambert law):

$$I(\tilde{\nu}) = I_0(\tilde{\nu}) \exp[- L \times c_A \times \sigma_A(\tilde{\nu}; T, P)] = I_0(\tilde{\nu}) \exp[-\tau(\tilde{\nu})]$$

In this equation, c_A is the concentration of species A (in molecule/cm^3) and the absorption cross section $\sigma_A(\tilde{\nu}; T, P)$ is expressed in cm^2/molecule. We can check the homogeneity of the optical depth $\tau(\tilde{\nu})$, that is, the argument of the exponential since [cm] × [molecule/cm^3] × [cm^2/molecule] is indeed a dimensionless quantity. The experimental determination of $\sigma_A(\tilde{\nu}; T, P)$ must be done (by adapting the resolution, gas concentration and path length conditions) to the spectral region of interest and selecting a set of (T,P) pairs that best cover the conditions encountered in the atmosphere. The corresponding tables (three dimensions) called look-up tables (LUTs) allow for the interpolation according to the spectral variable $\tilde{\nu}$ and the thermodynamic variables T and P of the medium.

For an isolated line under homogeneous conditions of pressure and temperature, we can make the link between the intensity of the line, the profile of the line and the cross section by writing as follows:

$$\tau(\tilde{\nu}; T, P) = L \times c_A \times S(T) \times \varphi_A(\tilde{\nu} - \tilde{\nu}_0; T, P) = L \times c_A \times \sigma_A(\tilde{\nu}; T, P)$$

So that in the vicinity of the isolated line of position $\tilde{\nu}_0$, we have:

$$\sigma_A(\tilde{\nu}; T, P) = S(T) \times \varphi_A(\tilde{\nu} - \tilde{\nu}_0; T, P)$$

The function $\varphi_A(\Delta\tilde{\nu}; T, P)$ (expressed in $1/\text{cm}^{-1}$) describes the normalized profile of the line such as:

$$\int_{-\infty}^{\infty} \phi_A(\Delta\tilde{\nu}; T, P) d\Delta\tilde{\nu} = 1$$

Dimensions are checked again since the product of the intensity of the line by the profile is expressed as:

$$[\text{cm}^{-1}/(\text{molecule cm}^{-2})] \times [1/\text{cm}^{-1}] = [\text{cm}^2/\text{molecule}]$$

The line profile depends on the pressure and temperature conditions as described below

A2.1.3. *Line profiles*

In the upper part of the atmosphere where the effect of collisions between the radiatively active molecular species A (which absorbs or emits the radiation) and other molecules is reduced, the profile of lines is determined by the molecular mass m_A and the thermal agitation, that is, the kinetic temperature T of the medium. It is therefore of the Doppler type:

$$\varphi_D(\Delta\tilde{\nu}) = (1/\gamma_D)\,(\ln(2)/\pi)^{1/2} \exp[-\ln(2)\,(\Delta\tilde{\nu}/\gamma_D)^2]$$

In this equation, $\Delta\tilde{\nu} = \tilde{\nu} - \tilde{\nu}_0$ is the deviation from the center of the line and we find $\varphi_D(\pm\gamma_D) = 1/2\ \varphi_D(0)$. The parameter γ_D is the half-width half-maximum Doppler (HWHM) of the line and is expressed as: $\gamma_D = \tilde{\nu}_0\ [2 \ln(2)\ k_B\ T/(m_A\ c^2)]^{1/2}$ or $\gamma_D \approx 3.58 \times 10^{-7}\ \tilde{\nu}_0\ (T/M_A)^{1/2}$ when T is expressed in K and the molar mass M_A is in g/mol. The constant k_B is the usual Boltzmann constant.

For the lower layers of the atmosphere, when collisions that contribute to the widening of the lines dominate, it is a Lorentz profile:

$$\varphi_L(\Delta\tilde{\nu}) = (1/\pi)\frac{\gamma_L}{\Delta\tilde{\nu}^2+\gamma_L{}^2}$$

In this equation, $\Delta\tilde{\nu} = \tilde{\nu} - (\tilde{\nu}_0 + \delta_L)$ is the deviation from the center of the line due to the shift δ_L under the effect of collisions and we find $\varphi_L(\pm\gamma_L) = 1/2\ \varphi_L(0)$. The parameter γ_L is the half-width half-maximum Lorentz (HWHM) of the line and is expressed as: $\gamma_L = (P_{air}/P_0)\ \gamma_{A\text{-}air}(T_{ref})\ (T_{ref}/T)^{nair}$ where $P_0 = 1$ atm =1013.25 hPa.

In the intermediate pressure domain (assuming the absence of correlation between the effects of widening by collisions and the effect of thermal agitation), we must use the Voigt profile which is written as the convolution product of the two profiles:

$$\varphi_V(\Delta\tilde{\nu}) = \varphi_D(\Delta\tilde{\nu}) \otimes \varphi_L(\Delta\tilde{\nu})$$

The Voigt profile is widely used in radiative transfer codes but for some molecules and under certain pressure conditions we may need (depending on the spectral resolution of satellite instruments) the use of more complicated line profiles that are discussed in Hartmann et al. (2008).

When looking at a particular species that has a dense spectrum of lines identified by the index k=[1,n_A], we can make the link between the line-by-line description and the description in terms of cross section by writing the contribution (of the radiatively active species A) to the optical depth for a homogeneous medium at temperature T and total pressure P (for a concentration c_A of the diluted absorbing gas in the N_2 buffer or in air) in the form of:

$$\tau_A(\tilde{\nu};T,P) = L \times c_A \times \sum_{k=1}^{k=n_A} S_k(T)\phi_k(\tilde{\nu}-\tilde{\nu}_k;T,P) = L \times c_A \times \sigma_A(\tilde{\nu};T,P)$$

$$\text{or } \sigma_A(\tilde{\nu};T,P) = \sum_{k=1}^{k=n_A} S_k(T)\phi_k(\tilde{\nu}-\tilde{\nu}_k;T,P)$$

The parameters $\gamma_{A\text{-}air}(T_{ref})$, n_{air} and $\delta_{A\text{-}air}(T_{ref})$ appear in the expression of the profile of the individual lines ϕ_k depending on whether we consider a Doppler, Lorentz or Voigt line profile.

Apart from the temperature dependence already introduced of the broadening coefficients (often measured experimentally), if we know the assignment of the line and therefore the lower level E” of the transition, the dependence of the line intensity as a function of temperature can be rigorously computed at the Boltzmann equilibrium for a medium at the kinetic temperature T by the formula:

$$S(T)/S(T_{ref}) = [Z(T_{ref})/Z(T)] \exp[-c_2 E'' (1/T-1/T_{ref})] \, rap(T,T_{ref})$$

The Z(T) function is the total partition function of the absorbing molecule. It is sometimes denoted TIPS for total internal partition sum, as can be found in Laraia et al. (2011).

The coefficient $rap(T,T_{ref}) = [1-\exp(-c_2 \tilde{\nu}_0 /T)]/[1-\exp(-c_2 \tilde{\nu}_0 /T_{ref})]\}$ translates the temperature dependence of the so-called induced emission term, and this term becomes very close to the unit when $c_2 \tilde{\nu}_0 >> T$ (i.e. the visible domain). But this term must be taken into account in the far-infrared or submillimeter range when we have $c_2 \tilde{\nu}_0 << T$. The constant c_2 used here (in connection with the use of wavenumber as the spectral variable) is called the "second radiation constant" and is equal to 1.438776877 K/cm^{-1}.

A2.1.4. *Continua and collision-induced spectra*

Numerous laboratory works and experimental measurements under well-documented atmospheric conditions have shown the limits of line modeling with a simple Voigt profile. In the case of water vapor, in particular, it is observed that the Voigt profile does not correctly reproduce the line profile far in the wings. A practical method which makes it possible, on the one hand, to limit the number of points of the profile to be calculated (gain in computer time) and which can use, on the other hand, atmospheric measurements and empirical modeling is to preserve the Voigt profile up to a certain distance from the center of the lines (often ± 25 cm^{-1}) and to attribute the absorption of the lines outside this interval to what is called the continuum (Clough et al. 2005). This continuum (which also exists for species other than H_2O) is modeled as a cross section by an empirical function of temperature, total pressure of the medium (and partial pressure of vapor which is in significant abundance in the lower layers of the troposphere). The latest version of the most used continuum is MT_CKD v3.5[2].

Other effects related to molecular collisions also require models that go beyond the line-by-line framework such as line-mixing effects or collision-induced absorption (CIA) spectra. All of the concepts and many useful details on the effect of collisions on molecular spectra are provided in Hartmann et al. (2005).

2. See: https://github.com/AER-RC/MT_CKD.

A2.1.5. *Line-by-line modeling of optical depth*

Since the atmosphere is a medium whose temperature and pressure conditions vary, it will always be necessary for radiative transfer calculations in this inhomogeneous medium to discretize the problem by carrying out what is called atmospheric layering (layers in which we will assume homogeneous conditions of temperature, pressure and concentrations). We must use an atmospheric layering adapted to the viewing conditions (nadir, limb, occultation, etc.) which will minimize numerical errors related to discretization. This operation is usually performed automatically by radiative transfer codes to find the right compromise between accuracy and calculation time. In any case, for a homogeneous layer (crossed over a length L by the atmospheric path) at temperature T, at pressure P, and containing n_e absorbing species $j = [1,n_e]$ of concentrations c_j having a spectrum of lines $k = [1,n_j]$, we can write (by generalizing to several species the formula given for species A):

$$\tau(\tilde{\nu};T,P) = \mathrm{L} \times \sum_{j=1}^{j=n_e} c_j \sum_{k=1}^{k=n_j} S_{k,j}(T)\phi_{k,j}(\tilde{\nu} - \tilde{\nu}_{k,j};T,P)$$

In this formula, the double sum relates to the species j and to the lines k of the species j, hence the use of the double index k,j for the intensities, positions, and profiles of lines. We see that when we have to consider a large number of absorbing species (generally of the order of a dozen), for each species we must take into account a large number of lines (in general of the order of a hundred) and that, for the spectral domain considered, we must calculate the optical depth for a large number of spectral samples $\tilde{\nu}$ (of the order of a thousand), the calculation of the optical depth for a single layer can be expensive in computing time, especially since the calculation of $\varphi_{k,j}$ profiles in the Voigt case (or for more complicated profiles) is itself the limiting factor. Once the contribution for a homogeneous layer (in which the path is L_i) has been calculated, we must combine the contributions of the different segments $i = [1,n_i]$ to the total atmospheric path by considering the optical depth $\tau_i(\tilde{\nu};T_i,P_i)$ for the values of T_i, P_i and $c_{j,i}$ specific to each homogeneous layer. To be complete and assuming that the diffusion phenomena are neglected (absorption and infrared emission without aerosol), the discretized equation of radiative transfer at local thermodynamic equilibrium for a ray propagating from layer 1 to layer n_j is written as:

$$\mathrm{R}_i(\tilde{\nu}) = \mathrm{R}_{i-1}(\tilde{\nu}) \exp[-\tau_i(\tilde{\nu};T_i,P_i)] + \{1- \exp[-\tau_i(\tilde{\nu};T_i,P_i)]\}\ \mathrm{B}(\tilde{\nu},T_i)$$

In this equation that gives the luminance (or radiance) at the exit of the layer i, the first term represents the transmission of radiation from the layer $i-1$ transmitted after absorption in layer i and the second term represents the proper emission of layer i. For the first layer and depending on the viewing geometry, the term $R_0(\tilde{\nu})$ represents either the surface term $\varepsilon(\tilde{\nu})\, B(\tilde{\nu}, T_{surf})$ for the nadir view of a surface of emissivity $\varepsilon(\tilde{\nu})$ and temperature T_{surf}, or the incident spectrum at the top of the atmosphere in the case of an external source (sun, moon, star).

Once the spectrum at the output of the atmospheric path has been calculated, it will remain to convolve the luminance $R(\tilde{\nu})$ calculated on the so-called "fine resolution" grid by the spectral response function of the instrument (we use the term ILS for instrument line shape or ISRF for spectral response function) which leads us to write the theoretical spectrum at the instrument output in the form:

$$R_{inst}(\tilde{\nu}) = \int_{-\infty}^{\infty} ILS(\tilde{\nu}' - \tilde{\nu}; \tilde{\nu}) R(\tilde{\nu}') d\tilde{\nu}$$

An instrumental function in the form of an integral (or kernel) has been introduced here to reflect the fact that some spectrometers may have a device function that depends on the spectral variable (off-axis detectors in the case of an interferometer or detector matrix in the case of a grating spectrometer).

A2.2. Spectroscopic databases

Several open-access spectroscopic databases (subject to identification) provide line-by-line spectroscopic parameters and absorption cross sections to be used as input to radiative transfer codes. Among the best known (and for the most recent versions of these databases) are HITRAN[3] (Gordon et al. 2016) and GEISA[4] (Jacquinet-Husson et al. 2016). These databases are multi-usage in the sense that they cover the entire electromagnetic spectrum from radio-frequencies to ultraviolet, through microwaves, submillimeter waves, far-infrared, all infrared, near-infrared and visible. Specialized databases are sometimes specific to each radiative transfer code.

A2.3. Atmospheric radiative transfer codes

A very large number of radiative transfer models (RTMs) have been developed for atmospheric applications and are more or less widely used. Dedicated codes have often been developed specifically for Earth observation and atmospheric sounding instruments and satellite missions. Wikipedia provides an interesting list of these

3. See: https://hitran.org.
4. See: https://www.sciencedirect.com/science/article/abs/pii/S0022285216301011?via%3Dihub.

codes in a fairly complete table[5]. Among the widely used codes, we can mention only the following:

– 4A/OP (Scott and Chedin 1981) is often used by CNES for IASI and IASI-NG[6];

– LBLRTM (Clough et al. 2005) is a successor to FASCODE and widely used by the North American community[7];

– RTTOV (Saunders et al. 1999) is widely accepted as an efficient tool by the ECMWF user community[8].

A2.4. References

Clough, S.A., Shephard, M.W., Mlawer, E.J., Delamere, J.S., Iacono, M.J., Cady-Pereira, K., Boukabara, S., Brown, P.D. (2005). Atmospheric radiative transfer modeling: A summary of the AER codes. *J. Quant. Spectrosc. Radiat. Transfer*, 91, 233–244. doi:10.1016/j.jqsrt.2004.05.058.

Flaud, J.-M., Camy-Peyret, C., Rinsland, C.P., Smith, M.A.H., Malathy Devi, V. (1990). *Atlas of Ozone Spectral Parameters from Microwave to Medium Infrared*. Academic Press, Cambridge.

Gordon, I.E., Rothman, L.S., Hill, C., Kochanov, R.V., Tan, Y., Bernath, P.F., Birk, M., Boudon, V., Campargue, A., Chance K.V. et al. (2017). The HITRAN2016 molecular spectroscopic database. *J. Quant. Spectrosc. Radiat. Transfer*, 203, 3–69. doi:/10.1016/j.jqsrt.2017.06.038.

Hartmann, J.-M., Boulet, C., Robert, D. (2008). *Collisional Effects on Molecular Spectra: Laboratory Experiments and Model Consequences for Applications*. Elsevier, Amsterdam.

Jacquinet-Husson, N., Armante, R., Scott, N.A., Chédin, A., Crépeau, L., Boutammine, C., Bouhdaoui, A., Crevoisier, C., Capelle, V., Boonne, C. et al. (2016). The 2015 edition of the GEISA spectroscopic database. *J. Mol. Spectrosc.*, 327(31–72). doi:10.1016/j.jms.2016.06.007.

Laraia, A.L., Gamache, R.R., Lamouroux, J., Gordon, I.E., Rothman, L.S. (2011). Total internal partition sums to support planetary remote sensing. *Icarus*, 215(391–400). doi:10.1016/j.icarus.2011.06.004.

5. See: https://en.wikipedia.org/wiki/Atmospheric_radiative_transfer_codes.

6. See: https://ara.lmd.polytechnique.fr.

7. See: http://rtweb.aer.com/lblrtm.html.

8. See: https://nwp-saf.eumetsat.int/site/software/rttov/download.

Saunders, R.W., Matricardi, M., Brunel, P. (1999). An improved fast radiative transfer model for assimilation of satellite radiance observations. *Quarterly J. Royal Meteorological Soc.*, 125, 1407–1425. doi:10.1256/smsqj.55614.

Scott, N.A. and Chedin, A. (1981). A fast line-by-line method for atmospheric absorption computations: The automatized atmospheric absorption atlas. *J. Appl. Meteorol.*, 20, 802–812. doi:10.1175/1520-0450.

Appendix 3

List of Sensors and Satellites Cited in the Book

Here, we list the names of the instruments cited in the book, as well as their platforms (or in some cases associated missions), and the chapters in which they are mentioned. The names of these instruments are generally acronyms that are developed here. The characteristics of these instruments and satellites are given accurately on the WMO OSCAR site[1].

Instrument	Acronym	Satellite	Chapter(s) (1) or (2) denotes volume number
3MI	Multi-viewing Multi-channel Multi-polarization Imager	MetOp-SG-A	**(1)** 6, 9
AATSR	Advanced Along-Track Scanning Radiometer	ENVISAT	**(2)** 3, 4, 6, 9, 10
ABI	Advanced Baseline Imager	GOES-16 to -18	**(1)** 1, 2 **(2)** 4, 6
ACE-FTS	Atmospheric Chemistry Experiment – Fourier Transform Spectrometer	SCISAT	**(2)** 11, 13
AGRI	Advanced Geostationary Radiation Imager	FY-4A and B	**(2)** 4, 6

1. https://space.oscar.wmo.int/.

Satellites for Atmospheric Sciences 1,
coordinated by Thierry PHULPIN, Didier RENAUT, Hervé ROQUET and Claude CAMY-PEYRET.

AHI	Advanced Himawari Imager	Himawari-8 and -9	**(1)** 1 **(2)** 4
AIRS	Atmospheric Infrared Sounder	Aqua	**(1)** 3, 5 **(2)** 1, 4, 7, 9, 13, 14
AMI	Active Microwave Imager	ERS-1 and -2	**(1)** 4, 5 **(2)** 3, 4, 6
AMSR	Advanced Scanning Microwave Radiometer	ADEOS-2	**(2)** 4, 14
AMSR-E	Advanced Scanning Radiometer for EOS	Aqua	**(1)** 3, 9 **(2)** 3, 4, 6, 14
AMSR2	Advanced Scanning Microwave Radiometer – 2	GCOM-W	**(2)** 3, 4, 6
AMSU (-A/-B)	Advanced Microwave Sounding Unit (-A/-B)	NOAA-15 to -19, MetOp-A to -C, Aqua	**(1)** 1, 2, 3, 4, 5, 9 **(2)** 1, 3, 4, 6
APT	Automatic Picture Transmission	ESSA-2, -4, -6, and -8, Nimbus-1 and -2, TIROS-8, NOAA-1	**(1)** 3
ASAR	Advanced Synthetic Aperture Radar	ENVISAT	**(1)** 4 **(2)** 6
ASCAT	Advanced SCATterometer	MetOp-A to C	**(1)** 4, 5 **(2)** 3, 4, 6
ASCENDS	Active Sensing of CO2 Emissions over Nights, Days, and Seasons		**(2)** 10
ASTER	Advanced Spaceborne Thermal Emission and Reflection radiometer	Terra	**(1)** 3
ATLID	ATmospheric LIDar	EarthCARE	**(2)** 8
ATMS	Advanced Technology Microwave Sounder	SNPP, NOAA-20	**(1)** 1, 2 **(2)** 4, 6, 9, 12
ATOVS	Advanced TIROS Operational Vertical Sounder: instrument suite consisting of HIRS and AMSU (-A/-B) or MHS instruments	NOAA-15 to -19, MetOp-A and B	**(1)** 2, 5
ATSR-2	Along-Track Scanning Radiometer – 2	ERS-2	**(1)** 4 **(2)** 8, 9, 10

AVCS	Advanced Vidicon Camera System	ESSA-3, ESSA-5, ESSA-7, ESSAA-9, Nimbus-1 and -2, NOAA-1	**(1)** 3
AVHRR	Advanced Very High Resolution Radiometer	TIROS-N, NOAA-6 to -19, MetOp-A to -C	**(1)** 1, 2, 4, 5, 7, 8 **(2)** 9, 10, 14
BUV	Backscatter Ultraviolet Spectrometer	Nimbus-4	**(1)** 3, 4
CALIOP	Cloud-Aerosol LIdar with Orthogonal Polarization	CALIPSO	**(1)** 9 **(2)** 9, 10, 14
CERES	Clouds and Earth's Radiant Energy System	TRMM, Terra, Aqua, SNPP, NOAA-20	**(1)** 2, 3 **(2)** 14
CLAES			**(2)** 11
CPR	Cloud Profiling Radar for CloudSat	CloudSat	**(2)** 6, 15
CO2M	Copernicus CO2 Monitoring	Sentinel 7	**(2)** 13
CrIS	Cross-track Infrared Sounder	SNPP, NOAA-20	**(1)** 1, 2, 4, 7, 9 **(2)** 1, 4, 7, 9, 10, 14
CZCS	Coastal Zone Color Scanner	Nimbus-7	**(1)** 1, 3
DDMI	Delay Doppler Mapping Instrument	CYGNSS	**(2)** 3, 6
DORIS	Doppler Orbitography and Radiopositioning Integrated by Satellite	Cryosat, Cryosat-2, HY-2A to D, JASON-1 to -3, SARAL, Sentinel-3A and B, SPOT-2 to -5, TOPEX-Poseidon, Sentinel-6A	**(1)** 3, 4
DPR	Dual-frequency Precipitation Radar	GPM Core Observatory	**(2)** 6, 15
EPIC	Earth Polychromatic Imaging Camera	DSCOVR	**(2)** 4, 9, 13, 14
ERB	Earth Radiation Budget	Nimbus-6 and -7	**(1)** 2, 3 **(2)** 14

ERBE	Earth Radiation Budget Experiment	NOAA-9 and -10	**(1)** 2, 3 **(2)** 14
ESMR	Electrically Scanning Microwave Radiometer	Nimbus-5 and -6	**(1)** 1, 3
ETM+	Enhanced Thematic Mapper+	Landsat-7	**(1)** 3
FCI	Flexible Combined Imager	MTG-I	**(1)** 1, 5
FLORIS	FLuORescence Imaging Spectrometer	FLEX	**(2)** 3
GEOCARB			
GEMS	Geostationary Environment Monitoring Spectrometer	GEO-KOMPSAT-2B	**(1)** 4, 6 **(2)** 8
GERB	Geostationary Earth Radiation Budget	Meteosat-8 to -11	**(1)** 4, 5 **(2)** 14
GIIRS	Geostationary Interferometric Infrared Sounder	FY-4A and B	**(1)** 2 **(2)** 1, 4
GLM	Geostationary Lightning Imager	GOES-16 to -18	**(1)** 1, 2
GMI	GPM Microwave Imager	GPM Core Observatory	**(2)** 3, 4, 14, 15
GOME	Global Ozone Monitoring Experiment	ERS-2	**(1)** 4, 9 **(2)** 7, 8, 9, 14
GOME-2	Global Ozone Monitoring Experiment – 2	MetOp-A to -C	**(1)** 4, 5, 9 **(2)** 7, 8, 9, 14
GOMOS	Global Ozone Monitoring by Occultation of Stars	ENVISAT	**(1)** 4 **(2)** 11
GPS/MET	Global Positioning System/Meteorology	Orbview-1	**(2)** 14
GRAS	GNSS Receiver for Atmospheric Sounding	MetOp-A to -C	**(1)** 4, 5
HALOE	HALogen Occultation Experiment	UARS	**(2)** 8, 11
HES			**(1)** 2
HIRS(/2, 3, and 4)	High Resolution Infrared Sounder	Nimbus-6, TIROS-N, NOAA-6 to -19, MetOp-A and -B	**(1)** 1, 2, 3, 4, 5, 9 **(2)** 4, 3, 14
HIS	Hyperspectral Imaging Sensor		**(2)** 4, 14

HRIR	High Resolution Infrared Radiometer	Nimbus-1 to -3	**(1)** 3
HSB	Humidity Sounder for Brazil	Aqua	**(1)** 3
HIRA	Chinese Infrared Sounder	FY-3D-3E	**(2)** 11
HIRDLS	HIgh Resolution Dynamics Limb Sounder	Aura	**(1)** 3
(H)SCAT	SCATterometer	HY-2A, B, and D	**(2)** 3, 4, 14
IASI	Infrared Atmospheric Sounding Interferometer	MetOp-A to -C	**(1)** 2, 4, 5, 6, 7, 9, 10 **(2)** 1, 4, 7, 8, 9, 10, 11, 13, 14
IASI-NG	Infrared Atmospheric Sounding Interferometer – New Generation	MetOp-SG-A	**(1)** 4, 5, 6, 9 **(2)** 1, 13
ICI	Ice Cloud Imager	MetOp-SG-B	**(2)** 3
IDCS	Image Dissector Camera System	Nimbus-3 and -4	**(1)** 3
IKPA	Russian Advanced Infrared ATmopsheric Sounder		**(2)** 1
IMG	Interferometric Monitor for Greenhouse	ADEOS	**(2)** 8
InSAR	Interferometric SAR		**(1)** 9
IPDA lidar	Integrated Path Differential-Absorption lidar	MERLIN	**(2)** 10
IRIS	Infra-Red Interferometer Spectrometer	Nimbus-3 and -4	**(1)** 1, 3 **(2)** 4
IRS	InfraRed Sounder	MTG-S	**(1)** 4, 5, 6, 9 **(2)** 1
ITPR	Infrared Temperature Profile Radiometer	Nimbus-5	**(1)** 3 **(2)** 4
JAMI	Japanese Advanced Meteorological Imager	Himawari-6	**(1)** 1 **(2)** 4
JMR	JASON Microwave Radiometer	JASON-1	**(1)** 3
LI	Lightning Imager	MTG-I	**(1)** 4, 5
LIMS	Limb Infrared Monitor of the Stratosphere	Nimbus-7	**(1)** 3 **(2)** 4
LITE	Lidar In-space Technology Experiment		**(2)** 14

LRA			**(1)** 3
LRIR	Limb Radiance Inversion Radiometer	Nimbus-6	**(1)** 1, 3 **(2)** 4
MERIS	MEdium Resolution Imaging Spectrometer	ENVISAT	**(1)** 4, 9
Merlin		CNES/DLR	**(2)** 10, 13
METimage	METeorological imager	MetOp-SG-A	**(1)** 5
MHS	Microwave Humidity Sounder	NOAA-18 and -19, MetOp-A to -C	**(1)** 2, 4, 5, 7, 9 **(2)** 1, 4, 6, 15
MI	Meteorological Imager	COMS	**(2)** 4
Microcarb	Micro satellite Carbone	CNES	**(2)** 10, 13
MIPAS	Michelson Interferometer for Passive Atmospheric Sounding	ENVISAT	**(1)** 4 **(2)** 11
MIRAS	Microwave Imaging Radiometer using Aperture Synthesis	SMOS	**(2)** 3, 6
MISR	Multi-angle Imaging Spectro-Radiometer	Terra	**(1)** 3 **(2)** 8, 9, 14
MLS	Microwave Limb Sounder	Aura	**(2)** 3, 8, 11
MODIS	MODerate resolution Imaging Spectroradiometer	Terra, Aqua	**(1)** 2, 3, 9 **(2)** 3, 8, 9, 10, 14
MOPITT	Measurements Of Pollution In The Troposphere	Terra	**(1)** 3, 4
MRIR	Medium Resolution Infrared Radiometer	Nimbus-2 and -3, TIROS-2 to -7	**(1)** 3
MSSCC	Multi-color Spin Scan Cloud Camera	ATS-3	**(1)** 1, 2
MSU	Microwave Sounding Unit	TIROS-N, NOAA-6 to -14	**(1)** 1, 2 **(2)** 4, 6
MTVZA-GY	Imaging/Sounding Microwave Radiometer – improved	Meteor-M N1, Meteor-M N2(-1 to -6)	**(2)** 4, 6
MUSE	Monitor of Ultraviolet Solar Energy	Nimbus-3 and -4	**(1)** 3
MVIRI	Meteosat Visible Infrared Imager	Meteosat-1 to -7	**(1)** 1, 4, 5, 6 **(2)** 14

MVISR	Multichannel Visible Infrared Scanning Radiometer	FY-1A to D	**(1)** 1
MWHS (-1/-2)	MicroWave Humidity Sounder (-1/-2)	FY-3A to E	**(2)** 4
MWR	MicroWave Radiometer	ENVISAT	**(1)** 4
MWRI	MicroWave Radiation Imager	FY-3A to D	**(2)** 4, 6
NEMS	Nimbus-E Microwave Sounder	Nimbus-5	**(1)** 3
NSCAT	NASA SCATterometer	ADEOS	**(2)** 3
OCO	Orbiting Carbon Observatory	OCO-2	**(1)** 9 **(2)** 9, 10, 13
OLS	Operational Linescan System	DMSP-F01 to 19	**(1)** 2
OMI	Ozone Monitoring Instrument	Aura	**(1)** 3, 4 **(2)** 7, 8, 9
OMPS-limb	Ozone Mapping and Profiler Suite	SNPP	**(2)** 9, 10, 14
OMPS-nadir	Ozone Mapping and Profiler Suite	SNPP, NOAA-20	**(1)** 1, 2 **(2)** 9, 10
OSCAT	Oceansat SCATterometer	OceanSat-2, ScatSat-1	**(2)** 3, 4, 6, 10
PARASOL	Polarization and Anisotropy of Reflectances at the top of the Atmosphere coupled with an Observation Satellite carrying a Lidar	CALIPSO	**(2)** 8
PMR	Pressure Modulator Radiometer	Nimbus-6	**(1)** 3 **(2)** 4
POLDER	Polarization and Directionality of Earth Reflectance	Adeos-I and Adeos-II, PARASOL	**(1)** 6, 9 **(2)** 8, 14
Poseidon (-2, -3, and -4)	Radar altimeter	TOPEX-Poseidon, JASON-1 to -3, Sentinel-6A and B	**(1)** 3
PR	Precipitation Radar	TRMM	**(2)** 6, 15
RA	Radar Altimeter	ERS-1 and -2	**(1)** 4
RapidScat	Rapid Scatterometer on the ISS	ISS	**(2)** 4

RFSCAT	Rotating Fan-beam SCATterometer	CFOSAT	**(2)** 3, 4
ROSA	Radio Occultation Sounder for Atmosphere	Megha-Tropiques	**(1)** 8
SAM II	Stratospheric Aerosol Measurement – II	Nimbus-7	**(1)** 3
SAGE	Stratospheric Aerosol and Gas Experiment		**(2)** 8, 9, 10
SAGE-II SAGE-II		ERBS	**(2)** 9, 10
SAGE-III/ISS SAGE-III		ISS	**(2)** 9, 10
SAMS	Stratospheric And Mesospheric Sounder	Nimbus-7	**(1)** 3 **(2)** 4
SAPHIR	Atmospheric Sounder of the Intertropical Moisture Profile by Radiometry	Megha-Tropiques	**(1)** 6, 8 **(2)** 4, 6, 9, 14, 15
SAR	Synthetic Aperture Radar	RadarSat-1 and -2	**(2)** 6
SASS	SeaSat-A Scatterometer System	SeaSat	**(2)** 3
SBUV	Solar Backscatter UltraViolet	Nimbus-7, NOAA-9 to -11, NOAA-13 and -14, NOAA-16 to -19	**(1)** 2, 3 **(2)** 8
SCAMS	SCAnning Microwave Spectrometer	Nimbus-6	**(1)** 3 **(2)** 4
ScaRaB	Scanner for Radiation Budget	Meteor-3-6, Resurs-O1-4, Megha-Tropiques	**(1)** 6 **(2)** 14
SCIAMACHY	Scanning Imaging Absorption spectrometer for Atmospheric CHartographY	ENVISAT	**(1)** 4, 5 **(2)** 4, 7, 8, 9, 10, 11, 13
SCMR	Surface Composition Mapping Radiometer	Nimbus-5	**(1)** 3
SCR	Selective Chopper Radiometer	Nimbus-4 and -5	**(1)** 3 **(2)** 4
SeaWIFS	Sea-viewing WIde Field-of-view Sensor	OrbView-2	**(1)** 3

SeaWinds	SeaWinds	ADEOS-2 QuikSCAT	**(1)** 3 **(2)** 4, 6
Sentinel-4	–	MTG-S	**(1)** 1, 4, 5, 6
Sentinel-5	–	MetOp-SG-A	**(1)** 4, 5
SEVIRI	Spin-stabilized Enhanced Visible and InfraRed Imager	Meteosat-8 to -11	**(1)** 1, 4, 5, 9 **(2)** 1, 3, 4, 9
SIFTI	Static Infrared Fourier Transform Interferometer	TRAQ (remained in draft form)	**(1)** 6
SIRAL	SAR Interferometer Radar ALtimeter	CryoSat, CryoSat-2	**(2)** 3
SIRS	Satellite Infra-Red Spectrometer	Nimbus-3 and -4	**(1)** 3 **(2)** 4
SLSTR	Sea and Land Surface Temperature Radiometer	Sentinel-3A and B	**(2)** 3, 8, 9, 10
SMAP	Soil Moisture Active-Passive	SMAP	**(2)** 3, 6
SMMR	Scanning Multichannel Microwave Radiometer	Nimbus-7, SeaSat	**(1)** 1, 3 **(2)** 3, 6
SSCC	Spin Scan Cloud Camera	ATS-1	**(1)** 1, 2
SSM (/T/T-2)	Special Sensor Microwave (Temperature/Humidity)	DMSP-F04, DMSP-F07 to 15	**(2)** 4
SSM/I	Special Sensor Microwave – Imager	DMSP-F08, DMSP-F10 to 15	**(2)** 3, 6, 14
SSMIS	Special Sensor Microwave Imager/Sounder	DMSP-F16 to 19	**(1)** 9 **(2)** 3, 4, 6
SSU	Stratospheric Sounding Unit	TIROS-N, NOAA-6 to -9, NOAA-11 and -14	**(1)** 1, 2 **(2)** 4
S-VISSR	Stretched Visible and Infrared Spin Scan Radiometer	FY-2A to H	**(1)** 1
TANSO-FTS	Thermal and Near Infrared Sensor for Carbon Observation – Fourier Transform Spectrometer	GOSAT 1, 2, 3	**(2)** 13

TEMPO			**(1)** 4 **(2)** 8
TES	Tropospheric Emission Spectrometer	Aura	**(1)** 3, 4 **(2)** 4, 8, 10, 13
THIR	Temperature-Humidity Infrared Radiometer	Nimbus-4 to -7	**(1)** 3
TMI	TRMM Microwave Imager	TRMM	**(2)** 14, 15
TOMS	Total Ozone Mapping Spectrometer	TOMS Earth Probe	**(1)** 1, 3 **(2)** 8, 9
TOVS	TIROS Operational Vertical Sounder: instrument suite consisting of HIRS, SSU and MSU	TIROS-N, NOAA-6 to -14	**(1)** 1, 2 **(2)** 4, 14
TROPOMI	TROPOspheric Monitoring Instrument	Sentinel-5 Precursor	**(1)** 4, 9 **(2)** 7, 9, 10, 13
TRSR	Turbo Rogue Space Receiver	GFO, ICESat, ICESat-2, JASON-1 to -3, Ørsted	**(1)** 3
UVNS	Ultraviolet Visible Near Infrared Spectrometer	MetOp SG-A 1, 2, 3	**(2)** 13
VAS	Vertical Atmospheric Sounder	GOES-4 to -7	**(1)** 1, 2
VHRR	Very High Resolution Radiometer	NOAA-2 to -5	**(1)** 1
VIIRS	Visible Infrared Imaging Radiometer Suite	SNPP, NOAA-20	**(1)** 1, 2 **(2)** 9, 10
VISSR	Visible Infrared Spin-Scan Radiometer	Himawari-1 to -4	**(1)** 1, 2 **(2)** 4
VTPR	Vertical Temperature Profiling Radiometer	NOAA-2 to -5	**(1)** 1 **(2)** 4
WindSat	WindSat	Coriolis	**(2)** 3, 6

List of satellites or missions in the field of the atmosphere

Satellite	Acronym	Agency
A-Train	Afternoon train of satellites	NASA
ADEOS	Advanced Earth Observing Satellite	JAXA

ADL-AEOLUS	Satellite mission for atmospheric wind	ESA
ALOS	Advanced Earth observation satellite	JAXA
AQUA	Afternoon train of satellites	NASA
ARKIKA-M1	Satellites to monitor Earth's atmosphere and surface in the Arctic	ROSCoSMOS
ATS (-1, -3, and -6)	Application Technology Satellite	NASA
BIOMASS	Seventh ESA Earth Explorer Mission (P-band radar)	ESA
CALIPSO	Cloud-Aerosol Lidar and Infrared Pathfinder Satellite Observation	NASA/CNES
CBERS	China–Brazil Earth Resources Satellite program	INPE
CFOSAT	Chinese-French Ocean Satellite	CMA/CNES
CLARREO	Climate Absolute Radiance and Refractivity Observatory	NASA
COSMIC	Constellation Observing System for Meteorology, Ionosphere and Climate	NOAA
CryoSat	Cryosphere Satellite	ESA
Earthcare	Earth observation satellite for polar sheet study	ESA
ENVISAT	Environmental Satellite	ESA
ERBS	Earth Radiation Budget Satellite	NASA
ERS	Earth Resources Satellite	ESA
FORMOSAT	Earth observation satellite	Taiwan
FORUM	Far-infrared Outgoing Radiation Understanding and Monitoring mission	ESA
FY Feng-Yun	CMA satellites	CMA
GCOM-W	Global Change Observation	JAXA
GeoCARB	Geostationary CARBon cycle observatory	NASA
GHGsat	GreenHouse Gas satellite	
GOES	Geostationary Operational Environmental Satellites	NOAA
GOSAT	Greenhouse gases Observing SATellite	JAXA
GOES	Geostationary Orbit Environmental Satellite-(East/West)	NOAA
GPM	Global Precipitation Mission	NASA
GRACE	Gravity Recovery And Climate Experiment	NASA/DLR

HY-XX Hai Yang XX	Chinese ocean satellite missions	CNSA
ICESat	Ice, Clouds and land Elevation Satellite	NASA
ISS	International Space Station	
JPSS	Joint Polar Satellite System	NOAA/NASA/DOD
LANDSAT	Earth observation satellites for high resolution imagery	NASA
MEGHA-TROPIQUES	Low inclination orbiting satellite to study tropical atmosphere	CNES/ISRO
MERLIN	Methane Remote sensing lidar mission	CNES/DLR
METEOSAT	Geostationary satellites for operational meteorology	EUMETSAT
METOP	Polar orbiting satellites for operational meteorology	EUMETSAT
METOP-SG	Polar orbiting satellites for operational meteorology (Second Generation)	EUMETSAT
Microcarb	Micro satellite for carbon dioxide monitoring	CNES
MSG	Meteosat Second Generation	EUMETSAT
MTG	Meteosat Third Generation	EUMETSAT
MTP	Meteosat Transition Programme	EUMETSAT
MTSAT	Multi-functional Transport Satellite	JMA
NIMBUS		NASA
NOAA-1 to -20		NOAA
NPOESS	National Polar-orbiting Operational Environmental Satellite System	NOAA
NPP	National Polar-orbiting Partnership	NOAA
OCO	Orbiting Carbon Observatory	NASA
PLEIADES	PLE	CNES
POES	Polar Orbiting Environmental Satellite	NOAA
QSAT	Water vapor (“q”) from Satellite	
ResursSat		ISRO
S5P	Sentinel 5 Precursor	ESA
S3	Sentinel 3	ESA
S4	Sentinel 4	ESA
S5	Sentinel 5	ESA
SCARBO	Space Carbon Observatory	
SPOT		CNES

Suomi-NPP		NOAA
TERRA		NASA
TIROS	Television InfraRed Observation Satellite	NOAA
TRMM	Tropical Rainfall Measuring Mission	NASA
UARS	Upper Atmosphere Research Satellite	NASA

Appendix 4

Microwave Sensors and Infrared Hyperspectral Sounders

The following tables present a non-exhaustive list of past, current or future instruments[1] in the microwave domain or hyperspectral sounders in the infrared domain.

A4.1. Non-exhaustive list of passive microwave instruments

The type 1 instrument is for across track scanning instruments and type 2 is for conical scanning instruments.

Acronym	Instrument	Agency	Type	Satellite	Use of data	
AMR	Advanced Microwave Radiometer	NASA	1	JASON-2 JASON-3	2008	2021
AMSU-A	Advanced Microwave Sounding Unit – A	NOAA	1	Aqua MetOp-A–19	1998	2027
AMSU-B	Advanced Microwave Sounding Unit – B	NASA	1	NOAA-15–17	1998	2021

1. An overview of satellites and instruments is available at: https://space.oscar. wmo.int/.

Satellites for Atmospheric Sciences 1,
coordinated by Thierry PHULPIN, Didier RENAUT, Hervé ROQUET and Claude CAMY-PEYRET.

Aquarius	Aquarius	NASA	1	SAC-D	2011	2015
ATMS	Advanced Technology Microwave Sounder	NASA	1	JPSS-2–4 NOAA-20 SNPP	2011	2039
HSB	Humidity Sounder for Brazil	INPE	1	Aqua	2002	2023
JMR	JASON Microwave Radiometer	NASA	1	JASON-1	2001	2013
MHS	Microwave Humidity Sounding	EUMETSAT	1	MetOp-A–C NOAA-18–19	2005	2027
MIRAS	Microwave Imaging Radiometer using Aperture Synthesis	ESA	1	SMOS	2009	2021
MSR	Microwave Scanning Radiometer	JAXA	1	MOS-1 MOS-1B	1987	1996
MSU	Microwave Sounding Unit	NOAA	1	NOAA-6–14	1978	2007
MW radiometer	MicroWave radiometer	NASA	1	SWOT	2022	2026
MWHS-1	MicroWave Humidity Sounder – 1	CMA	1	FY-3A FY-3B	2008	2021
MWHS-2	MicroWave Humidity Sounder – 2	CMA	1	FY-3C–I	2013	2030
MWR	MicroWave Radiometer	ESA	1	Sentinel-3A–3D	2016	2032
MWR	MicroWave Radiometer	CONAE	1	SAC-D	2011	2015
MWTS-1	MicroWave Temperature Sounder – 1	CMA	1	FY-3A FY-3B	2008	2021
MWTS-2	MicroWave Temperature Sounder – 2	CMA	1	FY-3C FY-3D	2013	2022

SAPHIR	*Sondeur Atmospherique du Profil d'Humidite Intertropicale par Radiometrie*	CNES	1	Megha-Tropiques	2011	2021
AMSR2	Advanced Microwave Scanning Radiometer – 2	JAXA	2	GCOM-W	2012	2022
AMSR-E	Advanced Microwave Scanning Radiometer for EOS	JAXA	2	Aqua	2002	2023
GMI	GPM Microwave Imager	NASA	2	GPM Core Observatory	2014	2026

A4.2. Non-exhaustive list of active microwave instruments

Acronym	Instrument	Agency	Type	Satellite	Use of data	
AMI-SCAT	Active Microwave Instrument – SCATterometer	ESA	Radar scatterometer	ERS-1 ERS-2	1991	2011
ASCAT	Advanced SCATterometer	ESA	Radar scatterometer	MetOp-A MetOp-B MetOp-C	2006	2027
NSCAT	NASA SCATterometer	NASA	Radar scatterometer	ADEOS	1996	1997
SCA	SCATterometer	ESA	Radar scatterometer	MetOp-SG-B1 MetOp-SG-B2 MetOp-SG-B3	2025	2045
OSCAT	OceanSat SCATterometer	ISRO	Radar scatterometer	OceanSat-2 OceanSat-3 OceanSat-3A ScatSat-1	2009	2028
SeaWinds	SeaWinds	NASA	Radar scatterometer	ADEOS-2 QuikSCAT	1999	2009
WindRAD	Wind RADar	CMA	Radar scatterometer	FY-3E FY-3I	2021	2030

ALT	Radar ALTimeter	NASA	Radar altimeter	SeaSat	1978	1978
ALT	Radar ALTimeter	NSOAS	Radar altimeter	HY-2C HY-2F HY-2G	2020	2030
AltiKa	Ka-band Altimeter	CNES	Radar altimeter	SARAL	2013	2021
Altimeter	Altimeter	NASA	Radar altimeter	SWOT	2022	2026
KaRIN	Ka-band Radar INterferometer	NASA	Radar altimeter	SWOT	2022	2026
SIRAL	SAR Interferometer Radar ALtimeter	ESA	Radar altimeter	CryoSat Cryosat-2	2005	2024
SRAL	Synthetic aperture Radar ALtimeter	ESA	Radar altimeter	Sentinel-3A Sentinel-3B Sentinel-3C Sentinel-3D	2016	2032
SWIM	Surface Waves Investigation and Monitoring instrument	CNES	Radar altimeter	CFOSAT CFOSAT follow-on	2018	2027
Poseidon-4	Poseidon-4	ESA	Radar altimeter	Sentinel-6A Sentinel-6B	2020	2032
CPR	Cloud Profiling Radar for Earth-CARE	JAXA	Cloud and precipitation radar	EarthCARE	2023	2026
DPR	Dual-frequency Precipitation Radar	JAXA	Cloud and precipitation radar	GPM Core Observatory	2014	2026
PR	Precipitation Radar	JAXA	Cloud and precipitation radar	TRMM	1997	2015
Rainradar	Rainradar	CMA	Cloud and precipitation radar	FY-3G FY-3J	2023	2030
RainCube	Radar In a CubeSat	NASA	Cloud and precipitation radar	ISS RainCube	2018	2020

A4.3. Non-exhaustive list of hyperspectral sounders

Name	Satellite	Agency	Pixel size in nadir	Spectral range in cm^{-1}	Spectral resolution in cm^{-1}	Radiometric noise NeDT @ 280 K
AIRS	Aqua	NASA	14 km	650–1,136 1,216–1,613 2,170–2,674	0.5 1.0 2.0	0.35 K 0.20 K 0.14 K
IASI	MetOp-A MetOp-B MetOp-C	CNES EUMETSAT	12 km	645–2,670	0.25 unmodified	0.2–0.5 (2) K
CrIS	SNPP NOAA-20 JPSS-2 JPSS-3 JPSS-4	NASA	14 km	650–1,095 1,210–1,750 2,155–2,550	0.625 1.25 2.5	0.24–0.39 K 0.19–0.28 K 0.18–0.21 K
IKFS	Meteor-M N2 Meteor-M N2-2 Meteor-M N2-3,4,5,6	Roscosmos	35 km	665–2,000	0.5	0.5 K
HIRAS	FY-3E FY-3F FY-3H FY-3I	CMA	16 km	650–1,136 1,210–1,750 2,155–2,550	0.625 1.25 2.5	0.2–0.8 K @ 250K 0.2–0.3 K @ 250 K 0.3–0.5 K @ 250 K
GIIRS	FY-4A FY-4B FY-4C, D, E, F, G	CMA	16 km 12 km	680–1,130 1,650–2,250	0.625 0.625	0.5 mW/m2/sr/cm-1 0.1 $mW/m^2/sr/cm^{-1}$
IRS	MTG-S-1 MTG-S-2	EUMETSAT	4 km	700–1,210 1,600–2,175	0.625	
IASI-NG	MetOp-SG-A1 MetOp-SG-A2 MetOp-SG-A3	CNES EUMETSAT	12 km	645–2,760	0.125	0.13–0.65 (1.43) K

Glossary

This glossary presents the main categories of instruments on board satellites and used in meteorology and climate science. The definitions are taken from the glossary established by Didier Renaut and published in *La Météorologie* in May 2017[1]. This glossary also offers definitions for some generic terms used to characterize satellite measurements.

Active sensor: a space instrument that irradiates a scene with a man-made electromagnetic radiation and measures part of the radiation reflected, backscattered or re-emitted by this same scene. Lidars and radars are examples of active sensors. Thanks to the emission of radiation in the form of pulses, the active sensor can provide a range-resolved measurement in the medium probed or an accurate measurement of the altitude of the surface.

Calibration: an operation to convert the magnitude of the output of a measuring instrument (e.g. the level of mercury in a thermometer or the detected backscatter power of a meteorological radar) in the magnitude of the physical input variable (e.g. the temperature or radar reflectivity) actuating that instrument.

Climate data record: a time series of measurements of sufficient length, consistency and continuity to determine climate variability and climate change.

1. https://lameteorologie.fr/issues/2017/97.

Satellites for Atmospheric Sciences 1,
coordinated by Thierry PHULPIN, Didier RENAUT, Hervé ROQUET and Claude CAMY-PEYRET.

Cloud fraction (or nebulosity): the Cloud cover or cloudiness defines the cloud-covered fraction of a geographic area. It is expressed as a percentage of the total area (between 0 and 100) or as a real number (between 0.0 and 1.0).

Column content (of a gas): number per unit area of molecular species or aerosol particles in a given atmospheric path. The column can be slant (SCD = slant column density) or vertical (VCD = vertical column density). A column can be partial (between two atmospheric levels) or total (between the surface and the top of the atmosphere or TOA).

The unit is molecule cm^{-2} or mol m^{-2}. For O_3 and SO_2, columns can also be expressed in Dobson unit (DU). 1 DU is equivalent to $2.68678011 \times 10^{20}$ molecule m^{-2}. The constant comes from the Loschmidt constant $2.68678011 \times 10^{25}$ molecule m^{-3} (exact number in SI), which is the number density of an ideal gas in standard conditions of temperature $T0 = 273.15$ K and pressure $P0 = 1013.25$ hPa.

The factor $10^{-5} = 10^{-3} \times 10^{-2}$ corresponds to an atmospheric path of one milli-atmosphere-cm. A standard vertical total column of ozone is 300 DU.

Constellation: in the spatial context, a constellation is a set of artificial satellites performing identical functions and distributed in such a way as to ensure their common mission at all times. A constellation can be made up of numerous orbiting satellites in low orbit or in medium orbit to carry out, for example, telecommunications, satellite navigation missions or observation. Observation of the atmosphere for meteorological purposes relies, among other things, on a constellation of operational geostationary satellites and on a constellation of polar-orbiting Sun-synchronous satellites.

Convective system: the cloudy zone of the atmosphere in which convective movements take place bringing warm and moist air from the surface to the top of the atmosphere. A mesoscale convective system (MCS) is a set of thunderstorms gathering together into clusters or lines over time.

Data assimilation: a mathematical method for defining the initial conditions of numerical prediction models by achieving an optimal combination (in the statistical sense) between observations and a previous short-range forecast.

Emissivity: the emissivity of the surface of a material is its effectiveness in emitting energy as thermal radiation.

ENSO (El Niño Southern Oscillation): the periodic change in the ocean–atmosphere system in the tropical Pacific that impacts weather patterns worldwide. This happens every 3–7 years (5 years on average) and usually lasts 6–18 months.

Essential climate variable (ECV): a physical, chemical or biological variable or a group of linked variables that contributes critically to the characterization of the Earth's climate.

Evapotranspiration: the sum of water evaporation and transpiration from a surface area (typically vegetated) into the atmosphere.

Footprint: the region of the ground observed (or illuminated if the instrument is active) instantaneously by a space instrument. The footprint is determined by the instantaneous field of view of the receiver for a passive instrument or by the divergence of the emission beam for an active instrument.

Geostationary Earth orbit (GEO): a circular orbit around the Earth characterized by zero orbital inclination (i.e. an orbit in the equatorial plane), and an orbital period equal to the Earth's rotational period. The sub-satellite point of a geostationary orbit always remains the same point on the equator. The geostationary orbit around the Earth is at an altitude of 35,786 km above the Earth's geoid; we commonly speak of satellites at 36,000 km.

GNSS radio occultation: a semi-passive space measurement technique based on the refraction by the atmosphere of radio waves emitted by satellites in high orbit in the GNSS system (Global Navigation Satellite System), which includes the American GPS (Global Positioning System), Russian Glonass and European Galileo, which represents in all more than 70 satellites. The receiver, placed on a low-orbiting satellite, detects and measures the radio waves from a GNSS satellite after refraction by their passage tangentially through the Earth's atmosphere. From the measurement of the refraction characterized by the angle of curvature followed by the wave, vertically resolved information is deduced about the temperature, humidity, and pressure of the atmospheric regions crossed, even for cloudy regions.

Ground track: known as a suborbital track, it is the vertical projection of the satellite's orbit onto the surface of the Earth.

Horizontal/vertical spatial resolution: as the measurements provided by a satellite instrument are generally very indirect, it is necessary to distinguish between the spatial resolution of the instrument and the spatial resolution of the measurement. The spatial (horizontal, vertical) resolution of the instrument is the smallest distance (horizontal distance, vertical distance) separating two neighboring targets that the instrument is able to distinguish. The spatial resolution of the measurement is the smallest distance (horizontal distance, vertical distance) separating two points in space for which the final measurements are calculated. For the same instrument, the spatial resolution of the measurement depends on the parameter to be restored. It can be affected by different spatial averaging operations.

Hydrometeor: any type of water or ice particles that have formed in the atmosphere or at the Earth's surface as a result of condensation or solidification. It is the form in which atmospheric water is condensed, suspended, falling or deposited on other objects: atmospheric hydrometeors are therefore cloud droplets, rain, drizzle, snow and hail. It can also refer to water deposits on the ground (dew or ice).

Hyperspectral: a spatial sensor is said to be hyperspectral if it makes its observations in many very narrow spectral bands. It restores the radiation spectrum (emitted, absorbed, scattered) with excellent spectral resolution.

Imaging radiometer: equipped with a scanning system, the radiometer becomes an imager. If its spatial resolution remains modest, it then covers a very large territory (dimensions of around a thousand kilometers). The scanning can be mechanical (in the case of a scanner), with a sequential reading of the image, or electronic thanks to matrices or strips of detectors (in the case of the push broom scanner), with a simultaneous reading of part of the image. In addition, the instrument has a number of spectral channels.

Imaging spectrometer: equipped with a scanning system, the spectrometer becomes a spectro-imager. High- or medium-spatial-resolution Earth observation instruments with multispectral analysis capabilities belong to this category, such as the Tropomi on Sentinel-5P or the OLCI instrument on Sentinel-3.

Intercalibration (of satellite instruments): involves systematically comparing the measurements acquired by these instruments with those acquired by reference instruments chosen for their accuracy, stability and durability and, if possible, traceable to the International System (SI) of units. The intercalibration of instruments is necessary to ensure the interoperability of their measurements and to allow for the analysis of long-term series of observations for the study of the climate.

Interoperability: the interoperability of observation systems characterizes their ability to deliver observations that can be combined in a coherent way even though they come from different sources. This implies that the sampling protocols are known and compatible, that the measurements are traceable to common references and that the data is processed and formatted according to common or compatible standards.

Kalman filter: a sequential data assimilation method for estimating an optimal state of the environment under study and its error covariance matrix, as well as their propagation over time using a linear model.

Leaf area index: the ratio between the surface area of a leaf and the surface area of the ground it covers. This index thus expresses the vertical density of the plant cover in m^2/m^2.

LEO, MEO, GEO (satellite categories): satellites in circular or near-circular orbit are classified into three categories according to their altitude, traditionally denoted LEO, MEO and GEO (respectively, for Low, Medium and Geostationary Earth Orbit):

– LEO: between 400 and 1,400 km (nowadays, it is instead between 600 and 900 km), with a period of revolution around 100 minutes.
– MEO: between 20,000 and 23,000 km, for navigation satellites (GPS-type), with a period around 12 hours.

– GEO: the altitude, nearly 36,000 km, is accurately determined so that the period of revolution of the satellite is equal to the period of rotation of the Earth, one sidereal day (a little less than 24 hours).

More than 99% of satellites in circular or near-circular orbit fall into one of these three categories. There are many gaps with regard to satellite altitudes. For example, we will not find a satellite with an altitude between 8,000 and 18,000 km.

Lidar: an active instrument using the emission and backscattering of light pulses emitted by a laser to locate and analyze atmospheric targets or the Earth's surface. There are various types of lidar: simple backscatter lidar, Doppler lidar, differential absorption lidar (DIAL), etc. Operating in the optical range, the lidar does not pass through the clouds once they have become slightly thick, unlike the radar (Chapter 13 of Volume 2).

Lidar altimeter: a lidar instrument (active) used for measuring the altitude of the Earth's surface located vertically below the satellite. A lidar altimeter provides topographic measurements of land masses, oceans and polar ice caps. Like all lidars, this altimeter is blind in the presence of thick clouds.

Low Earth orbit (LEO): a family of orbits around the Earth reaching an altitude of up to 2,000 km. Among these are the **polar orbit**, characterized by an orbital inclination close to 90°, allowing the two geographical poles to fly over at each revolution, and the **Sun-synchronous orbit**, whose altitude and inclination are such that the angle between the orbital plane and the direction of the Sun remains almost constant. A satellite placed in such an orbit passes over a given point on the Earth's surface at the same local solar time. As opposed to geostationary satellites, meteorologists often use the term **non-geostationary satellite** for a satellite in low orbit.

Mesoscale: the mesoscale corresponds to the intermediate spatial scale between the planetary scale (known as synoptic), which is that of lows and anticyclones on a continent (> 2,000 km) and the micro-scale which characterizes individual clouds (< a few km). Its horizontal extension extends between a few tens of kilometers up to a few hundred kilometers.

Metadata: information that characterizes data, for example, time and location of a satellite image.

Mixing ratio (of a gas): concentration of gas divided by the concentration of the air at the same place. It is a unitless number, expressed in parts per thousand, per million (ppm), or per billion (ppb).

Multidecadal Atlantic oscillation: periodic variation of the surface temperature of the North Atlantic with characteristic times of several decades, in relation to disturbances in atmospheric and oceanic circulations at the scale of the North Atlantic basin and beyond.

Numerical weather prediction model: computer-based software set of mathematical equations representing an approximation of the behavior of the atmosphere to obtain a simulation of its future states.

Operational: refers to a system or service that meets high performance and availability criteria to provide data almost continuously to critical applications. Distinguishes itself from R&D missions or services intended for technological or scientific demonstration where the exploitation of data can be exploratory and is not subject to real-time constraints.

Outgoing longwave radiation (OLR): radiation (or amount of radiation: flux) emitted into space in the infrared wavelengths at the top of the atmosphere.

Passive sensor: a space instrument that measures part of the electromagnetic radiation naturally reflected, backscattered, or emitted by a scene (reflected solar radiation, thermal infrared radiation emitted, etc.).

Phase 0: the initial phase of a space project during which the mission is expressed in terms of needs, the environment and its constraints are assessed and possible concepts are identified.

Phase 1: the second phase of a space project where the needs are refined and validated. The different possible concepts are explored, and the critical elements are identified and evaluated in order to estimate the associated risks.

Pixel: the pixel is the smallest homogeneous element of an image. Pixels are usually rectangular, sometimes square. We can distinguish the pixels of the detector matrix (of size fixed by construction) from the pixels of the restored image which may or may not be geo-rectified (taking into account the sphericity of the Earth).

Polarimeter: an instrument for measuring different elements of the Stokes vector of the incident radiation. We talk about spectro-polarimeter when it also does spectral analysis (PARASOL, 3MI).

Radar: an active space instrument using the emission and backscatter of microwave pulses to locate and analyze atmospheric targets or the Earth's surface. We often classify the frequency of radars in several bands (the limits of the bands depend somewhat on the sources).

Band L	400 MHz–1.5 GHz	20–75 cm
Band S	1.5–3.9 GHz	7.7–20 cm
Band C	3.9–6.2 GHz	4.8–7.7 cm
Band X	6.2–10.9 GHz	2.7–4.8 cm
Band K	10.9–36 GHz	0.83–2.7 cm
Band Ku	11–14 GHz	2.1–2.7 cm
Band Ka	20–30 GHz	1.0–1.5 cm
Millimetric	40–300 GHz	1–7.5 mm

Radar altimeter: a radar (active) instrument used primarily for measuring the altitude of the Earth's surface located vertically below the satellite. It provides topographic measurements of land and, above all, very precise measurements (within a few centimeters) of the topography of the oceans.

(Radar) Reflectivity: the amount of transmitted power returned to the radar receiver after hitting hydrometeors, compared to a reference power density at a distance of 1 m from the radar antenna.

Radio interferences: disturbances of the natural electromagnetic signals emitted by the Earth's surface and the atmosphere in the microwave range by anthropogenic emissions resulting from means of telecommunications such as mobile telephony.

Radiometer: a passive space instrument whose predominant function is the measurement of an electromagnetic flux. Its spatial resolution is generally modest (from a few hundred meters to a few tens of kilometers). Observation instruments with high spatial resolution (metric) such as

HRG on Spot-5 or the instrument of the Pléiades satellites are instead called "high-spatial-resolution imagers" or "cameras".

Revisit time (or period): the time interval separating two successive sightings of the instrument at the same point in space. This depends both on the orbit of the satellite and on the scanning and swath of the instrument. The inverse quantity, called the revisit frequency or rate, is also often used.

Roughness length: a horizontal scale characterizing surface heterogeneities. It influences the intensity of turbulent transfers between the surfaces and the atmosphere as well as the surface emissivity in the microwave range.

(Satellite) Data level: – level-0 products: telemetry as received from the instrument, after assembly with supplemental information needed for further analysis;

– level-1 products: measurements associated with a temporal and geospatial reference system, possibly with some corrections applied and/or calibration from processing of level-0;

– level-2 products: retrievals in geophysical space, relating to the environment, obtained from level-1;

– level-3 products: spatially and/or temporally complete retrieval data, resampled from level-2;

– level-4 products: indicators, products (as fluxes) or analyses, derived from level-3.

(Satellite) Platform: the subsystem of a satellite that provides services, i.e. all the generic functions necessary for its activity in orbit. The other subsystem is the payload which groups the instruments of the satellite.

Scan, field of view: remote sensing or meteorological satellites have instruments on board that observe the ground and most often take "images" in the visible, infrared or microwave domain. They are taken by the *scanning* of the instrument. In the microwave domain, the two main types of scanning are orthogonal scanning and conical scanning. In the infrared (longwave, shortwave, near-infrared) and VIS/UV, the images are acquired either by a pointing mirror to direct the LoS at the proper position on ground or by integrating (in a short time) the collected radiation along the swath using a 2D detector matrix to acquire a line of the image:

– orthogonal scanning: the instrument takes the images with a scan perpendicular to the ground track, generally symmetrically with respect to the central sight, known as the *nadir*. Seen from the satellite, the angle between aiming at nadir and aiming at a point P is called the *half-aperture angle* f. The extreme aiming angle (right or left), the angle of *maximum half-aperture* or f_0. The field of view is equal to $2\,f_0$;

– conical scanning: the scan of the instrument describes a cone whose central axis has a greater or lesser inclination with respect to the nadir. The rotational movement of the instrument describes a full revolution or fraction of a revolution. The width of the sweep's ground track, the *swath*, determines the field of view.

Scatterometer: a radar (active) instrument used to measure the backscatter coefficient of surfaces with very high precision. **Wind scatterometers** are the most common; with an appropriate choice of radar frequency and multi-directional (scanning) observation, they can measure wind speed and direction just above oceans with a wide swath.

Sounder: a space-based instrument capable of range-resolved measurements along the line of sight. The term "sounder" will be reserved here for passive instruments, even if active instruments such as lidar or radar are also sounders. A distinction is made between limb sounders that target a source of radiation (sun, star, Earth's atmosphere, etc.) tangentially to the Earth, which provides accurate measurements in the upper atmosphere, and vertical sounders, which target the Earth's surface at nadir and are more sensitive to the middle and lower atmosphere. Sounders are in principle spectrometers with very high spectral resolution in order to be able to analyze the absorption and emission lines of atmospheric constituents. It is from this very fine spectral analysis that the capability for altitude resolved measurements is achieved. Sounders are most often equipped with an "imager" function in order to produce maps.

Spectrometer: a passive space instrument whose predominant function is the measurement and analysis of part of the electromagnetic spectrum. There are several types of spectrometers. The most widely used are grating spectrometers or interferometers of different types: Michelson, Mertz, etc.

Specular reflection (sun glint): the property of a surface to redirect an incident radiation beam into a reflected beam at the same angle on the other side of the normal to the surface (like a mirror).

Sub-satellite point (SSP): the sub-satellite point is, at a given instant, the point on the Earth's surface where the straight line joining the centers of mass of the satellite and the Earth passes. It should be noted that this point generally differs from the vertical of the satellite (nadir), through which the straight line perpendicular to the surface of the Earth passes, passing through the satellite and whose continuation defines the track of the satellite.

Synoptic scale: there is a usual classification of atmospheric phenomena according to their spatio-temporal scale. The synoptic scale characterizes phenomena whose size is about a thousand kilometers and with a duration of a few days. For example, this is the typical scale for monitoring weather disturbances.

Synthetic aperture radar (SAR): an imaging (scanning) radar instrument capable of providing high-resolution images of the Earth's surface, day and night and through clouds. The computer processing of successive radar signals coming from the same target makes it possible to obtain a resolution equivalent to that of a very large "synthetic" antenna.

Time resolution: as with spatial resolution, a distinction must be made between the time resolution of the instrument and the time resolution of the measurement. The time resolution of the instrument is identical to the revisit period. The temporal resolution of the measurement is the smallest time interval separating two successive final measurements for the same point in space. For the same instrument, the temporal resolution of the measurement depends on the parameter to be retrieved. It can be affected by different time-averaging operations.

Uncertainty: the standard deviation of a sufficiently large number of measurements of the same quantity by the same instrument or method. Hence, the random part of the instrument reading or output; it represents the limit of measurement precision. The uncertainty of an instrument is caused by the unpredictable effects upon its performance of such factors as friction, backlash, electronic, photonic or detector noise (Chapter 12 of Volume 2).

Virtual constellation: in the context of the CEOS, a "virtual constellation" is a coordinated set of space-based and/or ground-based facilities contributed by different partners, which focus on the observation of a particular parameter or group of parameters of the Earth System.

List of Authors

Sid-Ahmed BOUKABARA
NOAA National Environmental Satellite
Data and Information Services (NESDIS)
Maryland and Wisconsin
USA

Claude CAMY-PEYRET
IPSL
Paris
France

Michel CAPDEROU
Laboratoire de Météorologie Dynamique
(LMD/IPSL)
Ecole polytechnique
Palaiseau
France

Carole DENIEL
CNES
Paris
France

Frank GALLAGHER
NOAA National Environmental Satellite
Data and Information Services (NESDIS)
Maryland and Wisconsin
USA

Mitch GOLDBERG
NOAA National Environmental Satellite
Data and Information Services (NESDIS)
Maryland and Wisconsin
USA

Andrew HEIDINGER
NOAA National Environmental Satellite
Data and Information Services (NESDIS)
Maryland and Wisconsin
USA

Ross N. HOFFMAN
NOAA National Environmental Satellite
Data and Information Services (NESDIS)
Maryland and Wisconsin
and
Cooperative Institute for Satellite
Earth System Studies (CISESS)
University of Maryland
USA

Paul INGMANN
Retired
European Space Agency's European Space Research and Technology Centre (ESA/ESTEC)
Noordwijk
The Netherlands

Satya KALLURI
NOAA National Environmental Satellite Data and Information Services (NESDIS)
Maryland and Wisconsin
USA

Fatima KARBOU
Météo-France
CNRM
Centre d'Etudes de la Neige
Saint-Martin-d'Hères
France

Jérôme LAFEUILLE
Retired
World Meteorology Office (WMO)
Geneva
Switzerland

Sylvain LE MOAL
Centre de météorologie spatiale
Météo-France
Lannion
France

François MONTAGNER
Retired
European Organisation for the Exploitation of Meteorological Satellites (EUMETSAT)
Darmstadt
Germany

Clémence PIERANGELO
CNES
Toulouse
France

Thierry PHULPIN
Retired from CNES and Météo-France
Toulouse
France

Didier RENAUT
Retired
Saint-Martin-de-Boscherville
France

Hervé ROQUET
Direction de l'Enseignement Supérieur et de la Recherche de Météo-France
Saint-Mandé
France

Timothy J. SCHMIT
NOAA National Environmental Satellite Data and Information Services (NESDIS)
Maryland and Wisconsin
USA

Michael SEABLOM
NASA Earth Science Technology Office
Science Mission Directorate
Washington, DC
USA

David SPENCER
NOAA National Environmental Satellite Data and Information Services (NESDIS)
Maryland and Wisconsin
USA

Pierre TABARY
Météo-France
Toulouse
France

Patricia WEIR
NOAA National Environmental Satellite
Data and Information Services (NESDIS)
Maryland and Wisconsin
USA

Index

A, C

A-Train, 87
absorption, 215, 217–219, 222, 227, 229, 230, 232, 235, 244, 245, 247, 248, 251, 252
aerosols, 143, 149, 151–155, 162
antenna, 5, 8, 23, 28
Aqua, 73, 76, 77, 82, 83
atmospheric
  composition, 133, 137, 147, 149, 157
  sounding, 118, 129
Aura, 73, 77–79, 82, 87
CALIPSO (Cloud-Aerosol Lidar and Infrared Pathfinder Satellite Observation), 81–83
camera, 3, 4, 6, 10, 18, 19
CEOS (Committee on Earth Observation Satellites), 165, 172, 173, 177, 178, 181, 182, 184, 185
CGMS (Coordination Group for Meteorological Satellites), 165, 167, 169–171, 173, 177, 178, 180, 182, 184, 185
clouds, 143, 149, 151–155, 159, 162
CloudSat, 81–83
Copernicus, 102–104, 111, 113, 118, 120, 121, 130, 133, 136–139
cosmodrome, 5, 18–20, 28
CubeSat, 85, 87

D, E

Degrees of Freedom for Signal (DOFS), 260
Earth
  Explorer missions, 97, 98, 114
  Watch missions, 102
environmental applications, 58
Envisat, 93–95, 104, 106, 114
ephemeris, 9
error transport, 259, 261
EUMETSAT (European Organisation for the Exploitation of Meteorological Satellites), 90, 95, 96, 100, 102, 103, 106
Europe, 117, 118, 122, 124, 130–132, 136
European
  Commission, 96, 103
  Remote-Sensing Satellite, 92–94, 96, 106, 114

F, G

field of view, 189, 200, 202–206
France, 141, 143, 145, 148–152, 154–160
geostationary orbit, 122, 127, 129
greenhouse gases, 142, 143, 150, 157
GSICS (Global Space-based Inter-Calibration System), 176, 177

H, I

hurricane, 7, 15, 19
image, 4, 5, 7–11, 13–20, 23–33
imagery, 118, 122, 126
infrared, 216, 217, 220, 223, 225–227, 230–232, 235, 237, 247, 249, 250
intercalibration, 164, 175–177, 185
interferometer, 237–239, 243
international cooperation, 143
interoperability, 164, 172, 184
inverse problem, 253, 256

J, K

Jacobian, 255, 256
Keplerian
  elements, 193, 195
  orbit, 190, 192, 194

L, M

lidar, 217, 239–246, 249–252
linearization, 254, 255
Look-Up Tables (LUTs), 257, 262, 263
meteorology, 122, 136, 137
Meteosat, 90–92, 95, 103
MetOp (Meteorology Operational Programme), 93, 95, 96, 102–104, 106, 107, 111, 112
microwave, 215–217, 219, 222, 223, 225, 226, 228, 230–232, 247–251

N, O

NASA, 67, 68, 71–77, 79, 82–86, 88
near real-time (NRT), 119, 127, 129, 131–133, 136, 138, 139
  oceanographic and atmospheric applications, 138
nephanalysis, 7–9
neural network, 257, 262, 264–266
next-generation space architecture, 38, 57, 59, 62
Nimbus, 67–72, 87
numerical weather forecasting, 130
operational, 118, 119, 122, 125–128, 130–132, 136–139
optimal estimation, 258, 260–262
OSCAR (Observing Systems Capability Analysis and Review) tool, 167, 174

P, R

parking longitude, 209, 210
perturbation theory (Lagrange), 195
perturbed orbit, 190
phase function, 221, 222, 228
photographs, 3, 4, 6, 8
planning, 164, 172, 174, 178, 184
platforms
  geostationary, 38, 39, 45
  polar-orbiting, 38
polar orbit, 118, 122, 135
precession
  apsidal, 195, 212
  nodal, 195
precipitation, 151, 155–157
radiance, 217, 229, 230
radiative transfer, 215, 224–227, 229, 232, 233, 250–252
radio occultation, 213
radiometer, 233–235, 247, 250
RainCube, 85–87
recurrence, 200, 201, 203, 204
remote sensing, 3
rocket, 3–6, 13, 17, 20, 21, 23, 28, 29

S, T

satellites
  collaboration, 208
  geostationary, 203, 210, 211
  recurrent, 200–203, 208, 211
  Sun-synchronous, 193, 194, 196–198, 200–206, 208
scan, 202, 203

scattering, 215, 218, 221, 222, 225–229, 232, 240, 242, 243, 245, 250–252
Sentinel missions, 103, 104, 107, 109, 111
space
  -based Earth observations, 37, 38, 45, 52, 58, 59, 62
  agency, 142, 143, 146, 151, 154, 159
spatio-temporal sampling, 200, 202, 205
spectrometer, 233, 235–238, 246, 250
spectroscopy, 215, 219
stationkeeping, 209, 210
Terra, 73–76

U, V, W

UARS (Upper Atmosphere Research Satellite), 72, 78
vision, 167–169, 171
weighting function, 227, 230, 231, 235, 245, 246, 260
WMO (World Meteorological Organization), 163–169, 171, 173–175, 178–180, 183–185

Summary of Volume 2

Acknowledgments

List of Acronyms

Introduction
Thierry PHULPIN

Part 1. Meteorology

Introduction to Part 1
Hervé ROQUET

Chapter 1. Operational Sounding of Thermodynamic Variables in the Atmosphere
Thomas AUGUST

1.1. Introduction
1.2. Operational use of TIR and MW sounders
1.2.1. Satisfying ever-more demanding users
1.2.2. Clouds: an obstacle to sounding and a very useful geophysical product
1.2.3. Demonstrating and maintaining product quality
1.2.4. Different operational algorithmic strategies
1.2.5. Application perspectives
1.3. Acknowledgments
1.4. References

Chapter 2. Wind Observations
Régis BORDE and Jean PAILLEUX

2.1. Introduction
2.2. AMVs

2.2.1. Extraction of AMVs
2.2.2. Current production and outlook
2.3. 3D winds derived from hyperspectral sounders
2.4. Measuring wind from space using Doppler lidar
2.4.1. Introduction
2.4.2. Measurements from ALADIN lidar onboard Aeolus
2.4.3. Culmination of a long process
2.4.4. Situation in 2022 and outlook
2.5. References

Chapter 3. Surface Variables
Jean-François MAHFOUF

3.1. Observation of the Earth's surface from space
3.2. Energy balances at the surface and at the top of the atmosphere
3.3. Ocean surfaces
3.3.1. Surface temperature
3.3.2. Surface wind
3.3.3. Sea ice
3.4. Continental surfaces
3.4.1. Surface temperature
3.4.2. Water content of soil
3.4.3. Surface albedo
3.4.4. Vegetation properties
3.5. Snow-covered surfaces
3.5.1. Spatial coverage and albedo
3.5.2. Equivalent water content
3.6. Expected changes
3.7. References

Chapter 4. The Assimilation of Satellite Data in Numerical Weather Prediction Systems
Bill BELL, Jean-Noël THÉPAUT and John EYRE

4.1. Introduction
4.2. Early meteorological satellites
4.3. Assimilation of satellite soundings 1970–2000
4.3.1. Early sounding instruments
4.3.2. Assimilation experience: 1970s
4.3.3. Assimilation experience: early 1980s
4.3.4. Problems arising in the late 1980s
4.4. Relevant aspects of data assimilation theory
4.5. The modern era (2000 to present)
4.5.1. Assimilation strategies

4.5.2. Advanced infrared sounders
4.5.3. Microwave sounders and imagers
4.5.4. Radiative transfer modeling
4.5.5. Observation uncertainties
4.5.6. Atmospheric motion vectors (AMVs)
4.5.7. Scatterometers
4.5.8. Radio occultation observations
4.5.9. Impacts
4.5.10. Reanalyses
4.6. Summary and conclusion
4.7. References

Chapter 5. Nowcasting
Thibaut MONTMERLE

5.1. Introduction
5.2. Satellite data for nowcasting
5.2.1. Polar-orbiting satellites
5.2.2. Geostationary satellites
5.3. Observed phenomena
5.3.1. Air mass instability
5.3.2. Convective systems
5.3.3. Characteristics of clouds
5.3.4. Hydrometeors
5.3.5. Wind
5.4. Nowcasting of detected phenomena
5.4.1. Method based on the tracking of structures
5.4.2. Method based on image extrapolation
5.4.3. Method based on artificial intelligence
5.4.4. Use of numerical forecasting
5.4.5. OBS-NWP fusion
5.4.6. Probabilistic forecast
5.5. Perspectives
5.6. References

Chapter 6. Observation and Monitoring of Tropical Cyclones from Space
Frank ROUX

6.1. Introduction
6.2. Visible and infrared imagery
6.3. Microwave imaging
6.4. Microwave sounding
6.5. Surface wind measurements

6.6. Ocean parameters
6.7. Climatology of cyclones
6.8. Conclusion
6.9. References

Part 2. Atmospheric Composition

Introduction to Part 2. Air Composition and the Contribution from Satellite Observations
Thierry PHULPIN and Claude CAMY-PEYRET

Chapter 7. Reactive Tropospheric Chemistry
Sarah SAFIEDDINE and Camille VIATTE

7.1. Introduction
7.2. Methane
7.3. Reactive organic species
7.3.1. Isoprene
7.3.2. Other non-methane volatile organic compounds
7.4. Reactive inorganic species
7.5. Conclusion
7.6. Acknowledgment
7.7. References

Chapter 8. Major Pollutants: Ozone and Fine Particulate Matter
Juan CUESTA and Gaëlle DUFOUR

8.1. Introduction
8.2. Tropospheric ozone
8.2.1. Beginnings of satellite-based tropospheric ozone observations
8.2.2. Current capabilities for tropospheric ozone monitoring
8.2.3. Multi-wavelength synergy for ozone pollution monitoring
8.3. Pollution aerosols
8.3.1. Optical thickness of pollution aerosols
8.3.2. Altitude of pollution aerosols
8.4. References

Chapter 9. Desert Dust
Juan CUESTA

9.1. Introduction
9.2. Qualitative satellite detection of desert dust
9.3. Satellite observation of the optical depth of desert dust
9.4. Vertical profiles of desert dust by spaceborne lidar
9.5. 3D distribution of desert dust by infrared spectrometer

9.6. Conclusion
9.7. References

Chapter 10. Species Emitted by Fires
Camille VIATTE and Pasquale SELLITTO

10.1. Introduction
10.2. Biomass burning gases
10.2.1. Greenhouses gases
10.2.2. Carbon monoxide (CO)
10.2.3. Volatile organic compounds (VOCs)
10.2.4. Ammonia (NH3)
10.2.5. Nitrous acid (HONO)
10.3. Biomass burning aerosols
10.3.1. AOD observations with nadir-viewing instruments
10.3.2. Extinction observations with limb-viewing instruments
10.3.3. Lidar profiles observations
10.4. Fire detection systems from space
10.5. Conclusion
10.6. Acknowledgments
10.7. References

Chapter 11. Stratospheric Chemistry
Claude CAMY-PEYRET and Sarah SAFIEDDINE

11.1. Introduction
11.2. Stratospheric ozone chemistry
11.2.1. Polar ozone depletion
11.2.2. Antarctic ozone distribution
11.2.3. Arctic ozone distribution
11.3. Stratospheric chemistry of other species
11.3.1. Chemistry of the stratosphere and models
11.3.2. Radical processes and cycles for the major families
11.3.3. The example of methane in the stratosphere
11.4. Satellite measurements of trace species in the stratosphere
11.5. Conclusion
11.6. Acknowledgments
11.7. References

Part 3. Atmosphere and climate

Introduction to Part 3 Atmosphere and Climate and the Contribution of Space
Paul POLI

Chapter 12. Climate Monitoring
Paul POLI and Jörg SCHULZ

12.1. General concepts about the climate
12.1.1. What is climate?
12.1.2. Is climate limited to atmospheric phenomena?
12.1.3. A question for Nobel Prize laureates: is the climate stable?
12.2. From space-based measurements to climate products
12.2.1. Sensing the environment
12.2.2. The role of space-based observations
12.2.3. The concept of essential climate variables
12.2.4. Observation-based products
12.2.5. Model-assisted climate products
12.3. Climate data records and uncertainty estimates
12.3.1. Why reprocessing?
12.3.2. Calibration
12.3.3. Uncertainty
12.4. The usage of climate data records in science and services
12.5. Looking ahead
12.6. References
12.7. References of the data sources cited in Figure 12.1

Chapter 13. Anthropogenic Greenhouse Gases: CO_2 and CH_4
Cyril CREVOISIER

13.1. Monitoring anthropogenic greenhouse gases
13.1.1. Biogeochemical cycles
13.1.2. Determination of gas sources and sinks
13.1.3. The global observation network
13.2. Contribution of spatial observation of greenhouse gases
13.2.1. Specificities of greenhouse gas observation
13.2.2. Particularly rich spatial programming
13.3. Measurement techniques
13.3.1. Passive observations in the infrared range
13.3.2. Passive observations by solar reflection
13.3.3. Passive observations by solar occultation
13.3.4. Active observations using lidar

13.4. From radiation measurement to gas flux at the surface
13.4.1. From radiation measurement to gas concentrations
13.4.2. From concentration to fluxes
13.4.3. Main limitations
13.5. Challenges for the future
13.5.1. Towards the observation of anthropogenic emissions by spatial imagery
13.5.2. Reducing spatio-temporal sampling biases
13.5.3. Towards an operational greenhouse gas monitoring service
13.6. References

Chapter 14. Clouds and Water Vapor
Hélène BROGNIEZ, Laurence PICON and Dominique BOUNIOL

14.1. Atmospheric water cycle and climate
14.2. Observations of water vapor
14.2.1. Passive sensors
14.2.2. Active sensors
14.2.3. Homogenization and intercomparison
14.3. Observation of cloud properties
14.3.1. Observations using passive instruments
14.3.2. Observations using active instruments
14.3.3. Multi-instrument synergy for the establishment of cloud climatologies
14.4. References

Chapter 15. Precipitation
Vincenzo LEVIZZANI and Christopher KIDD

15.1. Need for global precipitation measurements
15.2. Satellite observation of rainfall
15.2.1. Visible/Infrared
15.2.2. Passive microwave
15.2.3. Radar
15.2.4. Merged products
15.3. Observation of solid precipitation
15.4. Precipitation and the Earth water cycle
15.5. References

Appendices

Glossary

Printed and bound by CPI Group (UK) Ltd, Croydon, CR0 4YY
13/09/2026
14961068-0006